PHILOSOPHY AND TECHNOLOGY

BOSTON STUDIES IN THE PHILOSOPHY OF SCIENCE

EDITED BY ROBERT S. COHEN AND MARX W. WARTOFSKY

VOLUME 80

PHILOSOPHY AND TECHNOLOGY

Edited by

PAUL T. DURBIN
University of Delaware

and

FRIEDRICH RAPP
Technical University, Berlin

D. REIDEL PUBLISHING COMPANY

A MEMBER OF THE KLUWER ACADEMIC PUBLISHERS GROUP

DORDRECHT / BOSTON / LANCASTER

Library of Congress Cataloging in Publication Data

Main entry under title

Philosophy and technology.

(Boston studies in the philosophy of science ; v. 80)
"Proceedings . . . of a joint West German-North American conference on philosophy of technology, held at the Werner-Reimers Stiftung, Bad Homburg (near Frankfurt), West Germany, April 7–11, 1981"—Pref.
Includes index.
1. Technology—Philosophy—Congresses. I. Durbin, Paul T. II. Rapp, Friedrich. III. Werner-Reimers-Stiftung. IV. Series.
Q174.B67 vol. 80 [T14] 501s [601] 83–10936
ISBN 90–277–1576–9

Published by D. Reidel Publishing Company,
P.O. Box 17, 3300 AA Dordrecht, Holland.

Sold and distributed in the U.S.A. and Canada
by Kluwer Academic Publishers
190 Old Derby Street, Hingham, MA 02043, U.S.A.

In all other countries, sold and distributed
by Kluwer Academic Publishers Group,
P.O. Box 322, 3300 AH Dordrecht, Holland.

All Rights Reserved
© 1983 by D. Reidel Publishing Company, Dordrecht, Holland
No part of the material protected by this copyright notice may be reproduced or utilized in any form or by any means, electronic or mechanical, including photocopying, recording or by any information storage and retrieval system, without written permission from the copyright owner

Printed in The Netherlands

TABLE OF CONTENTS

PART III / RESPONSIBILITIES TOWARD NATURE

PART IV / METAPHYSICAL AND HISTORICAL ISSUES

PART V / DIRECTIONS FOR PHILOSOPHY OF TECHNOLOGY

ANALYTICAL TABLE OF CONTENTS

(*Note*: Where an author has not provided subheadings, these have been provided by the English-language editor and are enclosed in square brackets.)

PART II / TECHNOLOGY ASSESSMENT

PART III / RESPONSIBILITIES TOWARD NATURE

PART IV / METAPHYSICAL AND HISTORICAL ISSUE

PART V / DIRECTIONS FOR PHILOSOPHY OF TECHNOLOGY

PREFACE

Only recently has the phenomenon of technology become an object of interest for philosophers. The first attempts at a philosophy of technology date back scarcely a hundred years – a span of time extremely short when compared with the antiquity of philosophical reflections on nature, science, and society. Over that hundred-year span, speculative, critical, and empiricist approaches of various sorts have been put forward. Nevertheless, even now there remains a broad gap between the importance of technology in the real world and the sparse number of philosophical works dedicated to the understanding of modern technology.

As a result of the complex structure of modern technology, it can be dealt with in very different ways. These range from metaphysical exposition to efforts aimed at political consensus. Quite naturally, within such a broad range, certain national accents can be discovered; they are shaped by a common language, accepted philosophical traditions, and concrete problems requiring consideration. Even so, the worldwide impact of technology, its penetration into all spheres of individual, social, and cultural life, together with the urgency of the problems raised in this context – all these demand a joint philosophical discussion that transcends the barriers of language and cultural differences. The papers printed here are intended to exemplify such an effort at culture-transcending philosophical discussion.

The papers constitute the proceedings, amended and augmented, of a joint West German-North American conference on philosophy of technology held at the Werner-Reimers-Stiftung, Bad Homburg (near Frankfurt), West Germany, April 7–11, 1981. At the conference, barrier-transcending coincidences were as evident as national differences.

The idea for such a conference originated in a conversation involving two of the participants, Stanley Carpenter and Friedrich Rapp. That it could eventually take place – involving as it did the first international meeting devoted exclusively to the philosophy of technology in the West – is due to the active interest of K. von Krosigk, director of the Werner-Reimers-Stiftung. In addition to travel funds for the German participants, the Stiftung provided marvelous accommodations, excellent staff, and the most exacting consideration to individual detail; our warmest thanks go to the staff. Travel funds

Paul T. Durbin and Friedrich Rapp (eds.), Philosophy and Technology, xiii–xiv.
Copyright © 1983 *by D. Reidel Publishing Company.*

for the North American participants came from the U.S. National Science Foundation (thanks to Dr. Sidney Smith of the joint U.S.–West Germany program) and the Social Science and Humanities Research Council of Canada (for the one Canadian participant). Thanks for typing the entire final manuscript to Mrs. Bette Pierce and Mrs. Mary Imperatore of the Philosophy Department, University of Delaware; a number of the original papers for the conference were translated from English into German by Mrs. Joan Neikirk, formerly of the University of Delaware.

In addition to this English version of the proceedings, a German-language version (with some of the contributions reduced in length) was published in 1982, with the same editors, by the Vieweg Verlag, Braunschweig, West Germany, under the title, *Technikphilosophie in der Diskussion: Ergebnisse des deutsch-amerikanischen Symposiums in Bad Homburg* (*W. Reimers-Stiftung*), *7.–11. April 1981.*

PAUL T. DURBIN FRIEDRICH RAPP

PAUL T. DURBIN

INTRODUCTION: SOME QUESTIONS FOR PHILOSOPHY OF TECHNOLOGY

According to Carl Mitcham in his authoritative historical survey of the relationship between philosophy and technology,[1] there are two broad and partially overlapping types of questions that should be and have been addressed by writers in the field. Mitcham labels them "metaphysical/-epistemological" (others would just say "definitional") and "ethical/political."

One way of stating these questions precisely is to begin with a normative or "ought" question. Many scientists, engineers, and other technical people complain about interference or meddling by nontechnical people (government leaders, market-oriented industrial managers, etc.) in what they view as "technical decisions." Or, these same people complain about interference by citizens who do not know what they are talking about in decisions requiring expertise – e.g., nuclear-reactor safety-limit decisions. Here the question is: To what extent in an advanced industrial society ought the scientific/-technological community be allowed to make major techno-economic decisions without interference from non-technical people?

According to Mitcham and many traditional philosophers of technology, however, such a question (or questions) presupposes or at least involves another question: Just what, in fact, do we mean by "technology," a "technological community," or "technological society"? And even if one tends to view matters of definition as arbitrary or conventional (rather than, as Mitcham would have it, "metaphysical" or "epistemological"), there is a relatively factual question embedded in the "ought" question: How free, in fact, are technical people to make their own decisions in contemporary society?

Historically, this is related to the theoretical question of the distinguishing characteristics of the sciences (or cognitive skills more generally) and the practical (ethical or political) question of the proper hierarchy among the sciences or disciplines. According to Thomas Aquinas and the Aristotelian tradition, for instance, the intellectual disciplines – wisdom or metaphysics, the sciences, and the arts (productive skills broadly construed) – rank higher than moral virtues, and among these there is a further hierarchy: prudence (the intellectual moral virtue), justice, fortitude or bravery, and temperance or moderation of the appetites.[2] Such rankings tend to be strongly

Paul T. Durbin and Friedrich Rapp (eds.), Philosophy and Technology, 1–14.
Copyright © 1983 *by D. Reidel Publishing Company.*

culture-bound: Aquinas, in a Christian culture, ranked theological wisdom higher than natural or metaphysical wisdom in a way that his mentor Aristotle would not have; and a modern follower of Aristotle, J. H. Randall, Jr., claims that were Aristotle to have been writing in our modern (especially American) *praxis*-oriented culture, he would not "elevate knowing above practical action."[3]

For approximately the last hundred years, the hierarchy-of-the-sciences question, furthermore, has been closely related to questions of power, status, and support for academic departments in the universities and for the professional disciplines in society more generally.[4] There has also been – and this is even more closely related to our concerns here – the question of the relative ranking of the traditional cultural arts and humanities with respect to the sciences, and of the arts and humanities with respect to the professions. Here again the rankings are strongly culture-bound. Indeed, one way of restating our question might be: Which culture has the (ethically or socially or politically) *preferable* ranking of arts and humanities, sciences, and professions?

The departmentalization of the universities and the narrow specialization of the professions are part of a broader compartmentalism and specialism, according to many social theoreticians, that infects all of modern life.[5] Such theories suggest two further questions. Does our highly specialized society still have a place for vision, for the wisdom that would tell us what to do with all our specialized knowledge, with the products of our vast technological enterprise?[6] And, is our contemporary society markedly worse off either than earlier stages of our own culture or than other cultures?[7]

Given this set of questions and the related problematic aspect of modern culture – i.e., its specialism and fragmentation to such an extent that we are threatened with a lack of vision or wisdom – two of our German colleagues, Hans Lenk and Günter Ropohl, have raised a serious question for philosophers:

> The multidisciplinary and systems-like interlocking of techn(olog)ical problems requires . . . the interdisciplinary cooperation of social science experts and generalists . . . as well as the input of experts in engineering, . . . systems analysts and systems planners. . . . Philosophy has to accept the challenge of interdisciplinary efforts . . . . It has to step out of the ivory tower of restricted and strictly academic philosophy.[8]

What realistic possibility is there that philosophers will venture out of their academic ivory tower and contribute to interdisciplinary efforts to understand – let alone control – technological developments? In what follows,

I propose to survey the American academic scene in two areas, philosophy of science and general philosophy, in order to draw some lessons for philosophy of technology.

PHILOSOPHY OF SCIENCE

In recent years in the U.S., scientific decisionmakers have occasionally looked to philosophers for help. In these cases they have most often turned to philosophers of science: these philosophers have the mathematical skills to deal with technical people as equals; many have learned something of computer programming; their interests are presumably congenial with those of scientists (if not obviously of engineers); etc.

At the same time, there has been considerable loosening up in philosophy of science as people moved farther and farther away from traditional Logical Positivism. First came the work of Thomas Kuhn – especially in *The Structure of Scientific Revolutions* (1962) – and his allies and opponents. This social approach was paralleled by the historically-aware work of the Popperians, perhaps most notably that of Imre Lakatos. In my view, this broadening of focus in philosophy of science culminates in the work of Larry Laudan, in *Progress and Its Problems* (1977).[9]

Laudan focuses on *problem solving* as the key to understanding science. His theory of scientific growth takes Kuhn's concept of a "paradigm," combines it with features of Lakatos's "research programmes," and comes up with a new notion of "research traditions" as the key concept. I think the following gives the flavor of Laudan's approach:

> I am simply suggesting that we need a broadened notion of rationality which will show how the "intrusion" of seemingly "nonscientific" factors into scientific decision making is, or can be, an entirely rational process. Far from viewing the introduction of philosophical, religious and moral issues into science as the triumph of prejudice, superstition and irrationality, [my] model claims that the presence of such elements may be entirely rational [that is, if they contribute to the progress of science by aiding in its problem solving] (p. 132).

It would seem inviting to suggest that this broadened conception of scientific growth could be usefully applied to technological developments. But an obstacle – a "research tradition" (to use Laudan's phrase) of hostility toward interpreting science in social terms – stands in the way. This hostility can be traced back to Rudolf Carnap,[10] remains constant with Karl Popper,[11] and still infects Laudan's work.

Because Laudan's is the most openminded academic philosophy of science proffered so far (in the U.S., at least), his work can serve as a case study of the applicability of philosophy of science to technological decisionmaking.

In the last chapter of *Progress and Its Problems* – entitled "Rationality and the Sociology of Knowledge" – Laudan in effect proposes a test case for his problem-focused theory of scientific growth. Can it help us resolve the controversy between "externalist" (social) and "internalist" historians of science? He concludes, "We must understand what scientific rationality is before we can study its social background" (p. 222); and he is extremely caustic in his assessment of what he takes to be representative contributions from the social history and sociology of science.

Some samples:

"Despite decades of research on this issue, *cognitive sociologists have yet to produce a single general law which they are willing to invoke to explain the cognitive fortunes of any scientific theory*" (pp. 217–218; italics his).

"Nor should one be surprised at the exegetical bankruptcy of contemporary, cognitive sociology of science, for its current explanatory repertoire is far too crude to permit the kinds of discriminations that are called for" (p. 218).

"A judicious examination of the historical record seemingly undercuts the effort to link major scientific theories to any particular socio-economic group [in any historical period]. The Marxists are simply wrong in speaking of a specifically bourgeois mathematics; the followers of Weber have presented no convincing evidence for the existence of a specifically Puritan natural philosophy; contrary to fascist ideology, there is no distinctly Jewish physics" (p. 218).

"It is difficult to resist the *ad hominem* hypothesis that social historians are massively engaged in projecting their own disciplinary insecurities onto the history of science, convinced that scientists are as sensitive to questions of prestige as these historians evidently are" (p. 244, note 29).

Do these strictures suggest the cooperativeness that Lenk and Ropohl would require for one to participate in interdisciplinary projects with social historians, sociologists, scientific/technological managers?

I think there are three ways to reply to Laudan.

First, Laudan's mode of argumentation is flawed. While he begins the major argument in his discussion of sociology of knowledge with a clear cut distinction (p. 202) between explanations of scientific beliefs in terms of their "rational merits" and explanations in terms of social or historical causes, by the end of the section, he has backed off. "In distinguishing between the rational and the socially explicable as I have, I do not mean to suggest that there is nothing social about rationality or nothing rational about social structures" (p. 209). Even so, he adds immediately: "But this continuous interpenetration of 'rational' and 'social' factors should not hinder our capacity to invoke the arationality assumption" [to legitimate social explanations]; and he goes on to decry "that muddle-headed eclecticism which argues that intellectual and social factors can never be usefully distinguished" (pp. 209–210).

Laudan should have stuck with his *caveat*: it is not muddle-headed but sensible (even rational in his terms) to remind ourselves that there is constant interpenetration between rational and social factors. Indeed, the most useful distinction we can make is to recognize a spectrum, with some explanations leaning toward the social end, others toward a *relative* degree of independence from social factors. Science is inherently social, and as an institution it must depend on the larger society for legitimacy, not to mention economic support.

Second, Laudan's examples of "cognitive sociology" are not always well chosen. Having made an excessively rigid demand that social explanations of scientific beliefs be "arational," he must perforce look for extreme examples: Marxist putdowns of "bourgeois mathematics," ideological claims about "Jewish physics," etc. Laudan admits (e.g., p. 220) that there is far more to the sociology and social history of science than that – but even when he concedes this he tends to admit the legitimacy only of what he calls "non-cognitive sociology of science": explanations of why a particular scientific institution was founded when and where it was, etc. (p. 197).

This seems to do serious injustice to much sociology of science and social history of science that could be complementary to good philosophy of science – indeed, which admirably complements Laudan's own broadening of the concept of rationality in science. One example: even if Laudan might put down David Noble's *America by Design*[12] as a Marxist theory of "bourgeois American science" just as he puts down Marxist claims about "bourgeois mathematics," still there is much to learn in Noble's book about the particular way social institutions have influenced claims about scientific rationality in twentieth-century American scientific institutions. Suppose Laudan were to reply that he is willing to learn from Noble but insist that that does not mean we should accept Noble's (or Marx's) theoretical framework as supplying a "general law . . . to explain . . . any scientific theory" (pp. 217–218). In that case, we would have reason to retort that he is invoking an excessively demanding standard for what a social explanation of science ought to provide.

Third, Laudan, in his excessively high expectations with respect to the sociology and social history of science, seems to reflect an academic and disciplinary narrowness. This seems to show up in an offhand remark – that social historians suffer from "disciplinary insecurities" (p. 244, note 29) – which suggests that he feels secure in his own discipline. Laudan seems, throughout the chapter of his book under discussion here, quite confident that philosophers of science and "internal" historians of science can offer the sorts of explanations of scientific growth that "externalists" cannot. Even the standard he holds up to sociology of science, that it should offer explanatory laws capable of explaining particular scientific beliefs, suggests a Hempelian deductive-nomological model of what a good explanation ought to be. Philosophers of science of the old school (e.g., Ernest Nagel) have for years attempted to make some sense of historical explanations, usually finding them inadequate; meanwhile, historians (and historical sociologists of science) have gone right on offering explanations and criticizing one another's explanations as bad or good or outstanding – evidently using *standards different from those the philosophers would impose*.

No one should conclude that I have any fundamental disagreement with Laudan; his focus on problem solving as the key to understanding science is quite congenial to my American Pragmatist leanings. However, I doubt that he is likely to contribute to interdisciplinary or multidisciplinary solutions to technological problems unless he learns to be more open to social explanations and approaches.[13] And the same is true, indeed to a greater extent, for those who cling to older traditions in philosophy of science.

GENERAL PHILOSOPHY

In the U.S. outside narrow academic philosophy of science circles, the situation is not so gloomy – at least in one sense. For about a decade now, American philosophers – perhaps especially younger ones and those less well situated in academia – have been making a beginning in opening up to concerns of society outside the universities. Much of this might seem to be attributable to the anti-Vietnam War movement of the late 1960s, but there were other conditions as well.

In 1971, Marshall Cohen and others started the journal, *Philosophy & Public Affairs*. Although the tone and style of the journal were from the beginning and continue to be academic – perhaps because such a new venture needed to adhere to "the highest standards" to assure respectability for its contributors – the topics addressed have always been issues of concern to the larger society. Many articles that first appeared there have found their way into the several editions of James Rachels, ed., *Moral Problems* (1971, 1975, 1979).[14] (*The* U.S. academic work that has bridged the gap to other disciplines, of course, is John Rawls's *A Theory of Justice*, 1971.)[15]

At about the same time *Philosophy & Public Affairs* was founded, a philosopher-turned-journalist, Daniel Callahan, established the Institute for Society, Ethics and the Life Sciences at Hastings-on-Hudson, New York (the Hastings Center, for short). The *Hastings Center Reports* (beginning 1971), along with other publications emanating from the Center, have become a standard publishing outlet for philosophers interested in biomedical ethics – or, more briefly, bioethics. Numerous philosophers have contributed to the *Reports*, but none has been more consistent than Robert M. Veatch,[16] first an associate of the Center and then later with the Kennedy Institute for Bioethics at Georgetown University (Washington, D.C.). The Kennedy Institute, in turn, has produced the annual *Bibliography of Bioethics* (ed. L. Walters, beginning 1975)[17] and the *Encyclopedia of Bioethics* (ed. W. Reich, 4 vols., 1978).[18] Another philosopher recently at the Kennedy Institute, H. Tristram Engelhardt, is co-editor of the most influential book series in bioethics, *Philosophy & Medicine* (first volume, 1975).[19]

In the 1960s, several science, technology, and society centers were set up, and almost every one had a philosopher as part of the basic team of scholars and teachers.[20] The first major STS center was established by Emmanuel Mesthene, a philosopher, at Harvard University in 1964; that center produced an influential series of research reports.[21] In 1969, Cornell University's STS program came into existence, this time with a major figure in philosophy, Max Black, as a team member; Black, however, has not published any major work in philosophy of technology,[22] and many of the publications generated by the Cornell center have been curriculum/program surveys or politics-of-science volumes (especially the many publications of Dorothy Nelkin). Other centers have followed. For example, at Lehigh University, there is a program now headed by a philosopher, Steven L. Goldman; out of it has come an STS curriculum newsletter, as well as *Technology and Values in American Civilization* (a bibliography, 1980, ed. S. Cutcliffe, *et al.*).[23] Also, at the University of Delaware the Center for Science & Culture publishes the annual, *Research in Philosophy & Technology*.[24] The most influential U.S. journal in the field is *Science, Technology, & Human Values* (Wiley), but it has published only a very limited number of articles by philosophers.

In the 1970s two other movements have begun to take hold among U.S. philosophers that are relevant to our concerns here. One, professional ethics, began to apply philosophical (especially ethical) analysis to issues in the practice of professions other than medicine: nursing, law, business, engineering.[25] The other, called "applied philosophy," made the same sort of opening to the world outside academia as had characterized *Philosophy & Public Affairs*, bioethics, and professional ethics, but the approach was much more eclectic.

There are now at least two journals dealing with business and professional ethics (one edited by Deborah Johnson, from RPI, Troy, N.Y.; the other by Alex Michalos, Guelph, Canada). For applied philosophy, there is a center at Bowling Green State University in Ohio, which holds annual conferences and publishes a newsletter.

In most of this flurry of applied philosophy work, relatively little explicit attention has been given to problems specifically related to technology – although we should recall the Hastings Center work (some on medicine as influenced by technology), engineering ethics, and the STS contributions. There is another new journal, *Environmental Ethics* (University of New Mexico), which will almost inevitably include articles on technology-related problems. Kristin Shrader-Frechette has authored *Environmental Ethics,*[26] in addition to applying her skills as a philosopher of science and ethicist to technology assessment in *Nuclear Power and Public Policy.*[27] And Joseph Margolis[28] has stood out among general analytic philosophers, over a period of twenty years or so, in addressing himself to major concerns of our age, some of which are central to the concerns of philosophers of technology.

What has been the response of mainstream (U.S.) academic philosophers to these movements? Until very recently, it can be said that the attitude has been one of detached tolerance. Contributions to *Philosophy & Public Affairs* have generally been respected even in rather narrow academic departments, and the same has been true for many applied ethics papers published in other journals or anthologies. Bioethics papers for a number of years received less credit, and the Hastings Center's publications are still treated as popularized philosophy by some. Nor were contributors to science, technology, and society programs given much credit for explicit ventures into interdisciplinary work. By now, with the advent of the professional ethics and applied philosophy movements, the picture may have changed somewhat; still, it is too early to tell whether detached tolerance will turn into academic respectability (whether or not that is viewed as important).

And how about the openness of scientific/technological decisionmakers to contributions from applied philosophy? On this I have my doubts, on grounds that technical decisionmakers are often not open to non-technical meddling.[29] However, even assuming that some would be open, it is likely to be the case that general philosophers would not feel completely comfortable with technical details – so that any worries technical people might have about scientific laymen comprehending the technicalities of their decisions might be reinforced. Even so, at least these applied philosophers – in far greater numbers than philosophers of science – have tried; they thus might take the next step and make an extra effort to convince scientists that they (can learn to) know what they are talking about in areas of technical decisions.

SOME LESSONS FOR PHILOSOPHY OF TECHNOLOGY

At about the same time that philosophy of science opened up (however minutely) and that applied philosophy was beginning, philosophy of technology also entered the U.S. intellectual scene. At least three events conspired to bring this about. Philosophers, as much as other academics, experienced the repercussions of the anti-war, anti-technology, youth, and environmental movements of the 1960s. A limited number of philosophers got interested in artificial intelligence or systems theory (as well as STS programs). And, perhaps most important, a few philosophers less dominated by Anglo-American analytical philosophy and more open to European ideas began to introduce European philosophy of technology to U.S. audiences. Many of these philosophers eventually joined forces in the Society for Philosophy & Technology – and a representative sample attended the Bad Homburg/Reimers-Stiftung conference.

As the Society for Philosophy & Technology becomes more international, some lessons from U.S. academic philosophical experience seem appropriate. Given the questions raised at the outset here – How should technical and non-technical people relate to one another? Is there still a place in a high-technology culture for vision or wisdom? etc. – and recalling the plea of Lenk and Ropohl for philosophers to get out of their ivory towers and join interdisciplinary technology management teams, what is the message for philosophers of technology? First, that the traditional narrowness of U.S. academic philosophy of science would be an obstacle; but, second, the general – often interdisciplinary – movement labelled "applied philosophy" shows promise – *if* its practitioners become more interested in, and do their homework on, technological problems.

In academia in the U.S., as disciplinarity, departmentalization, and professionalization (even among philosophers) have become more and more entrenched,[30] the last vestiges of a generalist, anti-fragmentation approach have been relegated to interdisciplinary programs. STS centers and programs are one example, as were American Studies programs earlier and as have been applied humanities (including philosophy) programs more recently. What Lenk and Ropohl have called for goes a step beyond this: Get philosophers (along with such other generalists as are available) entirely out of academia and into the arena of technological decisionmaking. This, however, could backfire if they were not prepared for the complexities of technical decisions.

In this situation, it seems urgent not to turn philosophy of technology into another academic specialty – no matter the importance of going beyond "mere beginnings," as Elisabeth Ströker urges in her contribution to this volume.

What chance does an interdisciplinary approach have in a system massively oriented toward the disciplines and "professional" approaches? One critic asks: "What social consequences can we anticipate from what Paul Durbin called 'metascientific meddling'? What will become of the gadflies and meddlers [philosophers, historians, social theorists] inside the corridors of power?"[31] He seems to be implying that they will be swept aside if they get in the way.

As a result of the Bad Homburg conference and the move toward internationalization, the Society for Philosophy & Technology has a unique opportunity – and a unique challenge. Will the society retain its old loose-knit philosophy *and* technology orientation, or will it become a new philosophy *of* technology superdiscipline? It seems that the demands for a generalist approach, for interdisciplinarity, for vision or wisdom, would incline us toward the former alternative – even though the problems of generalists talking to specialists remain.

In almost all the papers in this collection, breadth, lack of narrowness, and interdisciplinarity are on display in one fashion or another.

The first symposium (Part I), "Can Technological Development be Regulated?," begins with Edmund Byrne's discussion of regulation, a bit of social commentary that makes no claim to be philosophical analysis of the academic sort. Robert S. Cohen's reflections on worldwide problems and opportunities for science and technology can also be classified as broad social commentary. Alois Huning, utilizing engineering codes of ethics and statements of basic human rights, maintains that technology can be guided by such statements of principle – if professionals work to live up to the ideals they profess. While Alex Michalos exercises his skills at traditional philosophical analysis, his purpose is to free science and technology regulators from what he sees as bondage to an indefensible fact-value distinction. Günter Ropohl also presents us with a technical exercise, but again his aim is practical: to free those who would control technological development, whether as managers or democratically, from the pessimism of those who say technology is out of control. Langdon Winner, author of a major study defending the technology-out-of-control thesis, completes the symposium with a history-of-ideas survey of the political ideal in an effort to get still another perspective on how, today, technology (in his view) dictates politics rather than the reverse.

The second symposium, since it focuses on technology assessment, could be expected to be narrowly technical. In fact, Stanley R. Carpenter opens the discussion by arguing that philosophers of technology need to look to so-called "alternative technology" more than technology assessment as it is currently practiced. (Hans Sachsse's comment questions whether "alternative technology" might not be a misnomer.) Friedrich Rapp, a strong advocate of a well-defined discipline of analytical philosophy of technology, here argues that specialized technology assessment is the best we have to offer as a means of controlling technology in pluralistic democratic societies; what we do not want (he thinks) is for this process either to become an apology for current practices or to float off in vague metaphysical speculations. Kristin Shrader-Frechette assumes that technology assessment will continue and will remain a matter mostly of narrow cost-benefit analysis; what she wants is to find ways to internalize within that narrow approach certain value concerns that cost-benefit people tend to ignore. Walther Zimmerli, who like Rapp espouses an analytical approach, nonetheless ends up agreeing with contemporary critics who want to slow technological development.

With the next two symposia, "Responsibilities toward Nature" and "Metaphysical and Historical Issues," we move almost completely outside philosophy of technology in anything like its recent sense.

Bernard Gendron's contribution to the nature symposium parallels work done by environmental ethicists in the U.S. Hans Lenk reflects on the responsible-for-nature views of Hans Jonas, and Klaus Michael Meyer-Abich makes an even stronger defense of the same view, arguing from a standpoint with roots in nineteenth-century German *Naturphilosophie*. All three contributions are clearly philosophical rather than interdisciplinary, but few would say the work is narrow, ivory-tower philosophy.

The case with the metaphysical/historical symposium is more complicated. Don Ihde and Wolfgang Schirmacher show a marked dependence on Heidegger, but both handle the material in imaginative and original ways.

Joseph Margolis, in his response at the end of the symposium, nonetheless questions whether their speculations are not needlessly "transcendent." In another way, Margolis's critique would apply to the efforts, by Reinhart Maurer and Carl Mitcham, to detect the historical roots of modern technology.

In the last symposium we have come back full circle, to the concerns of this introduction. Elisabeth Ströker's paper closes the volume defending a thesis exactly the opposite of that defended here; she thinks philosophy of technology ought to be an extension (with significant modification) of approaches developed in philosophy of science, and should become a recognized philosophical subdiscipline. In his contribution, Hans Heinz Holz counters with a *praxis*-oriented proposal, that if a philosophy of technology is to be developed, it must be as an anthropology of values, to orient technological development in the direction of socially desirable goals. Alwin Diemer opens this final symposium with a straightforward (though difficult) phenomenology of technology, but he thinks the future lies in applying such a philosophical approach to practical problems, e.g., Third World development.

NOTE: This introduction and the editing of the volume were done while working under a Sustained Development Award from the National Science Foundation and the National Endowment for the Humanities. While not specifically funded, work on the introduction and the conference was recognized by NSF as contributing to my SDA research project. The support of NSF and NEH is gratefully acknowledged. (P.T.D.)

NOTES

[1] Carl Mitcham, "Philosophy of Technology," in P. Durbin (ed.), *A Guide to the Culture of Science, Technology, and Medicine* (New York: Free Press, 1980), pp. 282–363.

[2] Cf. Thomas Aquinas, *Summa theologiae*, I–II, qu. 66; also, W. A. Wallace, *The Role of Demonstration in Moral Theology* (Washington, D.C.: Thomist Press, 1962), chapters 1–2.

[3] J. H. Randall, Jr., *Aristotle* (New York: Columbia University Press, 1960), p. 248.

[4] Alexandra Oleson and John Voss (eds.), *The Organization of Knowledge in Modern America, 1860–1920* (Baltimore: Johns Hopkins University Press, 1979); see also *Daedalus*, vol. 92 (1963), entire number devoted to the professions.

[5] The best available summary (with interpretation) of these views is that of Peter L. Berger (with others) in *The Homeless Mind: Modernization and Consciousness* (New York: Random House, 1973), but also in *The Social Construction of Reality* (Garden

City, N.Y.: Doubleday, 1966). For a more recent and more detailed view of the version of Herbert Marcuse, including the relation of his thought to that of Max Weber, see Morton Schoolman, *The Imaginary Witness: The Critical Theory of Herbert Marcuse* (New York: Free Press, 1980).

[6] Langdon Winner, *Autonomous Technology: Technics-out-of-Control as a Theme in Political Thought* (Cambridge, Mass.: MIT Press, 1977); and Jacques Ellul, *The Technological Society* (New York: Knopf, 1964; French original, 1954), and *Le Système technicien* (Paris: Calmann-Levy, 1977).

[7] E. F. Schumacher, *Small is Beautiful: Economics as if People Mattered* (New York: Harper & Row, 1977).

[8] Hans Lenk and Günter Ropohl, "Toward an Interdisciplinary and Pragmatic Philosophy of Technology," in P. Durbin, (ed.), *Research in Philosophy & Technology*, vol. 2 (Greenwich, Conn.: JAI Press, 1979), pp. 43–44 and 47.

[9] Larry Laudan, *Progress and Its Problems* (Berkeley: University of California Press, 1977); page references to this volume are cited in parentheses in the text.

[10] Rudolf Carnap, *Logical Foundations of Probability* (Chicago: University of Chicago Press, 1950), see p. 217 and *passim*.

[11] Karl Popper, "Normal Science and Its Dangers," in I. Lakatos and A. Musgrave (eds.), *Criticism and the Growth of Knowledge* (Cambridge: Cambridge University Press, 1970), p. 57; as with Carnap, many passages could be cited.

[12] David Noble, *America by Design: Science, Technology and the Rise of Corporate Capitalism* (New York: Knopf, 1977). Noble's interpretation can be usefully compared and contrasted with Alfred D. Chandler, Jr., *The Visible Hand: The Managerial Revolution in American Business* (Cambridge, Mass.: Belknap/Harvard University Press, 1977).

[13] A philosopher of science every bit as sophisticated as Laudan – and one perhaps a bit less open than he in other respects – who nonetheless manages to find a great deal of plausibility in cognitive sociology is Mary Hesse; see her *Revolutions and Reconstructions in the Philosophy of Science* (Bloomington: Indiana University Press, 1980), especially chapters 2 and 9.

[14] James Rachels (ed.), *Moral Problems* (New York: Harper & Row, 1st ed., 1971; 2nd, 1975; 3rd, 1979).

[15] John Rawls, *A Theory of Justice* (Cambridge, Mass.: Harvard University Press, 1971). See also Norman Daniels (ed.), *Reading Rawls: Critical Studies of "A Theory of Justice"* (New York: Basic Books, 1974).

[16] Robert M. Veatch, *Death, Dying, and the Biological Revolution* (New Haven: Yale University Press, 1976); *Case Studies in Medical Ethics* (Cambridge, Mass.: Harvard University Press, 1977); and, with Roy Branson, *Ethics and Health Policy* (Cambridge, Mass.: Ballinger, 1976).

[17] LeRoy Walters (ed.), *Bibliography of Bioethics* (an ongoing series; first volume, Detroit, Mich.: Gale, 1975).

[18] Warren T. Reich (ed.), *Encyclopedia of Bioethics*, 4 vols. (New York: Free Press, 1978).

[19] H. Tristram Engelhardt, Jr., and Stuart Spicker (eds.), *Evaluation and Explanation in the Biomedical Sciences* (an ongoing series; first volume, Dordrecht: Reidel, 1975).

[20] In 1969, two prominent U.S. philosophers, Kurt Baier and Nicholas Rescher, edited *Values and the Future: The Impact of Technological Change on American Values* (New York: Free Press, 1969); neither of them, however, was associated with a center.

[21] Harvard University Program on Technology and Society, *Research Reviews* (diverse authors): *Implications of Biomedical Technology; Technology and Work; Technology and Values; Technology and the Polity; Technology and the City; Technology and the Individual; Implications of Computer Technology*; and *Technology and Social History* (Cambridge, Mass.: Harvard University Press, 1968–1971). Mesthene also published a final report when the program was disbanded; it closely parallels the views of his *Technological Change: Its Impact on Man and Society* (Cambridge, Mass.: Harvard University Press, 1970).

[22] But see Black's "Are There Any Philosophically Interesting Questions in Technology?" in F. Suppe and P. Asquith (eds.), *PSA 1976*, vol. 2 (East Lansing, Mich.: Philosophy of Science Association, 1977), pp. 185–193.

[23] Stephen H. Cutcliffe, Judith A. Mistichelli, and Christine M. Roysdon, *Technology and Values in American Civilization* (Detroit, Mich.: Gale, 1980).

[24] Paul T. Durbin (ed.), *Research in Philosophy & Technology* (an ongoing series; Greenwich, Conn.: JAI Press, vol. 1, 1978; vol. 2, 1979; vol. 3, 1980; and vol. 4, 1981).

[25] Cf. Daniel Callahan and Sissela Bok (eds.), *Ethics Teaching in Higher Education* (New York: Plenum, 1980). This volume complements a set of nine volumes edited by Callahan and Bok (1980) and published directly by the Hastings Center; there is an overview, plus surveys of journalism, bioethics, business, social science, engineering and public policy ethics, as well as ethics in the undergraduate curriculum – each volume by a different author.

[26] Kristin S. Shrader-Frechette, *Environmental Ethics* (Pacific Grove, Calif.: Boxwood, 1981).

[27] Kristin Shrader-Frechette, *Nuclear Power and Public Policy* (Dordrecht: Reidel, 1979).

[28] Margolis also has at least one article to his credit explicitly on philosophy of technology: "Culture and Technology," in P. Durbin, ed., *Research in Philosophy & Technology*, vol. 1 (see note 24, above), pp. 25–37.

[29] Paul T. Durbin, "A Significant Limit on Applied History, Philosophy, and Sociology of Science and Technology," *Science, Technology, & Human Values*, No. 34 (Winter 1981), pp. 18–19. This is one of several comments on an applied history, philosophy, and sociology of science symposium; see *ST&HV*, Fall 1980.

[30] See Oleson and Voss, *The Organization of Knowledge ... 1860–1920* (note 4, above); a third volume in this solid series of historical studies is in preparation, and will bring the picture up to date.

[31] Sal Restivo, "Meeting Report: Notes and Queries on Science, Technology and Human Values," *Science, Technology, & Human Values*, No. 34 (Winter 1981), p. 21; Restivo's report refers to both the symposium in *ST&HV*, Fall 1980, and the comments, Winter 1981.

PART I

CAN TECHNOLOGICAL DEVELOPMENT BE REGULATED?

EDMUND BYRNE

CAN GOVERNMENT REGULATE TECHNOLOGY?

Recently, while I was in Dijon, France, I happened to be listening to BBC on the radio, and heard a Yorkshire consumer affairs official complaining that the law gave him power only to require that "push-chairs" (baby carriages, or prams) be stable and have reliable brakes, but not to require that their handles not come off. Reacting to an incident of local notoriety, he seemingly felt the need for broader power under law to see to it that Yorkshire mothers be able to use prams as nearly perfect as government regulation could make them.[1]

This tacit belief in the efficacy of government regulation, here expressed by an interested government regulator, is widely shared by people all over the world, especially by intellectuals who worry about the alternative risks of not controlling technology. Thus, for example, Victor Ferkiss sought in 1969 to determine the traits of the "technological man" whose skills would qualify him for the task of controlling technology. This task, according to Ferkiss, "may be the supreme test of our species' adulthood."[2] Nigel Calder, writing in the same year about the impact of technology on society, decided that "(m)oral and political efforts need . . . to be directed to compensating the totalitarian tendency (of technology), and to employing governments to encourage humane bonds of technology."[3] Garrett Hardin, in his well-known concern about the tragedy of the commons, has even gone so far as to propose that for every problem involving a system there be an appropriate government agency to regulate it.[4]

These altogether typical expressions of liberal confidence in government do not, of course, apply to just any government, but ordinarily to whatever the writer takes to be the supreme government – in the United States, this has usually been taken to mean the federal government. Thus, for example, when the movement for auto safety got underway in our country in the 1960s, proponents thereof warned against being sidetracked by the manufacturers into "the mare's nest of state legislation" and sought instead "national uniformity through the national representatives of the people, the U. S. Congress."[5] Daniel Moynihan supported the same movement in 1962 by claiming that autos and pipelines were "the only forms of interstate transportation . . . not regulated by the federal government for purposes

Paul T. Durbin and Friedrich Rapp (eds.), Philosophy and Technology, 17–33.
Copyright © 1983 *by D. Reidel Publishing Company.*

of safety," and that pipelines would soon be regulated.[6] Even Ralph Nader, who is hardly a blind devotee of government bureaucracy, insisted in 1966 that "(o)nly the federal government can undertake the critical task of stimulating and guiding public and private initiatives for safety" and meet "the urgent need for publicly defined and enforced standards of safety."[7]

The tendency thus exemplified to count on the federal government to keep technology within bounds far antedates the Corvair crisis and, of course, is with us still. In discussing the need for regulatory reform, an attorney recently indicated his acquiescence in the basic system by noting: "Federal regulation began in 1837 with the Steamboat Inspection Service, which was formed when a rash of steamboat explosions raised early questions about free enterprise."[8] In a report recently issued by the U. S. Office of Technology Assessment, itself a product of the federal fever under discussion, it was noted that, although both drugs and medical devices are subject to federal regulation regarding their safety and efficacy, "(s)urgical and other procedures that depend primarily on providers' techniques have not been subject to similar Federal controls."[9] So, not surprisingly, the report anticipates "increasing scrutiny" of "the use of medical technologies."[10] This increased interest is, it should be noted, justified on the grounds that the Federal Government (OTA's capitalization) is both "protector of the public" and a significant "developer and user of medical technology."[11]

There is, in short, a considerable amount of sympathy for the idea that the federal government is without question an appropriate instrument for effecting some desired level of control over technology. Why such sympathy should exist, however, is by no means obvious. For, there are countless examples of how inappropriately and ineffectively the federal government has been dealing with technology over the years. The record of failure here in question has occurred on every level and in every branch of government. Here it is sufficient to consider only the phases of legislation and regulation. First, some examples of ineffective legislation.

Controlling water pollution has long been a concern of the federal government, but it is arguable that this concern has done more harm than good. According to a study done by the U. S. General Accounting Office and published in November 1969, government programs have concentrated on municipal treatment systems and have either bypassed industry or encouraged overloading of municipal systems with industrial waste.[12] But this misplaced emphasis is itself a product of the way relevant legislation has been written.[13]

Casual recollections of federal government support for development of the Salk vaccine may be generally favorable. But a more detailed history would show that the Eisenhower administration very carefully avoided any program of widespread vaccination that would have taken business away from constituents of the American Medical Association. What resulted was not a public health program but privately paid fees for services with the single exception of the poor – potential carriers! – who would not otherwise have contributed to physicians' incomes.[14]

An even more glaring example of the inadequacy of legislation as a means of controlling technology is the fascinating process whereby the federal government came to intensify its involvement in the search for a cure for cancer through what came to be called the National Cancer Act of 1971. The federal government had long since come to be a funder of cancer research, and by 1971 some $64.7 million of the total National Cancer Institute budget of $78 million was spent on research.[15] However, a very powerful group of lobbyists decided that this was a woefully inadequate investment in such a serious problem, and that the solution was, in part, to pull the NCI out from under the jurisdiction of the National Institutes of Health. Senator Edward Kennedy, among others, expressed the opinion that heavy spending on a cure for cancer would nicely fill the vacuum left by federal disenchantment with the supersonic transport as an expensive technological project of dubious value.[16] This interest Kennedy was able to operationalize when, shortly thereafter, he became chairman of the Health Subcommittee of the Senate Committee on Labor and Public Welfare.[17] It was an election year, to be sure, and the incumbent president, Richard M. Nixon, was not to be outdone in concern for cancer victims and their families by a potential opponent. Thus it came about that the Nixon Administration also supported legislation of some sort in this area and allocated $100 million in its budget proposal for the ensuing fiscal year.[18] And accordingly the two political rivals were able to arrive at a generally supportive compromise.[19] Both public and private debate on the cancer legislation centered around whether and to what extent the NCI should become independent of NIH controls. The White House, hearing clearly the concerns expressed by the medical industry, generally opposed independence. Kennedy, in general agreement with the original proponents of the cancer legislation, favored a more autonomous agency. The end result, by way of a substitute bill introduced in the House by Congressman Paul G. Rogers, was essentially an agency that is operationally independent but nominally not.[20]

The key question that arises from all of this, of course, is what advances towards a cure for cancer did all this election year legislation produce. In terms of expenditures, the cancer control program alone, which calls for comprehensive cancer centers, had its budget increased from $5 million in 1973 to $34 million in 1974.[21] And the total NCI budget between 1972 and 1981 comes to a total of some $7–8 billion.[22] In terms of power distribution, a three-member President's Cancer Panel that was conceived of as an oversight group to assist the president has in effect become the executive committee of the board of directors. Its chairman, Benno Schmidt, has been wielding the most power; and he just happens to have been the chairman of the original "Citizens Committee for the Conquest of Cancer" (1969–1970) which became the Senate's "Panel of Consultants on the Conquest of Cancer" (1970–1971).[23] And, perhaps not entirely coincidentally, other programs administered by NIH have been getting comparatively less funding.[24] But when all is said and done, what is perhaps most revealing in this entire affair is that the NCI continues along the traditional lines of cellular– and molecular–oriented research and, abundant evidence as to chemical carcinogenesis notwithstanding, has not concerned itself notably with any sort of regulation of the industry at issue.[25]

This failure of relevant legislation to reach the front lines of an issue regarding technology is, at least in the judgment of theoretical pessimists, due to the inherent inequality of power arising respectively out of technology and out of government.[26] This thesis could be illustrated in many ways, e.g., by reviewing the history of legislation with regard to the railroads. It might also be illustrated by calling attention to an area where the limits of law are commonly recognized and worried about, namely, in international affairs. Inasmuch as the vast majority of significant technological developments impact far beyond any one nation's boundaries, laws adopted by any one nation will be of little use unless also adopted by other nations as well. This is patently true with regard to just about anything that might come under the heading of what in our country is called "interstate commerce." The vast field of telecommunications, including satellites and microwave communications, is an obvious case in point. But so is the troubled area of arms control. Recent political stances with regard to the proposed Salt II Treaty are in some respects just the tip of the iceberg. Consider also the long and still largely unsuccessful efforts of various international organizations to ban the use of all sorts of unpleasant means of warfare, notably chemical and biological agents.

Treaties, of course, are binding only on signatory nations, unless the contents of a treaty are generally held to have become "general principles of law recognized by civilized nations."[27] Yet the very failure of a nation to abstain from using a particular weapon itself becomes an indication that at least that nation does not recognize such usage to be in violation of any general principles.[28] And as a recent analysis of this problem asked, not without a certain note of discouragement: "In time of war can it ever be expected that the law of war will actually prevail over vital national or military interests?"[29] This query notwithstanding, the authors thereof conclude with what is essentially an optimistic view as to the capabilities of law. For, as they see it, what is mainly lacking is enough law to do the job. In their own words:

In a world distracted and disunited beyond human experience, science has placed in the hands of governments chemical and biological weapons systems far beyond the imagination of the founders of international law. But there has been no concomitant development of the legal process which should control and direct those in positions of supreme authority with reference to such weapons.[30]

It may be allowed, however, that just because legislation cannot resolve all problems does not mean it cannot resolve *some* problems. But not even this much can legislation accomplish without more. And the "more" that is here at issue is the manner in which the legislation is implemented – in a word, the regulatory system. Regardless of how well a legislative enactment may seem to address some problem involving technology control, the regulatory system that emanates from the legislation has a life force of its own, be it weak or strong.

For example, even when pushing for federal automobile safety standards in 1966, Ralph Nader had to acknowledge that the General Services Administration was not even exercising what powers it already had in this regard through federal procurements.[31]

The federal patent system, which obviously functions as a form of technology control, has never been entirely beyond suspicion. Acknowledging in 1939 that there were abuses, a knowledgeable writer on the subject insisted that "(t)hese abuses are not the fault of the patent law but are the fault of the Department of Justice in not enforcing the anti-trust laws."[32] Writing more recently, however, Irene Till reaches a notably more fatalistic conclusion. According to her,

(T)he patent has become a potent instrument for restraint of trade. . . . Particularly in industries where the technology is subject to change, (patents) are the basic weapons in corporate strategy for eliminating, subduing, or harassing competition.[33]

The same author who criticized the faults in federal water pollution laws notes in addition that even those laws that do exist have not been effectively enforced, in large part because physicians in the federal bureaucracy have tended to side with ineffectual medical colleagues at the state and local levels rather than with enforcement-minded lawyers in the federal service.[34]

In a recent study of the impact of FDA regulation of the pharmaceutical industry, it is asserted as "(a) consistent finding" "that regulation has had a significant negative effect on the rate of innovation."[35] The author goes on to claim that this negative impact gives U. S. firms "a strong incentive to develop and expand foreign production and other foreign operations."[36]

The Nader group, among others, has tended to attribute the weaknesses in our federal regulatory system to its having been coopted by the private interests which it was supposed to be regulating. Thus in 1966 Nader was identifying the Commerce Department as "the house of business" and warning against assigning any safety-oriented tasks to such a coopted agency.[37] Writing in 1973, Nader sees this process of cooptation as having taken on a still more troubling characteristic: "What is new," he says, "is the institutionalized fusion of corporate desires with public bureaucracy," resulting in what he calls "the industry-to-government-to-industry shuttle, where corporate risks and losses become taxpayer obligations."[38] Still thinking about such then contemporary interventions as those involving Lockheed and Litton, Nader might have been speaking of current government support of Chrysler and, perhaps still to come, the steel industry and others as well:

> Regulation that wastes scarce economic resources and inflates consumer prices, rewards inefficiency, and impairs service is a gross abuse. But when it assists in suppressing highly beneficial technology because the entrenched companies do not wish to displace their outdated capital, a new, worse, dimension of corporate socialism appears. In the last two decades, spectacular new technology has been developed that can throw the challenge of abundance to older technologies of scarcity – satellite communications vs. AT&T's cables, CATV vs. the traditional limited TV spectrum, mass transit systems callable on demand vs. buses and automobiles. These are a few of many developments which *the regulatory-industrial complex* has succeeded in blocking, limiting, or delaying.[39]

A condemnation of government regulation of technology that is in some ways more sweeping will be found in Lawless's study of one hundred cases of government intervention entitled *Technology and Social Shock*. Says Lawless by way of conclusion with regard to governmental action:

> One or more of the agencies or branches of the federal, state and local governments became involved in virtually all of the one hundred cases. [Fn.] In about half of the forty-five study cases [those more fully reported], one could say that a government agency took a definitive action . . . , but in fewer than ten of these cases could one describe the government action as prompt, and in two of these ten . . . the very speed

and timing of the government action added unnecessarily to the social shock. . . . In general, a delay in initiating remedial action (or in determining if it is needed) is relatively short (one to five years) when it results from a need for more technical information . . . , but much longer when it results from a conflict of powerful interests. . . .[40]

These by no means atypical criticisms of the role of government, especially the federal government, in regulating technology suggest the need, quite obviously, for improvements. Some of these improvements are of a purely technical nature, i.e., better equipment for monitoring problems subject to regulation.[41] Others, by way of corollary, are secondarily technical, i.e., the requirement that for any monitoring of a technology-based problem the monitoring agency be staffed with a sufficient number of appropriately trained *technical* specialists.[42]

Such an improvement in monitoring personnel (not, of course, easily achieved given the salary differential as between the public and the private sector) would contribute greatly to alleviating the problem, stressed by Mintz in his study of federal regulation of the pharmaceutical industry, of leaving the monitor dependent on technical data provided by the entity being monitored.[43]

Not even the upgrading of an agency's technical personnel will suffice, however, to disengage that agency from long-established subservience to the powerful interests it is supposed to monitor. This is well illustrated by the longstanding but ultimately futile effort of the Federal Communications Commission to reassess AT&T's rate structure. To no one's surprise, this effort could hardly be considered successful after over a decade of study. In the meantime, the FCC has entered some rulings which are not altogether favorable for AT&T, e.g., with regard to peripheral equipment and with regard to satellite communications, and this has been enough to cause Ma Bell to seek to induce Congress to emasculate the FCC and in effect reverse these unfavorable decisions.

The regulatory impasse thus laid bare has led one analyst to suggest an alternative approach that would include a "Natural Monopoly Contracting Agency," which would act as a kind of super-agency supposedly not subject to pressuring, at the federal level.[44] Others have sought to institutionalize impartial watchdogging in other ways, such as Nader's Public Citizen and John Gardner's Common Cause, not to mention a host of other more specialized public interest lobbyists and consultants. It is, however, precisely the plethora of watchdogs all of whom claim to represent "the public" that lends support to Lee Loevinger's onetime claim that we are wholly lacking any theory of regulation.[45] It is rather the case, I think, that we have many theories, none of which is conducive to uniformity. But in any event we are beginning to see studies of the impact of regulation, which hopefully will one day be sufficiently accurate that they can be required as a precondition to any significant proposed regulation. For example, when a ban on DES in cattle feed was proposed a few years ago, it was possible to estimate at least within broad parameters how much said regulation would affect the price of beef.[46] Similar estimates are proffered with proposals for tax increases, e.g., the recently proposed but politically doomed duty on imported oil.

The issue of the impact of regulation on technology is, however, much broader than just a question of prices. As the very idea of standardization suggests, regulation can affect the very nature of the technology developed. This is perhaps most clearly recognized by the general public with regard to automobile fuel consumption standards and how these have affected the structure and to some extent the operative technology of cars both current and yet to come. Of even more fundamental significance is the role regulation plays in the research and development of new technologies, most noticeably (at least after the fact) in the area of military technology.

What is perhaps little understood is the kind of limits a society's accepted ways of doing things imposes on any attempt to regulate a technology in one way rather than another. To cite some obvious examples, it would be the height of foolishness, however desirable otherwise, to attempt now in any finite period of time to require the English to drive their vehicles on the right side of the road, or to require people in most European countries to settle for a half-hour "lunch break" in the name of efficiency, productivity or whatever.[47] Problems of this kind have, of course, attracted considerable interest because of socio-cultural obstacles to technology transfer in Third World countries. But there is as much need for appropriate technology in developed countries. For example, the use in various European countries of automatic shut-off switches for corridor lights could and should be applied practically everywhere in the United States. Why is there no encouragement to do so? All it would take would be an appropriate government regulation that would so require! Nor would the obstacles – even if dignified as being "constitutional" – really be any more than an instance of *our accepted way of doing things*.

The kind of attention to the socioeconomic preconditions for a given technology and for the way that technology should be regulated is illustrated by default by a recent development in Switzerland. Like other European governments, not to mention some states in the United States, the Swiss parliament recently passed a law requiring everyone on a motorcycle to wear a helmet. However desirable such a practice might be, it was not made more desirable by federal fiat in a country whose bilingual survival has depended on recognition of many prerogatives for the individual cantons. As a result, opponents of centralization, of whom there are many in Switzerland, were able to activate provisions in the Swiss constitution to require a referendum on the helmet question in November, 1980.[48]

Inversely, a very wise assessment of socioeconomic limitations, albeit after the fact, will be found throughout Jacques Ellul's masterful study of the development of political institutions in France. Consider, for example, only his observations about the impact of the revolutionary government on French industry. In spite of the altogether expected emphasis on the guiding principle of laissez-faire, he points out, a government agency was made responsible for inventions, which it encouraged by means of prizes, industrial expositions, etc. Various measures were taken to facilitate domestic trade, including imprisonment for non-payment of debts, but commerce was hindered by a shortage of money and grossly inadequate means of communication. Attempts to improve the roads were not successful. The result was commercial decline, but at the same time economic stability was restored for the benefit of a large sector of the population.[49]

What Ellul's summary says, in more "in" terminology, is that post-revolutionary France was a pre-industrial economy not yet sufficiently developed for take-off. But, it should go without saying, the government at that time had to be as alert to built-in limitations as is any government in a country on the way to development or, for that matter, a country such as the United States until recently considered "developed" but in current jargon said to be in need of "reindustrialization."

The point of all this is that government policies with regard to regulation of technology cannot be any more appropriate than the limited insights of decisionmaking bureaucrats will allow. Accordingly, Ralph Nader's call for more highly specialized technicians in government agencies is helpful only to the extent that what these technicians would so efficiently regulate should in fact be so regulated or regulated at all. This is not necessarily a call for more "deregulation," so called, which is in its effects simply another form of regulation, just as a refusal to choose is itself a choice. Rather is this a recognition that any really important technological improvement (or, if you will, revolution) is almost by definition going to tax the competence of even the best trained technical experts in what Nader would presumably consider a well staffed agency. At work here, however, is not just the much criticized revolving door in the sense of crassly biased private career-conscious regulators. Rather is this a problem inherent in the very process of technological development. Who, after all, is appropriately competent to judge how if at all a new technology ought to be utilized in our society? I submit that probably the least able to judge is any expert narrowly trained in all the meticulous niceties of the technology or technologies which the new would simply render obsolete.

From this point of view, it is difficult to accept Garcia's ethical approach to a definition of "bureaucratization." As he wrote in 1971, "through internal corruption, the bureaucracies become self-serving and increasingly ineffectual in making decisions until they are entirely corrupt and immoral."[50] His call for "feedback" to "effectively destroy bureaucracy" is useful but commonplace when stated in more neutral terms. What I disagree with most about Garcia's analysis is that he sees the problem in terms of volition rather than in terms of knowledge. And since lack of adequate knowledge is so obviously the constant problem faced by regulatory agencies, any claim as to improper motivation must surely bear the burden of proof, as in a criminal trial.

Questions of guilt aside, however, the knowledge-gap in regard to regulation of technology is analogous in many ways to the problem of tenure in a university. The standard objection to tenure is that it allows individuals with outdated information to occupy positions which should be made available to those with up-to-date information to convey. Let us not be distracted here by an obvious pedagogical equivalent of the chicken-or-egg problem nor by economic considerations as to the comparative salary requirements of new versus experienced professors. Important here is the standard response to the anti-tenure argument, namely, that education in a discipline is vacuous if the "information" conveyed is not put into a complex context that only years of study enable one to understand to some extent. Call this, if you will, the Anti-Fad Principle. What is relevant about it here is that it applies to any allegedly well-trained would-be technocrat, the nature of whose training will almost certainly involve some form of faddism.[51] What society as a whole needs to appreciate even if our would-be technocrat cannot, is that there are more things in heaven and on earth than his or her slide-rule, or cost-benefit analysis, or MBO, or whatever tomorrow's trusted device will be, is able to generate for purposes of regulating technology.

An extraordinarily relevant case in point, although perhaps only indirectly concerned with control of technology, is that of a recent unsuccessful attempt on the part of federally funded technocrats to convince some people in Pennsylvania to allow the federal government to regulate the sale of land in the Brandywine watershed area. As one of the central protagonists observes, the technocrats had "a demonstrably 'good' plan."[52] But the demonstrably good plan did call for exercise of eminent domain, as required by the Ford Foundation and HUD.[53] All efforts to paint the program pretty were ultimately in vain, because "People already felt threatened by government, with the gas lines, the high tension line, the Marsh Creek project, and the proposed prison, and many saw the Brandywine Plan as another government program to force people to use their land in a certain way."[54]

In retrospect the same protagonist does acknowledge what is indeed basic in all such government intervention: "We *were* technocrats from outside, as they charged. Our lives, our future expectations of income, and our sense of personal control were not affected by the plan. Theirs were, and we owed them more than they received."[55] With these words is verified the observation of Peter Thompson writing about the Brandywine proposal in *Science* magazine: "(P)erhaps the lesson in human relations which come out of the confrontation between 'experts' and rural Americans is the most valuable piece of base-line data to emerge from the study."[56]

This matter is introduced here not in order to poison the well with regard to government regulation of technology.[57] Rather, it is meant to illustrate the proposition that there is more to regulation than putting the right words – or numbers – on paper. Nonetheless, if words and numbers are needed it is presumably better to have these generated by experts of one sort or another. To the extent that these experts are in some sense of the word "technicians" we are perhaps moving towards what Jacques Ellul calls a technological society. Whether this is good or bad is in part at least a matter of perspective. It is, however, also a matter of sorting out apples and oranges. For there are aspects of our technologies which should be regulated and are not, just as there are aspects which are regulated but were better not. And even if some form of regulation is appropriate, it is by no means obvious what form that should be. As noted by McCloy,

> The critical issue today is whether [regulation by such means as licensing, rate-setting, or imposition of standards] is always the most efficient approach or whether the provision of profit incentives, recourse to taxation, or increased reliance on the free market might be more effective.[58]

But even to speak of effectiveness assumes that one knows which of many possible goals the regulatory agency ought to be seeking to attain. And that is precisely where knowledge is lacking in government agencies as in the populace at large.

It is at this juncture that the increasingly common plea for some institutionalized arbitrator, such as a Science Court, comes to the fore. What has been lacking in proposals of this kind, however, is provision for the kind of humanistic input that cannot be obtained merely by a debate among technical experts. Assume, for example, that the introduction and regulation of new technologies were made subject to a judicial battle of experts. Leaving the standard problems of procedural obfuscation aside, let us assume that any given case would be argued on its merits. And leaving aside also the political realities that determine who will serve in decision-making roles, let us further assume that only our wisest citizens (whoever they may be) will ever sit as judges in a Science Court. To these assumptions add others to assure fundamental fairness, adequacy of available information, and efficacity of whatever decisions might be arrived at. Given all this, consider now a few not entirely hypothetical cases.

Case No. 1 A proposal to provide individual television receivers with a retrofitted device that would make possible satellite reception of any television broadcast from anywhere in the world, without the intervention of any "network" or any governmental agency responsible for "the public interest" in regard to the communication of information.

Case No. 2 A proposal to conduct all business transactions by a new system whereby a laser would read a potential buyer's pre-assigned identity number printed invisibly on the back of his or her hand and automatically debit his or her account the amount being charged, thereby eliminating the need not only for credit cards but for money as such.

Case No. 3 A proposal to ban all vehicles from the downtown area of any major city and provide mass transportation within that area by means of two ground-level conveyor-belt systems, one primarily for people and the other primarily for goods.

Case No. 4 A proposal to expand the capability of television to transmit information so that much work presently done at a business office could be done at home, and only key personnel would be required – permitted? – to journey to the centralized workplace.

Case No. 5 A proposal to substitute genetically-engineered nutriments for food as we now know it, thereby eliminating the need for agricultural production and distribution – in other words, no more grocery stores.

My purpose in formulating these cases is not to provide an exercise in futurology but to illustrate by way of suggestion how complex is the fabric of issues that arise in connection with any significant – and often even with comparatively insignificant – technological change. With regard to the "insignificant," and with the supposed advantage of hindsight, how would you have decided as a judge in a science court if asked shortly after World War II to approve a shift from a system of bottling products locally in glass to a system of packaging products regionally or even nationally in metal or plastic containers? Or, to return to the focus of this paper, how if at all should the federal government have responded to that proposal now seen to have entailed innumerable historically significant consequences? And how should the federal government have involved itself in the demise of the passenger train, or in the disregard of the bicycle as a convenient means of short– to medium–distance transportation? Or to the insistence that nuclear fission reactors offered the key to our future energy needs, or that women who wanted to avoid breast cancer ought to submit to mammography on a regular basis?

In a word, government as we know it is not in general able to regulate technology effectively or appropriately. Yet for the good of society technology ought to be regulated both effectively and appropriately. This can be done only imperfectly given the limited information available at any given time. But it should nonetheless be done as well as possible, and this means that it must be done on the basis of the most adequate information available. Such information will not, however, be adequate if it is only and narrowly technical. Included also must be information derived from many kinds of expertise on many levels and from many corners of reality. Along the way, of course, such a multifaceted search for appropriate choices may delay implementation of innovation. But the history of technology, including our own, is – or, perhaps, could be – filled with evidence that means do not necessarily lead to ends, and ends even if achieved may not constitute unadulterated progress.

Indiana University, Indianapolis

NOTES

[1] Subsequent investigation by the ever intrepid BBC produced the following information, disseminated to the public on August 14, 1980: The kind of accidents that had been occurring as a result of "buggy-handle failure" were only "of the second order," according to the Trade and Standards Office. To have qualified as failures of the first order such that the Minister would issue a warning, two conditions would have had to be met: (1) an accident must occur; (2) it must occur at a most critically inopportune time. Some 99,000 prams produced by the manufacturer (McLaren, Ltd.) from 1977 until a design change was introduced in September, 1979, were susceptible to the notorious handle failure. Some 106 accidents involving these prams had been reported, but

apparently none was "of the first order." In the meantime, the manufacturer has quietly notified retailers that the handle problem should be corrected on any of the suspect prams that are brought in for any reason whatsoever. For the August 14 broadcast BBC did a spot check of its own and discovered 6 prams with faulty handles. Pram owners seemed to know about the defect, according to BBC, but were hesitant to bring their prams in for replacement because a pram after use tends to be dirty, hence not, in their minds, deserving of replacement.

[2] Ferkiss 1969, p. 217.

[3] Calder 1969, p. 144.

[4] Hardin 1972, p. 134. Having called for this problem-specific regulatory structure, Hardin was at least realistic enough to point out that there is something of a second-order problem in figuring out how to watch the watchers (pp. 133–140). The appropriateness of this after-thought is well borne out by the long and arduous history of efforts in other countries, notably Western Europe, to develop institutions for precisely this purpose. See in this regard Chapman 1970, pp. 181–270. In Chapman's opinion, at least, the French approach via the Conseil d'Etat is by far the most advanced and most effective government watchdog to be found in Europe (p. 229). However, the areas most in need of technology assessment, namely, the nationalized industries (e.g., Electricité de France) are immune from the jurisdiction of the Conseil d'Etat on the grounds that they are essentially in the private sector, even though they exercise powers of eminent domain (p. 223; see also p. 54).

[5] O'Connell/Myers 1966, p. 208.

[6] Moynihan 1962, p. 266.

[7] Nader 1966, p. 249.

[8] Neustadt 1980, p. 129. See also Moynihan 1962, p. 266.

[9] OTA 1978, p. 90.

[10] *Ibid.*, p. 4. See also pp. 104, 110.

[11] *Ibid.*, p. 4.

[12] Ridgeway 1970, p. 107.

[13] *Ibid.*, p. 54.

[14] R. Carter 1967, pp. 290–327, 332–335. See also Lawless 1977, pp. 128–138; Rettig 1977, pp. 34–39, 313–315.

[15] Rettig 1977, pp. 42–46, 70. Compare pp. 30–34.

[16] *Ibid.*, p. 103.

[17] *Ibid.*, pp. 117–121.

[18] *Ibid.*, pp. 125–126. The precise amount which the Nixon Administration actually committed is not all that easy to determine. In the first place, the announced $100 million was in fact divided in half over two fiscal years (p. 183). And, secondly, they had actually cut some funds from cancer research just one year before (p. 210).

[19] *Ibid.*, pp. 184–193.

[20] *Ibid.*, p. 291: "The resolution of the legislative debate left the cancer crusade advocates with much of what they wanted, but gave the opponents the symbolic and material accomplishment of defeating the proposed separate agency recommendations."

[21] *Ibid.*, p. 303.

[22] *Ibid.*, p. 317.

[23] *Ibid.*, pp. 79–88, 278, 295–297.

[24] *Ibid.*, p. 310. It is for just this reason that proponents of heart research set about getting themselves comparable legislation providing for a categorical research program in 1972. *Ibid.*, p. 312.
[25] *Ibid.*, p. 305.
[26] See, for example, Ellul 1977, pp. 146–150, 292, 331.
[27] Van Wynen and Thomas 1970, pp. 188–236.
[28] *Ibid.*, pp. 137–187.
[29] *Ibid.*, p. 207.
[30] *Ibid.*, p. 249. See also preface, p. ix. A certain verbalistic legalism creeps out briefly at one point in the Thomas analysis. Commenting on the Hague Gas Declaration which signatory nations have been able to bypass rather handily, the authors suggest that use of "asphyxiating or deleterious gas" could have been avoided if instead of referring to such usage as the "sole object" the text had instead spoken of the "primary" or "main" object (p. 46; see in any event, p. 57).
[31] Nader 1966, pp. 226–227, 230.
[32] Toulmin 1939, p. 108.
[33] Till 1973, pp. 289–316.
[34] Ridgeway 1970, p. 54.
[35] Grabowski 1976, p. 37.
[36] *Ibid.*, p. 51.
[37] Nader 1966, pp. 241–242.
[38] Green 1973, p. x.
[39] *Ibid.*, p. xii. Italics added.
[40] Lawless 1977, pp. 508–509.
[41] *Ibid.*, pp. 58–60, 64, 67, 76, 80, 294.
[42] Green 1973, p. xii (Nader).
[43] Mintz 1965, pp. 410–415.
[44] Beverly C. Moore, Jr., in Green 1973, pp. 96–98.
[45] 11 Antitrust Bull. 101, 115 (1966).
[46] Campbell 1974, p. 50.
[47] At least one author, however, seems to feel that technology American-style will eventually bring about the demise of all of the cultural uniqueness of different peoples around the world. The end result, in his terminology, will be a race of "cybernanthropes" rather than human beings. These curious machine-men can be spotted by their liking for sleek offices and supermarkets. See Lefebvre 1971, pp. 201, 203–204, 209.
[48] As it turned out, the referendum upheld the helmet law in every canton. For a detailed analysis of Switzerland's constitutional mix of federal and canton control, see Chapman 1970, pp. 219–233. It is relevant to note that Chapman puts the Swiss on a par with the French as being the most distrustful of government (pp. 308–315).
[49] Ellul 1969, pp. 132–133.
[50] Garcia 1971, p. 99.
[51] See in this regard Stanley 1978, pp. 157–158; Lefebvre 1971, pp. 65–130.
[52] Strong 1971, p. 57; see also p. 117.
[53] *Ibid.*, pp. 125, 173.
[54] *Ibid.*, p. 117.
[55] *Ibid.*, pp. 198–199.

[56] *Science* March 14, 1969, quoted by Strong 1971, p. 174.
[57] This is done, in fact, by Jacques Ellul in his recent work. See Ellul 1977, pp. 119–120, 146–156, 330–334. See also my review of the English translation of this work, *Nature and System*, 3 (Sept. 1981): 184 ff.
[58] McCloy 1980, p. 462.

REFERENCES

Calder, N. 1969. *Technopolis: Social Control of the Uses of Science*. New York: Simon and Schuster.

Campbell, R. R. 1974. *Food Safety Regulation*. Stanford, Cal.: Hoover Institution on War, Revolution and Peace, and Washington, D.C.: American Énterprise Institute for Public Research.

Carter, R. 1967. 1966. *Breakthrough: The Saga of Jonas Salk*. New York: Pocket Books Cardinal.

Chapman, B. 1970, 1959. *Profession of Government*. London: Unwin University Books.

Crowe, B. L. 1969. "The Tragedy of the Commons Revisited." *Science* **166** 1103–1107 (1969), reprinted in Hardin and Baden 1977, pp. 53–65.

Ellul, J. 1969. *Histoire des institutions*, vol. 5. Paris: Presses Universitaires de France.

Ellul, J. 1977. *Le Système technicien*. Paris: Calmann-Levy.

Ferkiss, V. C. 1969. *Technological Man: The Myth and the Reality*. New York: NAL Mentor.

Garcia, J. D. 1971. *The Moral Society*. New York: Julian Press.

Grabowski, H. G. 1976. *Drug Regulation and Innovation*. Washington, D.C.: American Enterprise Institute for Public Policy Research.

Green, M. J. (ed.), 1973. *The Monopoly Makers*. New York: Grossman.

Hardin, G. 1972, 1968. *Exploring New Ethics for Survival*. Baltimore: Penguin.

Hardin, G. and Baden, J. (eds.). 1977. *Managing the Commons*. San Francisco: Freeman.

Lawless, E. W. 1977. *Technology and Social Shock*. New Brunswick, N.J.: Rutgers University Press.

Lefebvre, H. 1971, 1967. *Vers le Cybernanthrope*. Paris: Denoe/Genthier.

McCloy, J. J. 1980. "Federal Regulation: Roads to Reform," *ABA Journal* **66** 461–464.

Mintz, M. 1965. *The Therapeutic Nightmare*. Boston: Houghton Mifflin.

Moynihan, D. P. 1962. "The Legal Regulation of Automobile Design," in *Passenger Car Design and Highway Safety*, New York: Association for the Aid of Crippled Children, and Mt. Vernon, N.Y.: Consumers Union of U.S., Inc.

Nader, R. 1966. *Unsafe at any Speed*. New York: Pocket Books.

Neustadt, R. M. 1980. "The Administration's Regulatory Reform Program: An Overview." *Administrative Law Review* **32** 129–163.

Neustadt, R. E. and Fineberg, H. V. 1978. *The Swine Flu Affair: Decision-Making on a Slippery Disease*. Washington, D.C.: U.S. Dept. of Health, Education and Welfare.

O'Connell J. and Myers, A. 1966. *Safety Last: An Indictment of the Auto Industry*. New York: Random House.

OTA 1978. Congress of the United States. *Assessing the Efficacy and Safety of Medical Technologies*. Washington, D.C.: U.S. Government Printing Office.

Pennock, J. R. 1941. *Administration and the Rule of Law*. New York: Rinehart.

Rettig, R. A. 1977. *Cancer Crusade: The Story of the National Cancer Act of 1971*. Princeton, N.J.: Princeton Univ. Press.

Ridgeway, J. 1970. *The Politics of Ecology*. New York: Dutton.

Stanley, M. 1978. *The Technological Conscience.* New York: Macmillan, Free Press.

Strong, A. L. 1971. *Private Property and the Public Interest: The Brandywine Experience.* Baltimore and London: Johns Hopkins.

Till, I. 1973. 'The Legal Monopoly," in Green 1973, pp. 289–316.

Toulmin, H. A. Jr. 1939. *Patents and the Public Interest*. New York/London: Harper.

Van Wynen, A. and Thomas A. J. Jr. 1970. *Legal Limits on the Use of Chemical and Biological Weapons*, Dallas: Southern Methodist University.

Wilson, J. R. 1963. *Margin of Safety*. Garden City, N.Y.: Doubleday.

ROBERT S. COHEN

SOCIAL IMPLICATIONS OF RECENT TECHNOLOGICAL INNOVATIONS*

Science and technology are social phenomena. Like all social phenomena, they arise, develop, change – flourish or languish – even come to an end; and they do so in all continents, throughout all human civilizations at varying times of origin and with differing paces of development. But only tentatively and rarely did technology, with its deep roots in the craftsmanship of the earliest recorded times, or perhaps in the even earlier known techniques of the paleolithic hunters and artists, reach across barriers of social class occupations to join with science, and then only in periods of particular social requirements: we may think of Chinese pure mathematics and the practical calendars of military and political astrology; of Greek chemistry and the invention of liquid fire for naval defense (a science-technology link but very brief); of European stellar astronomy and practical sailor-navigators of the Renaissance. But the historically unique fusion of craft technology, raised to literacy and a new thoughtfulness in the Italian Renaissance – with its secularized philosophical effort to understand the order of nature, especially in the Galilean adaptation of Platonic mathematical idealism – was the achievement only of *post-feudal* Western Europe. Despite the immense sophistication of other civilizations and peoples – of the men and women of China, of India, of the high Arab and Persian culture of classical Islam, of Mayan and Aztec and Inca and others of the developed American societies – despite maturity of administration, maturity of the arts, literature, architecture, myths and religion, despite all this, technology and science were mainly apart in social function, and incapable of any thorough mutual fertilization.

Sooner or later, except for Western Europe, technological power over nature stagnated, along with the societies themselves or perhaps, first the societies reached a state of saturation: *either* stable, secure, culturally dull, without innovation and without entrepreneurs, and self-satisfied, without fundamental needs; *or* saturated but weak, hence ultimately unstable, ultimately based on value-schemes and epistemological ways with nature that were not viable, and hence open to internal decay or foreign conquest. With modern Europe since the late sixteenth century, however, which is to say and to emphasize, with the generation of an entire new class of men with their expanding role as entrepreneurial mercantile capitalists, both science

Paul T. Durbin and Friedrich Rapp (eds.), Philosophy and Technology, 35–47.
Copyright © 1983 *by D. Reidel Publishing Company.*

and technology were liberated from their separate traditional constraints of opportunities and conceptions. But the engineers and craftsmen, in one specialty after another during the subsequent four centuries, found progress impossible without science, without fundamental enlargement of their understanding of natural processes, which had to extend far beyond the subtlety of mere enriched skills, and unequivocally beyond the reach of the empiricism of old-fashioned trial-and-error exploration of nature. As the modern world opened, practical life was coming to terms with the explosive value of the new way of decoding nature (Galileo said, "read the book of Nature in mathematical language"): not by the seeking of more facts but by an imaginative and hypothetical curiosity to comprehend what the facts as such simply do not show. Science, in its modern revolutionary development, i.e., the Copernican revolution (and all the Copernican revolutions after it), mathematized and mechanized nature, tested and experimented upon nature, pushed and pressed far beyond what nature had ever spontaneously revealed; and with the resulting complex development of this new science, capitalist Europe not only increasingly mastered, but also learned to transform, nature.

The ultimate question for historians of science must be, why did modern science come into existence when and where it did, and why not where it did not? Was it made possible by this or that cultural quality, or perhaps by a religion whose metaphors of creation seem particularly attuned to explanations by causal laws, whose attitudes toward sexual relations of bodily love are those of domination, so that mathematically formulated laws and drastic experiments upon the eternally feminine Nature might be encouraged, even stimulated, when the socio-economic time was right? (Recall how Bacon and Boyle explained that the experimental method is to see natural events in artificial circumstances, to "torture Mother Nature.") And was the time right when the feudal order at last cracked open enough for the long-established but marginal merchants of the feudal towns to begin the march of urban society to social dominance? But then we would have to ask why capitalism came only to the Western Europeans in that dynamic form, or why their traditional religion evolved from its own complex tangle of metaphors and beliefs to the one form of Christianity that functioned so as to help science, in one major way, while hindering science in so many others.

Francis Bacon sensed all these issues when he, a giant among others who shared his views, called for deliberate social support of the scientific revolution by his Elizabethan England. Bacon, for England, initiated science policy; he initiated the social study of the human impact of scientific and technological innovations, and he did so with respect to the material practical realities of economic production and transportation and military power, and also with respect to the reality of a liberation of the human spirit from superstition, dogma, inherited errors of fact and inherited methods of thinking, from what he called idols and what we might call idolatry and fetishism. Bacon was a social optimist about science.

For us, looking at the world of the 1980s, the matter neither is, nor ever was, altogether as Bacon saw it. Nor as the French rationalist optimists of the eighteenth century believed, as they projected their Utopias that reflected their Voltairean adaptation of the Newtonian mode of thought to all technologies, sciences, educational structure, psychology, and plans for a new social order. In the attempt to understand the social impact of scientific technology, we must proceed simultaneously in two ways: first, in a far less sweeping and generalizing manner (is technology Good, or is technology Evil?) and second, in a far more self-critical and sceptical dialectical analysis (science gives life *and* death). We also should recognize the historical character of our attitude toward the social and human impact of science and technology within our own century: attitudes toward technology will differ depending on whose technology it is, on which specific technological advance we evaluate, on which portion of humankind is speaking or is represented, which class, which race, which tribe, which generation, which sex, at which cultural place the evaluator stands. And, even more assuredly, attitudes toward technology depend upon which technological advance, specifically, is at issue.

So generalization had best be avoided, or approached cautiously, even though in the end the entire situation of the human race will depend upon decisions regarding scientific technology. We are ourselves, then, in a new dialectic of specificity and universality; and this is due to the species-wide situation we confront, a situation which itself is due to the combined effect of the technology of the past two hundred years and to the world-wide political-economic market domination of the same period (briefly the mastery, so far as it goes, of nature and society by industrial, i.e., technological capitalism, and its aftermath).

What is all the more puzzling as we try to look over the range of issues that concern the social impact of scientific technology is whether the recent and prospective impact, taken as positive or as negative, taken as that of a single innovation or as that of the cumulative innovative currents of development, is so different from the singular or general impact of earlier strongly innovative technologies. The impact upon displaced rural persons in eighteenth- and nineteenth-century England, who were forced into their rootless, often work-less, often criminalized, normally "gin-soaked", morally anarchic, urban existence, was a drastic shock, a personal and cultural transformation whose details have become clear; the impact upon Japanese peasants under a quite differently arranged urban displacement a century later, was different; and now a third century later, we see vast metropolitan cities of millions (e.g., in India, Brazil, Mexico) who have been technologically displaced; and in this example, while the human misery is undoubted, and the scale of numbers so greatly raised, is the impact novel? Are the problems unique? This example leads to others that are linked to mass urbanization, and to the entire problematic of a world-wide, mass society that technology has made possible: technologically-induced mass unemployment, together with displacement of skills and of social relations, a phenomenon of commodity society, which seems endlessly drifting, or planning such shifts in human situations. Was colonial transformation, even partly without urban shifts, fundamentally different? We must ask whether the invention of mass society, early on, has entailed the continuity of social impacts that appear drastic to each generation, novel to each social observer, and then open to fresh analysis, which unhappily fails to take account of what has already been learned.

Mass war was new, partly a technological achievement, partly a political choice, but at any rate a novel extension of fighting not only to those obligated by the patriotic consciousness and by the legally powerful enforcement of a military draft but also to the civilian population, which itself was understood to be the substance of a country and a military source and resource. The American Civil War was the first to bring the mass war innovation to reality with an impact upon the United States of nearly unmeasurable disastrous quality. And all major wars since have continued this effect. Was total war, in the fifth decade of the twentieth century, technologically triumphant, of a different social impact?

To some extent, not quite clear, the social impact of technological innovation follows a continuous pattern, the continuity of innovation after innovation now reflecting, now reinforcing, now amplifying, a continuity of social transformation. These instances of mass population transfer through urbanization and of mass warfare may of course require discussion of specific technological advances: thus, city planning from ancient Rome to rebuilt revolutionary Paris, to the city of Washington, to new cities in, say, Siberia and Brazil and post-1945 Hiroshima, would have to be studied with respect to innovations in structural engineering, transport, sanitation, water supply, pollution of air and other environmental factors, educational techniques, and the social psychology of work and play, and health care. Are we faced with a radically changed impact of these new societies of today, or have the damaging social effects already (and always) come about wherever mass accumulation has existed?

Even when continuity might be established, the variations may be in themselves quite substantial. When we examine mass culture, we may at first recognize the expansion of everyday consciousness from rural constrictions to the larger community of city and state, to the democratization of mass schooling, popular literacy, large-scale newspapers and magazines from high-speed printing presses, magnificent and cheap reproductions of art works and musical performances through color photography and high fidelity recordings. (This was already subjected to pioneering analysis by Walter Benjamin in the 1930s.) But then we have also to reflect upon the negative inner development of that impact brought about by a new technology, television, which has profundly influenced mass consciousness, and must be characterized as a novel shift away from general literacy toward visual homogeneity and a stage of personal illiteracy of the written word. The problem of this particular social impact – let us say it is that of a passive, spectator culture, a nearly non-participant human practice – may not be new in human history (for we know of Roman circuses 2,000 years ago), but its extraordinarily rapid, and unusually pervasive and existential saturation, in the advanced industrial countries, testifies to an unprecedented and as yet insufficiently understood, technological (and socio-cultural) phenomenon.

When the social impacts of technological innovations are seen in this way, the continuity described in historical studies may be helpful to the social scientist who may hope to recognize parallels of causation, modification, response, as well as parallels of either implicit or explicitly deliberate social policy. Comparative historical studies, at their best, can shed such illumination upon present problems so that history becomes a heuristic for modern science and technology policy studies and practice. Hence the importance (recognized but as yet too little integrated within the advanced training of those in public administration and industrial leadership) of differential analysis of modernization across national or regional entities which have exhibited quite different social responses to their technological modernization: how different these are in England, Germany, the United States, Japan, the Soviet Union and China, for examples, as they are preceded or accompanied by differences in policy decisions of a social nature (whether of a market sort or of a politically planned socialist nature, or otherwise, needs to be sorted out in each case). The social impact of a technological innovation, then, is *underdetermined* by technology alone; although the constraints imposed upon human life by technology are severe at times, the range of possible social control and use of technology, which is a range from disaster in large or small ways to hope and fulfillment, seems still to be genuine.

Where the present time nevertheless seems to be fundamentally *discontinuous* with the past is, we may suppose, marked by two characteristics. *First*, certain quantitative increases have reached critical points at which, in a familiar phrase, quantity has changed into quality, into a new phase. The explosive power of nuclear bombs, the literally trans-human scale of data mastery in modern information technology, the biologically transformative and creative potential of biochemical genetic engineering, whereby new "natural" entities can be "invented," space science engineering, even efficient and accessible birth control techniques, all testify to the novel social potential of the scientific technology now available. But these technically original inventions and discoveries join the new phase of the impact of the older and continuous technologies suggested above, in what may be a *second* and vital characteristic of our time in the last quarter of the twentieth century. This is the world quality of social and technological problems; or, as the young Karl Marx might have described it, the "species nature" of these problems. For at this moment in human history, at last, the human species seems genuinely to be confronted by species-wide dangers and opportunities, superimposed upon and closely integrated with still urgent issues of local and regional sorts.

We may list the world-wide factors:

(1) Science and technology, despite their origins embedded within Western political economy and cultural sources, are now *world* science, and *world* technology.

(2) Production and distribution of goods and services comprise a *world market* system, despite some local autonomies and despite the variations of different mixtures of central planning, enterprise planning, and competitive mechanisms.

(3) Natural resources are abundant or scarce on a *planetary* accounting, despite local or national variations of the rich or the poor within nations of both the first and third worlds.

(4) Population problems are world-threatening, even while they continue to be locally or regionally of an immediate crisis nature.

(5) World-scale war, however restricted in specific national participation, is a species-wide ecological and genocidal threat, even while apparently local wars continue their own devastating technological and human impacts upon restricted portions of the earth.

(6) Religious solidarity, whether in the form of traditional and conservative

institutions or linked to innovative but anti-modernist cults, seems to be a world phenomenon, a protective response to the felt, or perceived, threatening impact of general technology and of impersonal mass urban peasant society; to use a phrase of Marx again, a world-wide religious tendency to provide the "heart of a heartless world."

It seems proper to suggest that these and perhaps other factors indicating the world-wide quality of the transformation should be followed by a list of world-wide failures:

(1) political and economic failure to utilize technology to eliminate poverty, within most of the advanced industrialized countries and – at its most degrading human quality of life – within the third world;

(2) failure of social scientific analysis, both of empirical studies in their historical and their current dynamical aspects, and of the appropriate rigorous methodology of the social sciences for this most practical of scientific tasks;

(3) failure of world-wide education *for these problems* in particular, and for a healthy and constructive understanding of science and its technology as a part of humanistic education in a scientific age, for the specialists and for the bulk of humankind alike (but specialists' education especially has been elitist);

(4) failure to solve the continuing need for accumulation of capital from present resources, either by adequate transfer from capitalist first world or Soviet sources or by extraction of surplus value from within developing countries (oil and certain mineral-rich lands excepted);

(5) failure within scientific and technological elites to transcend their social sources, their elitism, certain heroic exceptions aside (Pugwash and those within the WHO, for example) and, in particular, failure to identify and control ideology within science;

(6) fetishism of science matching the popular fetishism of consumer goods.

Now, in a survey of problems, I should briefly list a number of recent scientific developments and technological applications with serious social consequences, each of world significance.

(1) Nuclear explosives for military purposes, fission bombs in 1945, and decade after decade of further developments: fusion bombs, guided and programmed delivery systems, concealed launching sites on submarines or within deep-dug silos, multiple targeted bomb launcher rockets, psychologically misnamed "tactical" or "local-theater" nuclear artillery, the mysteries of military budgets for space stations in Earth orbits, and of the precision

destructiveness of laser technology. The nuclear novelty persists in military fact, a discontinuity in potential for worse disasters, but the conscious quality wears thin, is no longer shocking, no longer novel, a disturbing example of the adage that familiarity breeds contempt. But the truth is worse than ever, because the nuclear weapons have not been put under international agreement. Instead they have spread and have, as we have seen, worsened by continuous innovations. Along with the rest of so-called "military science" (no longer the study of battle tactics), they have received the bulk of available scientific and engineering financing. The story is incomplete, the technology too.

(2) Cybernetics, now an "old" innovative science (the extraordinary discovery of Norbert Wiener), a science of mechanisms with intelligence, is still incomplete; but robot labor, automated labor, artificial intelligence *in production*, exploration, quality control, all seem to be maturing. We still have no fully automated factory, but not for lack of scientific knowledge. The 1980s might be the robot decade, a time of robotics, of soaring relative productivity, decreasing general labor time, and increasing engineering labor elites. The situation now is that we have an incomplete *hardware* revolution.

(3) Information technology races on and on, with greater capacities, greater programming subtleties, shorter retrieval times, tinier mechanisms, spreading through all markets of production, distribution, social scientific research, transforming empirical research from the study of the cosmos to the investigation of hourly supermarket inventories, promising sensible decision capabilities for all sorts of complex planning, whether socialist or multi-national, in the gambling house or the tax revenue bureaus. There is no end in sight for this continuing *software* revolution.

We may say, to summarize these last two points, that automation to some extent promises and threatens to replace the blue-collar worker; and programmed computers, to replace the white-collar worker.

(4) Agricultural technology continues to transform food situations here and there, even while famines develop elsewhere. We have not yet come to terms with understanding the social impact of the green revolution, but we had better anticipate the probable success of saline water agriculture. Whereas once, only twenty years ago, nuclear power seemed likely to turn salt water to sweet by one of several workable power-based technologies but then became dangerous and too expensive, now a simpler chemistry, genetically engineered *within plants*, will perhaps cope with salt; perhaps then at last the desert will really bloom, world-wide. The social outcome is difficult to predict.

(5) Biological engineering, the application of the theoretical and experimental science of biochemical genetics, is now at the start of practicality. At every turn, social impacts appear: medical achievements by genetic manipulation of bacterial production, transforming the pharmaceutical industries; revival of imaginative projects of human genetic improvements, as dangerously idealistic as the older eugenics of the 1920s but far more practical; threats of biological warfare, cheap versions of expensive nuclear devastation; fantasies of animal food production and plant creation – indeed a directed evolution replacing the splendid statistical causation of Darwinian evolution. Perhaps the 1980s will also be a pioneering decade of applied biology, the decade of biological technology.

(6) Birth control, population control in general, at first mention hardly novel so accepting in the 1970s have we become of the pill, IUD, vasectomy; but it is still incomplete. And now resistance has arisen, in the West as well as the East. The incompleteness is everywhere evident, whether in technical efficiency, in male contraception research, in social psychology, or in the failure of sexual education. The social impact of frustrated birth control will itself be drastic.

(7) Mass communications, studied again and again by critics and by commodity managers since the twentieth century began, have advanced far beyond the skills of journalism, radio, and motion pictures, to the planetary outreach of manipulative and compulsive TV and the instant global contact of space satellite systems. This is the basis, in a technology linked to science, for the trivia, gossip, tensions and intimacies of the anticipated "global village," but also for a Utopian community analogous to the human support of the village.

(8) Twentieth-century medical technology, and its manifold resources in applied natural sciences spurring and motivating biology, chemistry, and physics, has opened up entire new special fields of investigation and competence. There are wholly novel problems of technological scarcity, priority, elitism, capital-intensive centers of research and therapy, and a renewed gap between excellence and poverty. For example, the impressive diagnostic tool of the CAT scan, which uses computer intelligence for calculation, simulation, and interpretation, and the external organ simulation in kidney dialysis, typify technology in life-assisting professional care, while the world-wide statistical and managerial skills of public health professions typify the possibility of applying interlinked medical and social-scientific understanding in dealing with species-wide issues of health and disease. These are far beyond the age of quarantine, close to the age of social (rather

than individual) medicine as a form of bio-engineered ecological management; we seem to have entered a time when medical ecology will seem the natural basis for understanding human health. *But how incomplete for actual life today*! Occupational and poverty hazards are still as grim and threatening as ever, in the industrial, in the mass urban, and in rural sectors of human life and work.

(9) How incomplete our modern "advanced" science remains, yet how promising too, seems most sharply shown by polymer studies. To understand polymers, and to produce them artificially, would be to initiate yet another age; the age of polymers would be the time of mastery of the principal building materials of living matter. This would be the chemical physics of life processes, especially of the information capabilities of material nerve fibers, and the extension and contraction dynamics of muscles. There is an unimagined enlargement that artificial polymer engineering would give to practical design, for muscle fibers directly convert chemical into mechanical energy. These "muscle motors" (as Kapitza terms them) are still, in the 1980s, the most extensive motor systems on Earth, with higher efficiency than engines, turbines, or other heat engines. How tempting to say that creating an artificial muscle fiber will be the stimulus to inventing an effective small mechanical motor, human size, and perhaps (as we now say) "appropriate."

(10) Scientific developments are partly autonomous, and so too are the many larger or smaller "revolutions" of science and technology. Nevertheless, these scientific and technological revolutions, and the great scientific and technological revolution of the mid-twentieth century, are socially revolutionary in a further sense which is distinctly non-autonomous: science and industrial production are now fused, mediated by technology (which motivates science even while it draws from discoveries of autonomous science), and mediated by the unusually subtle epistemic *praxis* that has invented the social system of science to produce knowledge; and thereby we have a technologically advanced analogue to ancient laboratory servants and assistants for the colossal experimental complexities of the knowledge industry. This industry, producing its commodity of knowledge, is only one element in the fusion of science and production processes. How incomplete the social sciences' treatment of the political economy of science is may be noted from the lack of any fully elaborated classical, Keynesian, or Marxist analysis.

(11) Within the network of world science, and of technological interdependence, resource exploitation is increasingly situated within regional and global markets, within the supra-national corporation and Communist

exchange coordination; in this matter we recognize a wholly natural and global situation of consumption patterns and production availabilities and of the potentialities of human lives within such consumption-production constraints. But the quality of incompleteness is also plain. Political units are barely rationalized and coordinated within themselves, much less interrelated among themselves by the global requirements; and international economic units, private and governmental, respond to goals derived from their internal values and autonomous functions. Controlled and central economic planning on regional if not planetary scales is mathematically, formally, possible, using high-speed artificial computer intelligence and cybernetic feedback control to distinguish the appropriateness of local from central authority, thereby accounting for the many variables (of multiple intrinsic values as well as instrumental values), related to social analysis of standards of living.

We see that these world-scale factors of problems, failures, and uncompleted technological innovative achievements all rest with elites, whether in countries with traditional political democratic institutions, including parliament or congress and a balance of responsible governing powers, or in those with a greater degree of centralized party-governing authority. In the historically evolved division of labor, technical elites have finally come to their own peculiar roles, their power deriving from specialized competence; they are partially insulated from other elites and from democratic decision-making by a scientific and technological sophistication which easily allows for esoteric secrecy (whether military or industrial). The advantages for cognitive achievements are evident in the scientific advances of the modern era. Unlike the specialized division of labor in the industrial work-force, modern scientific specialization has not tended to replace skilled by unskilled workers; rather we may see more highly specialized and skilled labor replacing less specialized and less skilled labor. But narrowness of scientific and cultural literacy may be similar to the quality of de-skilled workers in the factory, since there is no inherent requirement upon technological specialists to acquire or use an integrated outlook on either technological or social problems, and no inherent need for a humanist or broadly cultured education. Hence, threats to traditional cultural institutions, and to political or social democracy, have become ominous even while mastery of planetary resources by technologically incisive specialisms seems feasible.

These technological threats to human societies are broadly of three sorts: political, social, and ideological.

(1) *Political*: The threat due to elitism may outweigh the benefits of specialized learning and specialized practice, *first*, by undermining the competence of representative democracy or by distorting the procedure of electing representatives; *second*, by diverting or frustrating the development of self-management institutions (such as workers' control in the work-place, the market, or other production spaces of societies); *third*, by the overriding technological necessity of quick military response to security dangers with the consequent and accepted social necessity of hotline elitism; *fourth*, by linking populist counter-elitism to neo-Luddism.

(2) *Social*: Scientific and technological innovations, whether successes or failures, whether achievements or promises, threaten to undermine, dissolve, or drastically weaken the received qualities of cultural life and human consciousness, *first*, by challenging the power, validity, and even the presence of the literal as well as the figurative icons and rites of traditional religious and esthetic sensitivity in all their forms; *second*, by promoting the psychologically symbolical fetishism of science and technology, or of anti-science and irrationalism; *third*, by transforming human living relations through transforming the social relations of production, consumption, and communication; *fourth*, by transforming the social relations of pleasure and fulfillment, and, in the process, weakening the momentum of cultural traditions, leaving the individual increasingly without moorings, open prey for the immediacy and irrationality of quick-fix populist manipulation; *fifth*, technology of elite social planning is out of the individual's control and is felt by the individual human being as fragmentary rather than integrated, as chaotic, asymmetrical, short-term, as a matter of impersonal life-irrelevance rather than as a life-affirming achievement; *sixth*, the species-nature of global problems and planetary technological optimism then threatens the individual's life experience.

(3) *Ideological*: Technological society poses problems, sets criteria for explanations and solutions, provides resources of people and materials, creating but also distorting the cognitive culture along with daily life; the technological innovations produce their own political economy of culture along with a political economy of science. These are the objects of new work in the social sciences, and in turn they stimulate critics who then must consider whether science and technology are themselves partial, ideological, merely instrumental reasoning, whether seen from traditional (largely religious) premises or from humanist and other viewpoints.

Science and technology, innovative and global, cannot be neutral, whatever may be thought of their characterization by harsh critics as irresponsible instrumentality. They disturb, and with them modernization always threatens and displaces. We must inquire whether technology can be of human scale, whether science can be concerned with and directed by individual values, whether global threats ineluctably require elite technology, whether short- and medium-term benefits drag unpredicted secondary and tertiary penalties along with them, whether the perceived miracles of science can replace the miracles of religions, and the rhetoric of technology persuasively displace the rhetoric of religion; in fact, we ask whether social cohesion can be preserved.

The final report of the UNESCO Prague Symposium of 1976 pointed to the fundamental task of developing the theoretical foundations for the practical management of social processes under conditions which compel us to recognize that as technological progress accelerates, efforts toward solving social problems also have to be accelerated. This "gives rise to the key task of the social sciences as an irreplaceable instrument in helping to solve the great problems of our time" (see *The Social Implications of the Scientific and Technological Revolution*, Paris: UNESCO Press, 1981, pp. 365–370).

We are far from a satisfactory understanding of the joys and sorrows, the achievements and disappointments, of the technological saga of modern societies. There may yet be alternatives among technologies, and we must work as scientists, technologists, and philosophers to foresee dangers and opportunities – to *choose* with a sense of the still genuine possibility of fulfilling humane values.

Boston University

NOTE

* This is a modified version of a paper given at the UNESCO Symposium on Technological Innovations and Their Social Impacts, Bonn, November 1980.

ALOIS HUNING

TECHNOLOGY AND HUMAN RIGHTS

Technology has been essential for man's life at every moment of history; even under the most favorable climatic and geographic conditions man needs at least some techniques, if only to take into his possession and use the goods of nature. Man is a being of lacking ("Mängelwesen"), who for mere survival cannot but make use of his technological capabilities; he does not possess the same natural equipment that enables animals to live immediately on natural resources; the advantage of man and the condition of continuous progress in comparison to the rather firmly fixed animal behavior are man's reason and his capability of verbal communication, which allow him to pass the tradition of what has once been learnt from generation to generation.

Initially the needs of securing his existence, then his wants that arose from the knowledge of possibilities of utilizing nature and natural forces, and finally the development of new possibilities urged by more developed wants and demands, pushed man to develop technology to such a degree and extent that it became like a new body for him, a body without which he is unable to survive in this world. (An extreme parallel can be seen in sick people whose anti-infection system is too weak or not functioning at all so that they constantly have to live in a sort of protective tent in order to survive at all.) It was Occidental rationalism that made possible such a formation and construction of the world by man, that now the whole earth bears its sign. In this century we can speak of a global civilization and of a beginning of a world society that has become possible by universal communication, both material and intellectual. It is not only technological potential that becomes universally communicable and available; the same is true about the universal articulation of needs and wants: every member of mankind claims to have the same right to lead a life worthy of a human being, and this to the same degree and extent.

We know that economic connections and technological possibilities have brought about world-wide information and communication, which leads to a quick global implementation of new technologies. "Technology is ... not only part of special cultures, but it also contributes essentially to the constitution of a modern culture in the entire world.... Science

Paul T. Durbin and Friedrich Rapp (eds.), Philosophy and Technology, 49–57.
Copyright © 1983 by D. Reidel Publishing Company.

and technology are a common heritage of mankind. Technology is the basis of intercultural communication and comprehension. Modern means of transport and communication have brought the different cultures closer together, have made our world smaller."[1]

Our entire life is embodied in a technical surrounding and depends on it: "Technology penetrates our whole life and is indispensable for assuring our future."[2] "Technology must be a means to make sure that we can lead a life worthy of a human being and a means of improving the quality of our lives. Thus human, cultural, social progress becomes the regulative factor for our technological decisions and actions. This true progress is not possible without or in opposition to technology; it can only be achieved by a technology that has been changed in several aspects."[3] "Science and technology thus can first create the presuppositions for the realization of human rights; and only by means of science and technology are we able to observe the situation in different countries."[4]

"In spite of a lot of ideological effort, no political system will now be able to escape uncensored electronic communication forever. Our world has become small and observable. The formula of defense, 'non-interference, non-intervention,' which served to hide all the violations of human rights throughout history, has become obsolete through technology. . . . Without exaggerating one may say that human rights have formed an existential union with technology."[5] "In that sense human rights and technology are united in one"; both can only be realized together.[6]

It is only since technology has become a global phenomenon that a universal society or community of communication is developing (which surely does not yet correspond exactly to the ideal community of communication according to the ideas of Habermas and Apel), which nobody can evade forever and to which everybody has to respond for justification. Technology assures the material basis of this universal community of communication and also presents important aspects of the intellectual discussion about what belongs to the "true" needs and the inalienable claims of man and of the human society.[7]

The primordial human right is the right to life and assurance of life. This basis of all the other rights already shows the close relation to technology. The technological equipment of medicine contributes to saving many human lives which otherwise would not overcome critical periods. But this very basic right also shows the other side of technology, which brings about a qualitative and quantitative increase of the potential of killing and destroying parallel to the desired progress.

Since the determination of goals for technology does not belong to technology in the strict sense, but is prior to it (this is logically correct even when an existing technical object or a known process is combined with a new purpose), the ethics of human rights is decisive for the use of existing technology and for its further development.

This is recognized by representative organizations of scientists and engineers. Scientists of many nations expressly acknowledge that their work is bound to serve human rights. Thus until the early seventies, the Association of German Engineers (VDI – Verein Deutscher Ingenieure) required a "profession of the engineer" (the last formulation of which dates from May 1950) of every new member. Here we read: "The engineer has to fulfill his professional work in the service of mankind. . . . The engineer has to work with respect to the dignity of human life and in fulfillment of his service to his neighbor. . . . The engineer may not give way to those who do not fully respect the rights of a human being and who abuse the true essence of technology; he must be a faithful collaborator of human morality and culture. . . . " It was the same motivation that led to the foundation of a centre of work and research in the domain of technology assessment, or as it is called in a broader sense "evaluation of technology."

The Carmel declaration on technology and moral responsibility of December 1974 is another example; it says: "Every technological enterprise has to recognize the fundamental human rights and has to respect the dignity of man. . . . Every new technological development has to be judged according to whether or not it contributes to the development of man as a truly free and creative person. . . . Absolute preference should be given to the diminution of human suffering, to the extinction of hunger and sickness, to the fight against social injustice and to the struggle for lasting peace."

By human rights we understand the rights that every human being has by birth and which cannot be given or taken away, which belong to every individual independently of recognition by the public authority and of his nationality simply because he is a human being.[8]

While civil rights in their concrete formulation are given or recognized by an institution that is above the individual, human rights exist by the fact that a human being exists. To speak of human rights therefore means to accept the existence of rights that are given by the nature of a being; it means to recognize natural law or natural rights. There is something like a nature common to all human beings which has to be recognized by everybody who wants to keep and protect his own nature as a good or property that nobody is allowed to attack or destroy. In different formulations this natural rights claim has been expressed in many national constitutions and fundamental laws as well as in international declarations and treaties.

One presupposition of this development was the acknowledgment of the fact that all human beings as such are fundamentally equal. No longer are different functions in a society understood as belonging to different natures, as was the case with Aristotle and still in the Middle Ages.

In history, this is most evident in the case of the human right of freedom. Aristotle distinguished between citizens, who were born free, and others, who were born as slaves – and he considered this fact a natural or essential expression of being – and of obligation ("is" means "ought"). In Christianity all human beings were equally free – before God, or more precisely, only before God. The French Revolution made the fundamentally equal freedom of everybody a political claim, which especially in our century has led to concrete formulations in international law declarations, regulations and treaties. One bench mark may be seen in the Nürnberg War Crimes Tribunal, where people who had obeyed the positive laws of their country were condemned with international approval, and the condemnation could be justified only by claiming unwritten rights and obligations based on human nature as such.[9]

From the philosophical tradition, Hegel can be cited as one who declared freedom the substance of law, and who saw in the system of law the "realm of realized freedom" in which every man is a person and respects all others as persons. Everything that is external to this person or to this fundamental freedom, Hegel considers as a thing, and he lists "my body, my life" among these things.[10] But it seems that Hegel did not yet take as human everybody whom we now consider as a human being; he certainly did not accept that women have equal rights with men.

Some of Hegel's remarks may show his understanding of natural law: "To state the absolute injustice of slavery means to maintain the idea of man as mind (*Geist*), which is free of itself; this idea takes man as free by his very nature."[11] "All the goods or rather the substantial determinations which constitute my very person and the most general essence of my consciousness of myself – like my being a person in general, my general freedom of will, my morality, and religion – are inalienable."[12] "All the other determinations of our existence, age, profession, etc., are transitory and changeable; only the determination of freedom remains. It is my most inner being, my essence, my category, that I cannot be a slave; slavery is a contradiction of my consciousness."[13] "I am free only as far as I state the freedom of others and as far as I am acknowledged as being free by others. Real freedom presupposes that many are free. Freedom is only real if it exists in a plurality of persons."[14] "In the oriental world there was only one free being. . . ; in the Greek world only a few were free. It is a completely different situation for us to say now: man as such is free. Here the determination of freedom is quite general. The subject as such is thought to be free; and this determination is valid for everybody."[15] "Man's substance is freedom."[16]

This is a historical attainment from which there is no way back; historical processes of apprehension and culture are irreversible, since they have proved to be the way to the true "nature"; this way became possible only through the progress of science and technology.[17]

In the socio-anthropological interpretation of the Hegelian dialectics of master and slave, Marxism declares the realization of the principle of freedom for everybody to be the intended end of the natural direction of the development of society. Thus H. Hörz can define Marxist humanism as the theoretical and practical formation of society, by use of the possibilities of the economic and technical basis, for the purpose of bringing about the highest possible degree of freedom for freely associated persons.[18]

Realization of this freedom has become a possibility for everybody in modern times because it is only now that, through technology, everybody acquires the material, economic, and technical potential to make free use of his capabilities. But it is also true, now, that while, on one hand, we cannot think of the realization of human rights for an important part of the world population without the help of science and technology, on the other hand science and technology could lead to a total collapse if they are not directed by the ethics of human rights.[19]

Above all in socialist programs the complete development of the personality is clearly proclaimed as the aim and right of all human beings. In fact, however, it seems that the so-called "capitalist" societies are still ahead in this regard; one cause certainly lies in the better developed economic and technical conditions.

Without being too optimistic, we can now register universal consensus about the necessity of specifying human rights as the implementation of the goals of technological development. But so far this has only been acknowledged as a guiding principle, which means that the acknowledgment remains rather abstract.

A good example again is presented by the Association of German Engineers, which is the largest scientific and technical professional organization in Europe. Here, the leitmotif of the engineer's work is expressed as follows: "The aim of all engineers is the improvement of the standard of living of all mankind by the development and sensible use of technological means." It would not be right to expect technocratic proposals here; there are no purely technical solutions to essentially ethical questions and problems: "We refuse technocratic ideas of society." "Engineers are capable of showing what is technologically possible by the development of technology; and for that we can take responsibility." "Engineers must take into account the dimensions of societal aims and claims, together with the merely technological solutions, when they fulfill their task."[20] That this is quite difficult in practice can clearly be seen in the case of information technology, which contains the problem of procedural safeguards to protect personal data from abuse.

It is therefore necessary to use all the possibilities of steering technological progress by rational social means, to make them known to the public, to work for their acceptance in the process of the formation of the public opinion, and to create the proper institutions for their realization. There can be no doubt about what has to be the first aim that should lead this effort. This is the ideal of humanity, of a life worthy of man for all mankind. This ideal can be found in all the accepted value systems, although in different concrete interpretations.

A negative description of this ideal might point out that a human world should contain as little pain, poverty, and privation, as little violence and aggression, and as little restriction of personal freedom as possible.

A positive description of humanity will have to list quite a series of notions (which in addition are interpreted and given concrete applications in different ways in different cultures): freedom, peace, justice, equality of rights, health, sufficient nutrition, safeguards for the private sphere of the person, protection of life, right to material, physical, emotional, and intellectual enjoyment and satisfaction, free exercise of religion, etc. – with this list neither providing a hierarchy of values nor claiming to be exhaustive or complete.

It is not without presuppositions that one can seek to achieve global consensus about the values that should guide the progress of technology, although these presuppositions are not always clearly perceived or expressed. Such presuppositions might include every individual's will to survive, or the conviction of the value of mankind's surviving; or also, fear of ruin, reverence for life, Christian love of neighbor, and all kinds of manifest or hidden interests, hopes, and fears of the sort that are expressed in the traditional moral rule: *Quod tibi fieri non vis, id alteri ne feceris* (Do not do to your neighbor what you would not want him to do to you).[21] "Technological science which aims at changing the world is justified by the service that it renders to the individual and to mankind. We cannot say that progress has gone too far as long as many people, even whole nations, live in conditions that are humiliating and unworthy of man, but which can be improved by means of scientific and technological knowledge" (Pope John Paul II). Here there are still great tasks awaiting us, especially in our technologically highly developed societies, which can no longer maintain the exclusivity of their advantages.

Marxism is right in emphasizing the "societal determination of technology"; but this also means that technology has to take its orientation from human rights, since they are the limits of self-determination of different societies. This question nowadays is discussed under the title of "scientific-technological progress and humanism" and the dialogue crosses political and ideological borders.[22]

Personally, I continue to defend a theory of convergence which I do not consider utopian.[23] Even Karl Marx thinks that naturalism or humanism is the truth that can unite idealism and materialism.[24] The ideal of humanism can only be formulated in an abstract way, and there are different levels of abstraction. Concrete consequences of these abstract formulations of human rights depend on historical conditions and their application remains a continuing task for every moment of history. These concretions cannot be isolated from the given historical social-economic-political-technological context or system.

In the process of uniting the development of technology with human rights claims, the decisive question seems to be the institutionalization of means for the effective realization of these human rights. Both Heraclitus and Hegel are right, when they see war, not peace, at the beginning of mankind, in man's "natural" and undeveloped state. Our highly developed cultures are very much products of our own formation of the world; our nature thus is culture. Man can only fulfill his nature as a cultural being, as has been explained by Ortega y Gasset and Gehlen; it is only in culture that man reaches the peak of his true nature.

Societal and cultural conditions constitute the framework in which science and technology develop; here, therefore, the means and institutions have to be created, which bind the progress of science and technology to human rights. But this binding must always be open to a new future since the concretion of the contents of human rights claims can and must change together with the historically increasing possibilities which are due in part to scientific and technological progress. We must create and safeguard possibilities for claiming and realizing human rights and assuring their observance.

The minimum moral obligation of every scientist and engineer includes the disapproval of silent acceptance of the application of scientific knowledge for suppression and exploitation of man, for the destruction of conditions of human life and of life in general.[25]

Further discussion is needed on one problem: What kind of institution is best suited to ensure the regulative and guiding role of human rights for science and technology? For this we need interdisciplinary research on technology, and such research is unfortunately quite underdeveloped – one reason for this poor state being the lack of institutional means. Another is the fact that the cultural sciences concerned are far from presenting the reasonable means of moral guidance, at the level of instrumental reason, achieved in technology.

University of Düsseldorf

NOTES

[1] VDI [Association of German Engineers] Study Group, *Der Ingenieur in Beruf und Gesellschaft* (ed.), *VDI: Zukünftige Aufgaben* (Düsseldorf, 1980), p. 29.
[2] *Ibid.*, p. 20.
[3] *Ibid.*, p. 34.
[4] W. Kraus, "Menschenrechte und Technik," *Neue Zürcher Zeitung*, 30 August, 1980.
[5] H. Höcherl, "Menschenrechte und Technik sind eng miteinander verbunden," *Ingenieur-Karriere, Sonderteil der VDI-Nachrichten*, 44 (31 October, 1980), p. 50.
[6] W. Kraus (see note 4, above).
[7] On the discussion of needs, see S. Moser, G. Ropohl, and W. C. Zimmerli (eds.), *Die "wahren" Bedürfnisse oder: wissen wir, was wir brauchen*? (Basel and Stuttgart, 1978; includes bibliography); K. Lederer and R. Mackensen, *Gesellschaftliche Bedürfnislagen* (Göttingen, 1975); P. Kmieciak, *Wertstrukturen und Wertwandel in der Bundesrepublik Deutschland* (Göttingen, 1976); K. M. Meyer-Abich and D. Birnbacher, *Was braucht der Mensch, um glücklich zu sein? Bedürfnisforschung und Konsumkritik* (Munich, 1979; includes bibliography).
[8] Definition from the Brockhaus Encyclopedia, vol. 12 (1971), p. 411; see also W. Heidelmeyer (ed.), *Die Menschenrechte: Erklärungen, Verfassungsartikel, Internationale Abkommen* (Paderborn, 1972).
[9] The text here should have made clear, with respect to the notion of nature, that this idea does not refer to an unchanging static fact; it includes the historicity of human *praxis* in its evolutionary – or even revolutionary – development. Therefore a nature-based or natural law ethics does not fall under the verdict of dogmatic foundationalism; it is compatible with the moderate relativism of many American pragmatists – whose view I think I may be allowed to interpret as a kind of realism or naturalism.
[10] G. W. F. Hegel, *Grundlinien der Philosophie des Rechts* (Hamburg, 1955), § 4 (p. 28), § 36 (p. 52), and § 40 (p. 54).

[11] *Ibid.*, § 57 (p. 66).
[12] *Ibid.*, § 66 (p. 72).
[13] Hegel, *Einleitung in die Geschichte der Philosophie* (Hamburg, 1959), p. 234.
[14] *Ibid.*
[15] *Ibid.*, p. 245.
[16] *Ibid.*, p. 106.
[17] Cf., on this, F. Böckle, "Resultate der Moralphilosophie," in S. Moser and A. Huning (eds.), *Werte und Wertordnungen in Technik und Gesellschaft* (2d ed.; Düsseldorf, 1978), p. 107.
[18] See my "Ein Ost-West-Dialog," in *VDI-Nachrichten*, 4 (23 January, 1981), p. 17. Hörz has since published an enlarged version of his contribution to the dialogue: "Wissenschaftlich-technischer Fortschritt und sozialistischer Humanismus," *Deutsche Zeitschrift für Philosophie* 29 3–4 (1981); 343–356.
[19] W. Kraus (see note 4, above).
[20] VDI (see note 1, above, pp. 3, 20, 5, and 6).
[21] See A. Huning, "Zur Frage der Begründung sittlicher Werte bei Bonaventura," in I. Venderheyden (ed.), *Bonaventura: Studien zu seiner Wirkungsgeschichte* (Werl, 1976), pp. 104–112.
[22] Examples are the Summer conferences involving the so-called "*praxis* philosophers" in Yugoslavia and the symposia at Deutschlandsberg in Austria. See also S. Wollgast and G. Banse, *Philosophie und Technik: Zur Geschichte und Kritik, zu den Voraussetzungen und Funktionen bürgerlicher "Technikphilosophie"* (Berlin, 1979), p. 18.
[23] This puts me at odds with Wollgast and Banse (see previous note), p. 236.
[24] K. Marx, *Kritik der Hegelschen Dialektik und Philosophie überhaupt* (Nationalökonomie und Philosophie Ms. 6); see I. Fetscher (ed.), K. Marx and F. Engels, *Studienausgabe*, 4 vols. (Frankfurt, 1966), vol. 1, p. 70.
[25] H. Hörz, "Naturerkenntnis und Ethik," *Deutsche Zeitschrift für Philosophie*, special issue, "Mensch, Wissenschaft und Technik im Sozialismus" (1973), pp. 84–103 (see p. 98).

ALEX C. MICHALOS

TECHNOLOGY ASSESSMENT, FACTS AND VALUES*

1. INTRODUCTION

The central thesis of this paper may be put thus: There are good theoretical and practical reasons to avoid any appeals to a fact-value distinction in the assessment of technology. Alternatively, one could say that in the assessment of technology any appeals to a fact-value distinction will be at least useless and at most dangerous and self-defeating.

The structure of my defense of this thesis is as follows. In Section 2 I draw an ontological distinction between facts and values, and in Section 3 an epistemological distinction between facts (truths) and falsehoods is drawn. In Section 4 I present five illustrations of types of arguments that are frequently used to draw erroneous conclusions as a result of confusing the two distinctions drawn in Sections 2 and 3. In Section 5 I consider possible, but by no means necessary, methodological advantages of distinguishing facts and values (i.e., making the ontological distinction). Then I briefly outline six theories of value (in Section 6) in order to show that one cannot simply talk about values as if everyone meant the same thing. The idea of a value, like the idea of a fact, is a theoretical idea, and a very controversial theoretical idea at that. In the light of the discussion in Sections 5 and 6, I believe it is fair to say (in Section 7) that appeals to a fact-value distinction in the assessment of any technology will be useless because of the controversial status of all theories of facts and values. What's more, such appeals will be practically or strategically dangerous or self-defeating because (a) the confusion in Section 4 will probably not be avoided; (b) the appeals usually involve unwarranted appeals to authority (credentialing); (c) the appeals encourage belief in an epistemological handicap for all evaluations which is only warranted with some theories of value; (d) the appeals encourage belief in an epistemological robustness for all alleged factual judgments which is only warranted for some; (e) the appeals encourage unwarranted self-certification of allegedly factual claims. For all these reasons, then, I recommend that in all decision-making, including all decision-making involving technology assessment or any policy-making, one should avoid any appeals to an (ontological) fact-value distinction.

Paul T. Durbin and Friedrich Rapp (eds.), Philosophy and Technology, 59–81.
Copyright © 1983 *by D. Reidel Publishing Company.*

2. FACT AND VALUE: ONTOLOGY

Sometimes people talk about questions of fact, factual issues, or factual matters in order to classify the questions, issues, or matters prior to attempting any answers or other assessments. For example, among all the issues related to the development and use of energy from alternative sources (e.g., coal, oil, gas, uranium, water, biomass, wind, etc.) there are supposed to be some straightforward factual matters such as how much energy is obtainable from a given amount of one of these sources, using a certain technology in a certain period of time. Even those of us who don't know the answers to such questions are willing to grant that these sorts of questions are factual, if indeed *any* questions are factual.

To say that a question is factual, involves an issue or matter of fact, is obviously not to say that the question is true. It is logically impossible for questions to have truth values, to be true or false. Normally, but not always, the same is true of exclamations (e.g., "Whoopie!") and commands (e.g., "Close the door.") I say "normally, but not always" because exclamations like "Fire!" may be regarded as doing at least double and possibly triple duty. One who yells "Fire!" may be regarded as asserting that there is a fire, expressing fear and deep concern, and urging listeners to do something – run for their lives, call the fire department or whatever. Only a linguistic purist would insist that "Fire!" cannot do all these jobs at once, and I doubt that many people would believe the purist.

It is not some sort of defect in questions, exclamations, and commands that they don't or normally don't have truth values. They have their own unique linguistic jobs to do. Declarative sentences do have truth values, and this is their distinguishing feature. Nevertheless, to say that a declarative sentence involves a factual matter or an issue of fact is not to say that the sentence is true. While it is logically possible for such a sentence to be true, saying that the sentence involves an issue of fact is not the same as saying that it's true. It is merely to classify the ontological category (ball park, pigeon hole) of the sentence. For example, the sentence "Canada has seven operating nuclear reactors" involves a factual matter but the sentence may be false. Similarly it may be false that the United States has fifty-four nuclear reactors, although the issue of the existence or non-existence of the reactors is a factual issue.

At least since Hempel's classic paper "Problems and Changes in the Empiricist Criterion of Meaning" in 1950 it has been known that there is not and is not likely to be any generally acceptable criterion of empirical meaningfulness. In other words, there is not and is not likely to be any rigorous criterion to apply that will allow one to decisively categorize sentences as empirically factual, sentences about matters of empirical fact or, briefly, factual sentences. We need not review Hempel's review of the demise of the verifiability, falsifiability, confirmability, etc., criteria. Before most of us, Hempel too was a teenage logical positivist. It troubled him as much as anyone to discover that nothing worked, that finally we had intuitions and hunches separating factual matters from metaphysics, theology, ethics and other things that frightened logical positivists. Still, that's where that story ends. You will not see a criterion of empirical fact. You will only see examples of sentences about empirical (sometimes logical) facts or factual matters. Here, as elsewhere, we can often agree on what or what not to do without being able to agree on why or reasons for doing it.

One may also talk about questions of value, evaluative issues, or evaluative matters in order to classify the questions, issues, or matters prior to attempting any answers or other assessments. For example, one may raise the question, "Is thermal energy from nuclear fission or burning coal more desirable from the point of view of site attractiveness?" That is, are coal-burning facilities more or less esthetically attractive than nuclear reactors?

Again, to say that a question is evaluative, involves an issue or matter of value, is obviously not to say that the question is true. It is merely to classify the ontological category of the question. A declarative sentence like "Nuclear reactors are esthetically more attractive than coal furnaces" involves an evaluative matter, but again the truth or falsity of the sentence doesn't follow from the mere fact that it involves an evaluative issue. Some people believe that such sentences are pseudo claims, that although the sentence is formally declarative, it cannot function as a declarative sentence (cannot have a truth value at all). However, even on this extreme view, to say that a sentence is evaluative, involves a matter of value or an evaluative issue, is still not to say that the sentence is true or false. Indeed, on this extreme view, it is logically impossible for the sentence to be true or false.

In short, according to some common usage, issues of fact may be contrasted with issues of value in a purely ontological classificatory sense. In this sense, truth or falsity is not at issue. Classifying a matter in this sense is not asserting the truth or falsity of the matter. Classifying an issue in this sense is logically prior to the question of truth or falsity, and the latter question may not even arise. For example, the question will not arise if one is only interested in getting a question classified as factual or evaluative, and the question will not arise when an evaluative declarative sentence is asserted and its assessor holds the extreme view that such sorts of sentences cannot have truth values.

3. FACT AND FALSEHOOD: EPISTEMOLOGY

Sometimes people use the term "fact" and its cognates as a synonym for "true." For example, people say things like "As a matter of fact there are several operating nuclear reactors in Ontario"; "It is a fact that there are nuclear reactors in Ontario"; and "In fact there are nuclear reactors in Ontario." The last sentence, for example, just means "In truth there are nuclear reactors in Ontario." The two before that may be expressed as "It is true that there are nuclear reactors in Ontario." Such usage of the term "fact" is not ontological but epistemological. The term is used not to designate an ontological category, but to designate an epistemological assessment or appraisal.

In the epistemological usage of "fact" it would be self-contradictory to talk about factual matters that were false. Factual matters in this epistemological sense are true, i.e., the sentences about the matters are true. For example, it would be self-contradictory for someone to say "In fact there are nuclear reactors in Ontario and it's false that there are nuclear reactors in Ontario." If in fact there are nuclear reactors in Ontario then it is true that there are such reactors there and it cannot also be false. Similarly, if in fact John is five feet tall then it would be self-contradictory to say "In fact John is five feet tall but he's not five feet tall."

Thus, in the epistemological sense of "fact," facts are contrasted with falsehoods, or, more precisely, factual sentences are contrasted with false sentences because "factual" is regarded as synonymous with "true." In this sense of "fact" to say "In fact nuclear reactors are visually more attractive than coal furnaces" is to affirm the truth of the claim that nuclear reactors are visually more attractive than coal furnaces. Clearly, those who believe that evaluative sentences cannot have truth values would insist that a sentence like "In fact nuclear reactors are visually more attractive than coal furnaces" must be self-contradictory. That is, if "in fact" is regarded as synonymous with "in truth" and evaluative sentences cannot be true or false, then it is self-contradictory to claim that it's true that nuclear reactors are more attractive than coal furnaces *and* it's not true or false. If it's not true or false, then it's not true.

It is the epistemological sense of "fact" that is intended when people talk about gathering relevant facts or undertaking a fact-finding investigation.

If, for example, a journalist is sent out to dig up all the facts he can find about Cleveland's mayor, no one will expect the journalist to return with old shoes, chairs, and hubcaps. Nor would anyone expect to see the journalist bring a pick and shovel to work. What would be expected is a search for true relevant propositions or sentences about the mayor. In an ontological sense of the term, the mayor's shoes, chairs, and auto hubcaps are facts in his life, literally the furniture of his world. But gathering the facts about his life does not imply gathering up such furniture. (Anyone who wants to make something out of the difference between "facts in" and "facts about" is free to do so.)

It would be nice if we had a generally acceptable criterion of truth. There seems to be general agreement that Tarski's formal definition is sound: i.e., a sentence represented by "p" is true if and only if p; e.g., "John is sick" is a true sentence if and only if John is sick. As a meaning analysis of a common notion of truth, this seems to be unexceptionable. But it's obviously worthless as a criterion for determining which sentences are true and which are false. From the meaning analysis we don't find out how to prove that John is or isn't sick. (The problem is roughly analogous to knowing what it means to bake a cake, but not knowing how to do it.) To solve the criteriological problem of truth, I think one must appeal to some notion of coherence. Nicholas Rescher's *The Coherence Theory of Truth* is the most recent attempt to carry out such a program. It's not the sort of program one can expect to just carry out as one carries out the garbage. However, Rescher has taken significant steps forward.

4. FROM ONTOLOGY TO EPISTEMOLOGY

Given the two common ways of talking about factual matters, an ontological way in which facts are contrasted with values and an epistemological way in which facts are contrasted with falsehoods, the stage is set for unwarranted inferences. It is easy to move from the assumption that an issue is a factual matter in the ontological sense of the term "factual" to the quite different assumption that the issue is factual in the epistemological sense of the term. In such cases, one starts with an issue that is ontologically factual and converts it into an epistemological fact, i.e., a truth or true claim, or a falsehood. Whether the conversion is intentional or not, the fallacy of equivocation is committed. One logically cannot derive epistemological appraisals of truth or falsity from mere ontological categorization.

As Edgar Bergen might have asked Mortimer Snerd, "How could anyone be so stupid?" And as Mortimer might have replied, "Wal, it's a lot easier in a crowd!" The fact (truth) is that several common ways of talking and thinking (in English at least) conspire to drive people to equivocate on the ontological and epistemological uses of the term "fact." For example, it often happens that when a group meets to discuss a problem, say, the development of an energy source, one or more members think it will be helpful to share some factual matters. If there is a thin edge of the wedge leading to complete consensus in a group decision-making situation, so the assumption goes, it must be at or near the factual side of things. So, by all means, let us lay out some of the facts as soon as possible. The move from factual matters to facts here can be a move from ontology to epistemology. Because factual matters (ontology) tend to be relatively unproblematic (relative to evaluative matters), the assumption goes, and factual matters are going to be discussed first, facts (epistemology) are going to be introduced first. The mere description of a claim as factual in the ontological sense tends to soften the path toward its acceptance as factual in the epistemological sense.

As a second example of how the equivocation arises, consider the case of expert advice given by scientists. Suppose it is granted that there really are experts on matters of scientific fact, e.g., chemical facts, physical facts, economic facts, etc. At a minimum that means that there is a body of factual (ontology) claims about which so-called experts know more than non-experts. To be an expert in, say, chemistry is just to know more about the factual claims of chemistry than most people know. Apart from the question of the truth or falsity of any factual claim of chemistry (or any chemical issue or problem for chemistry or matter of chemistry), most people I guess would be willing to grant that it is usually possible to distinguish such claims from many others. If I say, for example, that sugar is water-soluble, no one is going to quarrel about describing that as a fact of chemistry rather than a fact of sociology, economics, or arithmetic. Granted that these disciplines have fuzzy edges and no knock-down criteria of identification, it's not likely that the water-solubility matter is going to be resisted as a matter of chemistry. However, because there are some factual matters that clearly are in the province of chemistry (an ontological point) and there are experts on such matters, it is possible for the expert chemist to claim to assert chemical facts in the epistemological sense of "fact" and get away with it. The *non sequitur* runs: Smith is a chemical expert. x is a matter of chemical fact (ontologically speaking). So, if Smith claims that x is a chemical fact (ontologically yet) then x is a fact (epistemologically speaking, true). In short, Smith's credentials are doing the job (invalidly) that Smith's evidence should be doing (validly). We often tell students of introductory logic courses to beware of people with credentials in one field making claims in another field. We should also tell them to beware of people with credentials using those credentials instead of evidence to warrant claims in the field where the credentials are held.

There is another way in which an equivocation on the term "fact" occurs related to judgments that something is a matter of scientific fact. When one judges that some issue is scientific or is a matter of science, one might base the judgment on one's knowledge that there is a body of scientific literature about it. For example, one might think that the question of the thermal energy of coal is a scientific question because there is a recognized body of physical science literature dealing with precisely this question and related questions. The same person might think that the question of the visual attractiveness of a coal furnace is not scientific because there is (as far as he knows) no recognized body of science dealing with precisely this and related questions. Now he can reason invalidly: Science deals with facts (ontology). There is no science of visual attractiveness. So, there are no scientific facts (epistemology) of visual attractiveness. Thus, whatever one says about visual attractiveness cannot have the status of a scientific fact (truth). Whatever claims about visual attractiveness are, they cannot be true.

Besides noticing that the argument just presented involves an equivocation, it ought to be noticed that the criterion of scientific significance used in the preceding paragraph is far from rigorous or decisive. There could be no new sciences if the criterion of scientific significance required every allegedly scientific claim to have a recognized body of scientific literature devoted to it already. The scientific enterprise could never have begun if that had been the criterion of selecting scientific claims. Nevertheless, I don't doubt at all that many people do use such a criterion as a rough and ready guide to judgments about what is scientific. Since I have already said that there are no generally accepted, decisive criteria of scientific significance, I have no objections to people using rough and ready criteria. I only object if people use such criteria to pronounce all evaluative claims as in principle unscientific and hence untrue. After all, in such cases the principle is known to be defective and the inference is invalid.

A fourth way in which an equivocation on the term "fact" occurs is related to a defective argument for distinguishing empirical facts from values. It is sometimes argued that facts must be different from values because it is possible to get complete agreement on the facts of a situation (event, person, thing) but still have disagreement about its value. For example, we would be able to get complete agreement from virtually everyone watching me present this paper that I have a nose. The fact that I have a nose is literally as plain as the nose on my face. We could get agreement on its size, color, shape, physiological structure and so on. Nevertheless, the argument runs, it would still be possible to have disagreement about the physical attractiveness of my nose. Some people would say it's a nice nose, others that it's ugly, or plain, superb, classical Greco-Roman, and so on. In short, unlike the fact of my nose, the esthetic value of my nose is open to a variety of opinions. The fact (truth now) that I have a nose can be established beyond reasonable doubt, but the esthetic value of my nose remains doubtful forever. The preceding sentence is crucial. What is being contrasted in that sentence is not two ontologically distinct types of things. Rather, it is the epistemological status of factual claims that is being contrasted with evaluative claims. The issue is now one of truth versus something less than the truth.

To appreciate the weakness of the argument in the preceding paragraph, consider the following analogue. Values must be different from facts because it is possible to get complete agreement on the value of something but still have disagreement about factual aspects of it. For example, long before there was agreement on the facts about things like rainbows, sunsets, mountains, waterfalls and clouds, most people found such things not only attractive but often awe-inspiring. It would have been possible for two ancient Greeks (or two contemporary folks) to agree that a particular rainbow was beautiful while disagreeing on its origin, color, size, and shape. Anyone familiar with the history of science can probably multiply such cases easily. Anyone at all can probably think of many more cases in which complete agreement in evaluative issues comes at least as easily as agreement on facts. For example, we would be able to get complete agreement from virtually everyone watching me present this paper that I don't have a moral obligation to cut off my nose during the presentation. Similarly, nobody in his right mind would think that I would be physically more attractive at this minute with a bloody hole in the middle of my face instead of a nose. Nobody thinks I would look pretty with three noses, or four, and so on. Clearly there are as many evaluations or judgments of value as there are judgments of fact about which one can get virtually complete agreement. In principle there must be an infinite number; e.g., everyone will agree that I don't have two noses, three, etc., and everyone will agree that I would not be prettier with two noses, three, etc. There does not seem to be any ontological lesson to be learned about facts and values from counting heads regarding things people agree or disagree on readily. With a little thought one can cook examples to suit oneself.

While I am on this subject of agreement, I should take a few lines to address the frequently made assumption that if people cannot readily identify a good reason for some judgment, agreement or disagreement, then there must be an evaluation involved. In brief, people often argue: There is apparently no good reason to assert that p is the case. Therefore, the assertion that p is the case must involve an evaluation, a value, a value judgment or some such thing.

Instead of leaping to such a conclusion, one should at least consider the following seven other possibilities. The judgment that p is the case may involve (a) a perception that is not clearly identifiable; (b) an idea that is vague or open-textured; (c) principles or rules of inference that are obscure; (d) so many complicating factors to aggregate that, although one can intuit an appropriate judgment, one cannot clearly support it; (e) some premises that one cannot identify; (f) a mistake about some rule or premise; (g) a disagreement about facts difficult to discern.

A fifth and final way in which an equivocation on the term "fact" occurs is related to standard textbook treatments of rational decision-making or action on the Bayesian or estimated utility model. Briefly, according to this view decision-makers are confronted with a set of mutually exclusive and exhaustive possible future states of affairs and courses of action. Each combination of state of affairs and action yields an event with a particular value or utility. By quantifying these values, multiplying these quantities by the probabilities of the events that may occur and summing the products of every event attached to every action, one can obtain the estimated value of performing that action. A rational actor then is supposed to perform the available action that has the highest estimated value.

What seems to be built into the Bayesian procedure as just described is the assumption of a fact-value dichotomy in the form of a probability-value dichotomy. Rational decision-making and action on this model requires an amalgamation of two apparently ontologically distinct kinds of entities, namely, values and probabilistic facts or factual probabilities of some sort. But if probabilities have some sort of factual status and values do not, then there is a tendency to regard values as epistemological nonstarters. Probability assessments, the assumption goes, may be more or less accurate or true, but evaluations are simply given or not. The latter is especially the case when the values used in the model are based on revealed preferences which are uncritically given and accepted. Again, then, we have the unwarranted move from an ontological to an epistemological use of "fact."

On top of the objectionable equivocation just described, it should be noticed that it is a mistake to assume that probability assessments do not require evaluations and that evaluations do not involve probability (factual) assessments. I have discussed these connections in other places and will therefore be very brief. In order to measure any probabilities one must make decisions concerning one's universe of discourse, its appropriate description, the appropriate subset of the total universe to be used for any given problem situation, and the appropriate methods to be used to make all these decisions. These and other decisions involve assessments of the benefits and costs attached to alternative decisions; i.e., the decisions involve evaluations. Similarly, the evaluations required on the other side of the Bayesian model require epistemological assessments. For example, in order to form a preference for apples over oranges one must know something about both. Without building some sort of information or knowledge component into evaluation, it becomes impossible to distinguish the proverbial Fool's Paradise from Real Paradise. Conceptually, the difference between the two is either that the fool does not have any good reasons for valuing or preferring his Paradise while someone in a Real Paradise would, or that there are good reasons for believing the Fool's Paradise is not what he thinks it is; it is not, that is, the Real Paradise he thinks it is. For example, heavy smokers who believe they can go on smoking without damaging their lungs can only have the pleasure of a Fool's Paradise. In short, uninformed evaluations are worthless.

5. FROM ONTOLOGY TO METHODOLOGY

I think that one of the reasons people believe that the so-called fact-value distinction is very important is that they have not distinguished the fact-value distinction from the fact-falsehood distinction. They believe that by determining the correct ontological status of an issue, as factual or evaluative, the truth status of the issue has also been settled. Given the particular views that some people have about evaluative issues (e.g., the people I characterized as having extreme views), evaluative sentences don't have any truth values. Hence, for people with such views, the ontological categorization of an issue as evaluative has the epistemological consequence that the concepts of truth and falsity will not be applicable at all. Against such extremists, it is pointless to try to establish the truth or falsity of an evaluative sentence. It would make as much sense as trying to establish the truth or falsity of a sneeze or a glass of beer. For these folks evaluative sentences, by definition, logically cannot have truth values. The sentences are epistemological non-starters and that's that.

If one is not an extremist on evaluative issues, then – unless one is a metaphysician interested in determining the ontological structure of the world – there is probably little of interest in the fact-value distinction. On the other hand, the fact-falsity, or better, the truth-falsity distinction is fundamental. If one can't distinguish truth from falsity, true sentences from false ones, then one runs the risk of eating stones, swimming in molten lead and making love to ostriches. Being able to distinguish truth from falsity really does have survival value in ways that being able to distinguish facts from values doesn't. So why fool around with the metaphysical question?

One answer is that it might be the case that if we got an issue correctly categorized ontologically, then we could address it with the right method. At any rate, we might be able to eliminate some methods as useless. For example, if one holds an extreme view of evaluative sentences, then once one knows that a sentence is evaluative, one knows that there is no method that can establish the truth or falsity of the sentence. This is indeed an extreme case, but it illustrates how one can use ontological knowledge to make methodological decisions. In this extreme case, the decision is simply to admit the inapplicability of every method. Given one's definition of evaluative sentences, no method is relevant to establishing their truth or falsity.

Suppose that you do not hold an extreme view. That is, suppose you believe that categorizing a sentence as factual or evaluative leaves the truth status of the sentence undecided but not undecidable. The question is: Can it help your epistemological task to know that the sentence involves an ontologically factual or evaluative claim? It might. It depends on the particular claim, your methodological resources (time, energy, strategy, money, logico-mathematical sophistication, etc.), your epistemological beliefs, and your value theory, to mention only a few salient variables. I wish there were a short answer. The only relevant short answer I can think of now is that correct ontological classification into the categories of matter of fact or value is no panacea. No one should imagine that a solution to the ontological-metaphysical problem would constitute a major epistemological breakthrough.

Consider some examples in support of the previous sentence. (1) The ultimate matter of the world is wave-like rather than particle-like. (2) Every innocent human infant should be tortured to death before the age of four months. (3) A human being wrote these sentences. (4) A morally good human being wrote these sentences. It seems to me, though this is not a knock-down demonstration, that sentence (1) is a factual claim whose truth is extremely difficult if not impossible to decide; (2) is an evaluative claim whose falsity will be granted by all but extremists and crazies; (3) is a factual claim as uncontroversial as (2); and (4) is more controversial than (3), less controversial than (2) and probably not as controversial as (1).

All I want to do with these four sentences (1)–(4) is illustrate and lend some support to my claim that solving the ontological status problem may leave one with enormous epistemological problems, depending on the particular claim to be assessed. I also said it depends on your methodological resources. For example, consider (5) It is wrong to torture all living babies to death and (6) The negation of a conjunction is logically equivalent to the disjunction of two negations. (5) is an evaluative sentence whose truth will be granted by virtually everyone who reads it, while (6) is a logical factual sentence (a sentence stating a logical fact or logical truth) whose truth will not be recognizable by everyone or even most people. Depending on one's training and methodological resources, the evaluative (5) and the factual (6) will be more or less easy to assess epistemologically.

Epistemological beliefs have a crucial role to play in determining how useful correct ontological categorization can be. I can't review the history of epistemological skepticism here. Suffice it to say that some folks believe there can be no certainty concerning the truth or falsity of empirical sentences. They may rest their case on so-called arguments from dreams, from illusions, from limited sense perception, from peculiar definitions of complete verification, and so on. If they are thoroughgoing skeptics, then there's no hope. They will regard empirical claims as inherently unjustifiable and hence doubtful. Alternatively, they may be less-than-thoroughgoing skeptics, holding that one might have complete certainty concerning empirical claims although such claims are inherently fallible. For example, they may say that sentences like "There's only one doctor in this town" can be known to be true with complete certainty, but they are still liable to be falsified. They are still fallible or corrigible but not doubtful. Although I find this view satisfactory, some folks don't. It depends on how one defines key epistemological terms like "knowledge," "certainty," "complete verification," and so on.

The last important variable I mentioned that determines how useful correct ontological categorization would be in determining the truth status of any sentence is one's theory of value. With some theories of value, the truth or falsity of evaluative sentences is a routine matter, provided that one is not a thoroughgoing epistemological skeptic. With other theories, as I have already suggested, "truth" and "falsity" just don't apply. In the next section I will review some basic alternative views of the nature of value.

6. THEORIES OF VALUE

Before I begin my review of value theories, a couple of preliminary remarks will be helpful. In the first place, one must distinguish *things that have value* from *the value that things have*. Although virtually anything may be regarded as a thing that has value (or, briefly, a thing of value or a valuable thing), many people (especially social scientists) seem to focus on things like personal freedom, political efficacy, honesty, and self-esteem as examples. That's okay, as long as one remembers that strictly speaking sticks and stones and telephones, and anything else, may be regarded as valuable things in some sense of "value."

Just as theories of molecules, electricity, and democracy have as their subject matter molecules, electricity, and democracy, respectively, so a theory of value has value as its subject matter. Just as in molecular theory one gets meaning analyses of basic terms, measurement procedures, working principles and lawlike statements, a theory of value may be expected to have a similar battery of features. I have reviewed alternative views of the structure of scientific theories elsewhere. So here it is enough to say that there are over half a dozen views about what scientific theories ought to look like, ranging from mere unrestricted universal generalizations to axiomatic systems with logical and empirical vocabularies, rules of formation and transformation, and so on. Depending on what one means by the terms "theory" and "science," there have been more or less fully developed scientific theories of value. Perhaps the most fully developed theory we have had is that of Ralph Barton Perry in his classic, *General Theory of Value* (1926).

The alternative views of the nature of value that will be outlined below should be regarded as central features of possible (and more or less plausible) theories of value. Generally speaking, I describe the sort of meaning analysis that might be adopted for each view, and briefly suggest some of its salient strengths and weaknesses. I can't, of course, provide a thorough analysis and appraisal of any theory. Moreover, I do not try to press my own preference for any of the views. For the points to be made in the following section, it does not matter what theory of value one adopts. It only matters that one appreciates the variety of theories available.

When people say that something is valuable or good, just how should their remarks be understood? For example, how should one interpret remarks like "Jesus was a good man," "Clean air is valuable," and "There is some value in friendship"? We could, of course, retreat to a handy dictionary and get a standard lexicographer's definition of the key terms "good," "value," "valuable," and so on. But that would be an inappropriate response to the question. What we want to have is not merely correct English usage, but a more or less systematic or theoretical account of the meanings of these key words, and an account of the things referred to by these words – if there are such things.

Generally speaking, seven distinct kinds of answers have been offered to our question, each one suggesting a different view of the nature, essential being, or *ontological status* of value. The names used by different authors to refer to these seven different views vary a great deal, to put it mildly. Hence, it is vitally important that as you read this section, you must try to get each view fixed in your mind without becoming dogmatic about the labels used for the view. You should also not be surprised to find some people combining some of these views or elements of some of them to obtain hybrid forms.

The relations between the seven views are illustrated in the following diagram, which will be discussed in the remaining paragraphs of this section.

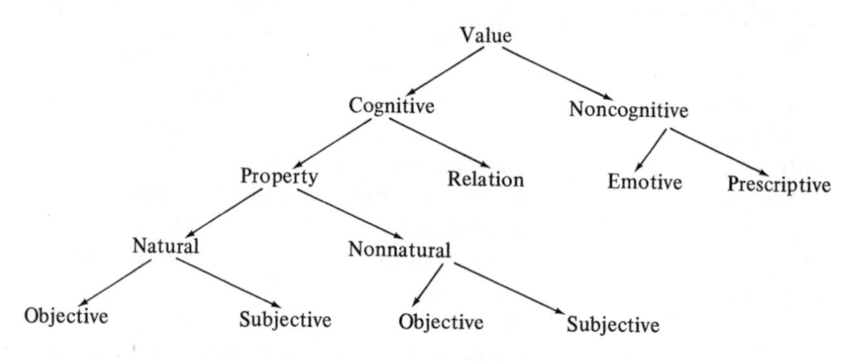

The first distinction to be drawn is that between cognitivism and non-cognitivism. In the *cognitivist* view the terms "value" and "good" designate something which is identifiable and namable, but not necessarily describable. Things that have value or goodness are thus similar to things that are yellow or sweet insofar as one can identify and name the qualities they possess, but cannot necessarily describe them in other words. In the *noncognitivist* view the terms "value" and "good" do not *designate* anything at all. Things do not have value or goodness on this view any more than "Such" is the name of something, or "Help me, please" describes any state of affairs.

As you can see from the diagram above, there are two species each of cognitivist and noncognitivist views of the nature of value. Since the division of noncognitivist views takes us to the *infimae species* or rockbottom of the righthand branch of our family tree, while the division of cognitivist views does not, I will consider the latter next.

Supposing that it is granted that "value" and "good" do designate something which is identifiable and namable, one might ask: Is value or goodness a property or a relation? From an ontological point of view, the difference between properties and relations is important. Consider, for example, the sentence "John is standing to the left of Frank." John and Frank do not seem to exist the way the relation designated by "is standing to the left of" exists. The *particular* relation seems to obtain (exist) when and only when John is standing to the left of Frank. Some people think that the *universal*, abstract entity, type or sort of relation designated by "is standing to the left of" has its own mode of existence, which Bertrand Russell called "subsistence" to distinguish it from ordinary existence.

I can't settle the old philosophical problem of the existence of abstract entities (Platonic forms, if you like – almost). But on a *relational* view, value may be thought of as a relation obtaining between at least two things (usually a person and a valued thing) *and* the relation may be regarded as a concrete particular or an abstract type of entity. I have not put two branches below "relation" on the diagram above in order to simplify the diagram. Strictly speaking, however, two branches could be drawn, one for the view that "value" designates particular relations and one for the view that "value" designates an abstract entity or universal relation. Similar divisions could be made for the view that "value" designates a property.

Whether value is regarded as particular or universal, insofar as it's regarded as a relation one will not be able to pick it out the way one picks out the terms of the relation. For example, suppose "Clean water is valuable" is taken to mean something like "There is a relation between clean water and people in virtue of which it is correct to say that clean water is valuable." We may be able to observe the clean water and the people (the two terms of the relation), but we can't observe the relation. Moreover, on this view, it is precisely that relation that provides the warrant for the attribution of value to clean water. Obviously, then, people who hold that value is a relation have special problems of characterizing the relation and giving criteria for identifying it. Here the ghost of David Hume looms large, because causal relations are prime candidates to figure in relational theories of value and Hume argued forcefully that such relations are nothing more than human artifacts, products of human imagination. Thus, relational theorists will have special problems establishing the truth or falsity of evaluative sentences. If they happen to be committed to abstract entities, they are even worse off. Or, turning things upside down, if one were an epistemological nominalist (believing there are no abstract entities) and if one became convinced that "value" designated an abstract entity, a relation-type, then one would be stuck with the view that value doesn't exist. For such people sentences like "Clean water is valuable" might be false or maybe meaningless, but certainly not true.

In contrast to a relational view, one might hold a quality or *property* view according to which "value" and "good" designate some sort of a property. On this view, value belongs to things roughly the way other properties belong to things, e.g., the way height, weight, triangularity, etc., belong to things.

Supposing it is granted that "value" and "good" designate a property of things that is identifiable and namable, one might then ask: Is value or goodness a natural property of the world or not? On the *naturalistic* view of value or goodness, this property or quality is as natural as water, mountains, headaches, and itches. Just as one does not have to design or make trees and toothaches, one does not have to design or make value or goodness. It just grows naturally, the way some tadpoles naturally become frogs and people naturally grow old, if all goes well.

On the *nonnatural* view ("supernatural" would do as well) of value or goodness, these qualities do not just emerge or evolve like flowers and snowflakes. They are somehow outside, above, or beyond the natural stream of things. They have roughly the same status relative to the world of nature that God and other supernatural spirits are supposed to have. One need not believe that there are any nonnatural or supernatural things in order to understand the point of view, I think. Nonnaturalists just happen to believe that value or goodness is not finally rooted in or grounded on anything in the physical, natural, or material universe.

There are two species of naturalism and nonnaturalism (supernaturalism) that merit attention. I will consider each in turn.

Naturalistic Objectivism

On what may be called a *naturalistic objectivist* view of the nature of value, to say that something is good or valuable is to say that something has some natural property in virtue of which the predicates "is good" or "is valuable" may be appropriately applied to it. For example, suppose one says that it is good to be kind to people, that kindness is worth a lot, or that kindness is a valuable characteristic for a person to possess. On the naturalistic objectivist view, such remarks have the status of genuine truth claims, i.e., they must be true or false. Moreover, they are true if and only if whatever is referred to by the subject of each remark has some property or attribute named by the predicate of the remark. Thus, for example, it *is* good to be kind to people if and only if kindness has some special property called "goodness." Similarly, Jesus was a good man, on this view, if and only if he possessed the particular property of being a good man. Accordingly, being good, having value or worth, on this view, is precisely analogous to being triangular or having the property of triangularity. Just as polygons can be triangular whether or not anyone knows or cares about it, someone or something can be good or valuable whether or not anyone knows or cares about it. Again, just as mountains can have snow and teeth can have cavities whether or not anyone knows or cares about it, someone or something can be good or valuable whether or not anyone knows or cares about it. Furthermore, it is true to say that something is triangular or good just in case that thing has the special properties named by "is triangular" and "is good," namely, the properties of being triangular and being good.

Naturalistic Subjectivism

On what may be called a *naturalistic subjectivist* view of the nature of value, to say that something is good or valuable is to say that somebody has some natural attribute, characteristic, or feeling in virtue of which it is appropriate to apply the predicates "is good" or "is valuable" to something. The term "somebody" might designate a single person, a small group, a culture, a nation, everyone in Western Civilization, or just human beings in general. As one moves from a single individual (i.e., naive egoism) to human beings in general, one's naturalistic subjectivism tends to become more plausible, but that is a long story that need not detain us now. The attributes, characteristics, or feelings possessed by the subject of one's naturalistic subjectivism might be, say, feelings of approval, satisfaction, pleasure, or favorable interest of some sort. Hence, for example, to say that kindness is worth a lot is to affirm something like the fact that Canadians approve of kindness, most people in the world approve of being kind, or that everyone you know approves of kindness. Supposing that one can recognize acts of kindness when one sees them, it would be true to say that they are good or valuable just in case Canadians do approve of them, most people approve of them, or everyone you know approves of them. It depends on how you specify the subject for your naturalistic subjectivism. If goodness or value is determined by, for example, ninety percent of Canadians having an empirically discoverable feeling of approval toward something, then that thing is good or valuable when and only when ninety percent of Canadians have such a feeling about it.

In short, then, naturalistic subjectivism and objectivism are alike insofar as both views hold that judgments of goodness, value, or worth are *genuine truth claims*, i.e., they are in principle determinably true or false. They are different views insofar as subjectivists locate the properties that justify evaluations in people while objectivists locate the justifying properties in whatever is claimed to be good or valuable. One way to keep the two views separate is to think of them as roughly analogous to two properties of sugar, namely, mass and sweetness. When one says that sugar has mass (ignoring relativity theory), one is talking directly about a property located in sugar. On the other hand, when one says that sugar is sweet, one is talking directly about the sensation of sweetness that is in people having the sensation, rather than in the sugar. "Sugar is sweet" is an indirect or roundabout way of referring to how sugar is perceived by people, while "Sugar has mass" is a direct way of referring to how sugar is. Just so with values. For naturalistic objectivists, "Sugar is valuable" is a direct way of referring to how sugar is, while for naturalistic subjectivists it's an indirect way of referring to how sugar is perceived or apprehended.

Nonnaturalistic Objectivism

On what may be called a *nonnaturalistic objectivist* view ("supernaturalistic objectivist" would do as well) of the nature of value, to say that something is good or valuable is to say that something has some nonnatural property in virtue of which the predicates "is good" or "is valuable" may be appropriately applied to it. The nonnaturalistic objectivist accepts everything that the naturalistic objectivist accepts about value judgments *except* their justification in natural phenomena. Things really do have goodness or value in them, but these are nonnatural or supernatural qualities inexplicably existing or, perhaps, existing by the will of God. The qualities do not just emerge like snowflakes and they are not natural the way my appetite for apple pie is natural. They are *something else*, and that's all there is to it!

Nonnaturalistic Subjectivism

On what may be called a *nonnaturalistic subjectivist view* of the nature of value, to say that something is good or valuable is to say that somebody has some nonnatural attribute in virtue of which it is appropriate to apply the predicates "is good" or "is valuable" to something. The nonnaturalistic subjectivist accepts everything that the naturalistic subjectivist accepts about value judgments *except* their justification in natural phenomena. It is the spiritual aspect of human beings that is the locus, ground, or origin of value in the world. Value exists insofar as people have feelings, attitudes, likes and dislikes, but human feelings, attitudes, likes and dislikes are not just natural phenomena like the pains or itches of cats and dogs. Human beings, on this view, have something in them that is not quite natural. Although the view does not logically entail theism, those who have spoken of the image of God in man have shared this view of the extraordinary status of human beings.

Again, nonnaturalistic objectivism and subjectivism are alike insofar as both views hold (1) that value judgments are genuine truth claims, and (2) that values are not grounded in natural phenomena. They are unlike with respect to where they locate the nonnatural properties that justify evaluations.

Before moving on to a discussion of the two kinds of non-cognitive views of value, some comments are in order about the strengths and weaknesses of the four views just considered. First, if objectivist views could be sustained then there would be a universal, objective, or absolute basis for evaluations. That would be marvelous indeed. It would mean, for example, that when people say things like "Abortion is wrong" or "Jesus was a good man" that anyone who doesn't agree with these judgments is seriously out of step with the structure of the world, either the natural or the supernatural world. Such a person would be mistaken in roughly the same way that someone who ate stones for nourishment would be mistaken. If one were a nonnaturalistic objectivist, then one would have special problems of characterizing the property designated by "value" and explaining how natural beings can have access to it. If one were a naturalistic objectivist, then one would have special problems explaining the fact that different people at different places or the same people at different times have quite different views about what is or is not valuable. (Recall the discussion in section 4 above.)

If subjectivist views could be sustained, life could be better or worse, depending on the views. Naive egoism makes every person his or her own court of last appeal, which would probably make life lonely, nasty, brutish, and short. It is also a consequence of this view that two people would not be affirming the same proposition if each judges that something is good or valuable, and they would not be affirming contradictory propositions if one judged that something was valuable while the other judged that it was not valuable.

G. E. Moore's famous "open question argument" was intended to undermine all naturalistic theories of value, but it could also be used against all forms of subjectivism. In either case, as Frankena showed some years ago, the argument is question-begging.

Emotivism

On what may be called an *emotivist* view of the nature of value, to say that something is good or valuable is not to affirm anything at all, but merely to express a feeling or attitude. Emotivists depart radically from all cognitivists because on the emotivist view assertions of goodness, value, or worth are not genuine truth claims, i.e., they cannot (logically) be true or false. Moreover, they cannot be true or false because they have the logical status of exclamations, not of declarative sentences.

Let us consider the view more carefully with the help of an analogy. Suppose I step on your toe. You might say "You're hurting me," or "Ouch," or you might just pull your foot out from under mine. If you just pull your foot back then you have made no claim that might be true or false. Similarly, if you just scream "Ouch!" or "Eeaah!" you have made no claim. On the other hand, the declarative sentence "You're hurting me" must be true or false, though for our purposes it does not matter which. What the emotivist view amounts to then, is just this: Sentences that appear *formally* as declaratives in which some property, namely, goodness or value, is being affirmed of something are *functionally* exclamations in which nothing at all is being *affirmed*, though something is being *expressed*. Thus, for example, when one says "Clean air is valuable," one should be understood exactly as if one were exclaiming "Hooray for clean air!" or "Whoopie! Clean air!" – neither of which is true or false. Therefore, and most importantly, it would be foolish to try to find any property in virtue of which the predicate "is valuable" may be appropriately applied because there is no such property. Since the sentence "Clean air is valuable" is not functioning as a declarative sentence, the predicate "is valuable" is not functioning as the name of some attribute in people or air or anything else. The fundamental assumption of all cognitivists is, according to the emotivists, just plain wrongheaded. Value judgments are not genuine truth claims at all; so it is pointless to ask why some are true and some are not.

Prescriptivism

On what may be called a *prescriptivist* view of the nature of value, to say that something is good or valuable is not to affirm anything at all, but merely to recommend or prescribe something. Like the emotivists, prescriptivists also depart radically from cognitivists because on the prescriptivist view assertions of value are not genuine truth claims. On this view such assertions have the logical status of commands, not of declarative sentences. Just as the prescription "Take two aspirins and go to bed" does not have a truth value, the sentences "It is good to be healthy" and "Better to be healthy than sick" are supposed to be without truth values. Apparent declaratives like "It is good to be healthy" should be interpreted functionally as prescriptions like "Promote healthiness," "Take care of your health," "Prefer health to illness." Again then, as is the case for emotivists, according to prescriptivists the fundamental assumption for all cognitivists is just wrongheaded. Value judgments are not genuine truth claims at all; so it is pointless to ask why some are true and some are false.

The nice thing about noncognitivist views is that they provide an explanation for the apparent fact that people have not been able to discover and display the properties in virtue of which things are said to have or lack value. If "value" and "good" are not the names of any properties, there were none to be found. Noncognitivist theories also have the virtue of reminding us that evaluations often if not always are used as expressions of emotion and as prescriptions. Even if one rejects pure noncognitivism, it's difficult to reject these insights of the noncognitivist position. Evaluative sentences frequently seem to do double or even triple duty.

7. RATIONAL DECISION-MAKING

Two sections ago I raised the question: Can it help your epistemological task (the task of deciding which sentences are true and which are false) to know that a sentence in question involves an ontologically factual or evaluative claim? I answered that it might, and spent some time unpacking that answer. It might have been comforting to be able to say straightaway that yes, by God, once we know if we're confronted by facts or values we know where we are. At the very least some methodological roads are closed. Some kinds of answers will never be forthcoming, and so on. It might have been comforting, but it would not have been true.

The truth is that for some epistemological and evaluative views, the fact-value distinction is problematic and the gap between facts and values is logically unbridgeable. For thoroughgoing epistemological skeptics, the gap, insofar as it is recognized at all, is as serious as a pimple on a malignant tumor. For people holding a naturalistic theory of value, whether it is of an objectivist or subjectivist variety, there is no ontological distinction between facts and values. There are merely formally and functionally different ways to talk about and deal with facts, empirical or logical facts.

What I want to do in this final section is to suggest a useful procedural rule of rational decision-making. The rule is not, as far as I can tell, derivable in a strict sense from the preceding discussion. But it has emerged fairly directly in my own mind as a consequence of that discussion. The rule is just this: Forget about the fact-value distinction. It's almost always a red herring.

Whether the question at issue is epistemological, ontological, or axiological (i.e., a question of truth, being, or value) one is almost always wiser to tackle the question directly rather than to try to cut one's ontological pie into facts and values. Martin Luther King said that the trouble with resorting to violence is that then the issue becomes violence. I think that the trouble with resorting to metaphysics is that then the issue becomes metaphysical. Instead of fighting battles about whether something is true, good, or worth doing, one finds oneself fighting battles about credentials, either the ontological credentials of the issue before one (Is it a matter of truth or goodness or action?) or the credentials of the disputants. Instead of pressing the question "Are there good reasons for believing, preferring, or doing this or that?" the red herring question is raised "Is this a matter of values or facts?" The red herring question can have the effect of silencing voices and eliminating points of view that might happen to be right, or giving too much ground to those who are wrong. Neither error has to be risked if one insists that debate must focus on the issue of truth or falsity, or on what is to be done. Either one can substantiate one's case to satisfy most people or not. How one classifies the type of argumentation used is, I think, quite beside the point.

In the area of energy development decision-making in particular, I think that what we have been witnessing for the past few years is an attempt by some folks to have a greater voice in such decision-making. If one describes this effort as an attempt to get some values into energy policy development, then one may be drawing an imaginary line that some people will never be able to cross. On one side will be hardheads with the facts and on the other side will be softheads with values. It would be strategically wiser to describe the effort as an attempt to get a wider spectrum of issues considered or voices heard. Given our fuzzy understanding of facts and values as ontological categories, we have a stronger epistemological and moral warrant to describe the effort in this way. There will be facts and values (whatever both of these are) that everyone regards as relevant, some that everyone regards as irrelevant, and some about which people will disagree. If we can get agreement on what is true, good, or ought to be done, who but a metaphysician cares if we're agreeing on something factual or valuable or both? On the other hand, if we can agree that something is factual or valuable or both but can't agree on whether it's true, good, or ought to be done, then who but a metaphysician cares about that? In short, it seems to be a good idea for everyone but metaphysicians to forget about the fact-value distinction and get on with the pursuit of truth and the performance of good deeds.

8. CONCLUSION

In a roundabout way I think that all I have been arguing for here is humility in the interests of rational decision-making. I don't expect anyone to stop talking about facts and values as if I had laid these beasts of burden to rest. Indeed, I don't expect to stop talking about facts and values either. If I have managed to make it more difficult for some people – whether technology assessors, science policy analysts, or others – to use the terms glibly, and more difficult for them to be certain of the strengths of their own positions and the weaknesses of other people's positions, that will be enough.

University of Guelph, Ontario

NOTE

* Earlier versions of this paper were presented to the Research Group on Expertise and Lay Participation in Health Policy Decisions at the Institute of Society, Ethics and the Life Sciences, Hastings-on-Hudson, New York; the Morris Colloquium on Morality, Rationality and Environmental Crises at the University of Colorado, Boulder; and the Policy Studies Organization Workshop on Social Values and Public Policy, Washington, D. C.; as well as this German-American conference on the philosophy of technology. Obviously I have had plenty of good suggestions in all these places, including the Bad Homburg conference – for which I am grateful.

GÜNTER ROPOHL

A CRITIQUE OF TECHNOLOGICAL DETERMINISM

1. THE TECHNOPOLITICAL CHALLENGE

During recent decades all industrialized countries have established political institutions and procedures to influence technical development. It would take several pages to list all the ministries of technology, planning authorities, committees on nuclear energy, commissions for astronautics, national boards for automation, offices of technology assessment, and so on, which are expected to control technical development according to social requirements and political goals. To say it in a word: While in former times technics had been reserved to private enterprise, it now ranks among public affairs.

Roughly speaking, there are two main tendencies in technopolitics, each contradicting the other in some sense. One approach tends to promote technical development as much as possible; here it will be called the innovation policy. The other trend, on the contrary, which tends to restrain technical development, is connected with the concept of technology assessment. Innovation policy is, so to say, an accelerating force, whereas technology assessment may be regarded as a decelerating factor with respect to technical development.

Innovation policy is exclusively oriented to economic factors. It starts from the assumption that the introduction of new products and new production processes is the main source of economic growth, with all its benefits for the individual standard of life and the social welfare of the nation. So this policy tries to increase the rate of product and process innovation at any price and hardly bothers about possible non-economic impacts of those innovations; sometimes it does not even care whether those innovations really meet actual needs; and, in case of doubt, it does not hesitate to produce new needs together with the new products. Innovation policy continues that significant trait of capitalism called "creative destruction" by J. A. Schumpeter (1942), but it is a continuation by different means: No longer can the private entrepreneur, on his own, manage large-scale innovations; wherever this type of business leader still exists, he has to be supported by a network of governmental and quasi-public supports, both financial and informational. Technology transfer[1] is the key word for all the processes

Paul T. Durbin and Friedrich Rapp (eds.), Philosophy and Technology, 83–96.
Copyright © 1983 *by D. Reidel Publishing Company.*

that constitute innovation; it means especially (i) the transfer of scientific and technological knowledge from pure science to technical application, (ii) the transfer of technical know-how from one field of application to another, and (iii) the transfer of knowledge and know-how from the industrialized to the developing countries. So the objective of innovation policy is to facilitate and to accelerate technology transfer by any means.

On the other hand there has arisen a strong movement against the abuses of technical development. The growing scarcity of natural resources, the increasing damage technics inflicts on the environment, and its ambiguous impacts on the mental and social situation of men have forced the question, whether anything that can be done actually ought to be realized. "Can does not imply ought" is the slogan of this criticism, which, as a consequence, leads to the establishment of institutions for technology assessment. Technology assessment may be defined as a field of research which, in an anticipatory way, analyzes technical innovations, calculates their consequences for the environment and for society, checks the expected results with regard to relevant values, and makes recommendations based on these studies to the responsible decision makers in business and politics. Thus, technology assessment is designed not to prevent all innovation, but to select out of the abundance of feasible innovations, the actually desirable ones and to slow down technical development wherever it could become dangerous.

These technopolitical activities are in fact taking place and are financed by considerable sums of public money – no matter whether their effectiveness can be justified theoretically. This, to be precise, is the technopolitical challenge to philosophy of technology, namely, that there appears to be an enormous lack of theory to justify an unreflective practice.

2. THE BASIC PROBLEM

Although innovation policy and technology assessment are rather different in intention, they agree on a fundamental assumption. Both take for granted that technical development, whether as a whole or at least in detail, can be controlled. But this hypothesis is precisely the problem. The assumption that technical development is controllable has not been examined thoroughly enough up to now, especially since the contradictory assumption – in the guise of technocracy suspicion or the even stronger fear of technological determinism – is an obvious commonplace in philosophy of technology. Incidentally, it could be asked, moreover, whether the control of technical development, even if it were possible, is desirable. (This is a question which liberals, e.g., W. Becker, 1978, customarily deny; but this ideological aspect has to be left for another discussion.)

Obviously it could be argued that this problem is, in the end, an empirical one: If and inasmuch as either innovation policy or technology assessment were to succeed, the controllability of technical development could be taken as proved. But, unfortunately, both examples in favor and examples against the hypothesis can be found; and some other cases seem rather undecided. Take the supersonic passenger airplane as an instance: Technology assessment was successful insofar as the development of that transportation system was stopped in the United States; but was it really a success, since, notwithstanding, development in Great Britain and France was not stopped at all and the Concorde frequents New York? At the moment, it seems as if supersonic transportation might not be finished because of technology assessment activities but, quite simply, because it is a bad bargain for the aircraft companies.

Anyway, as we know from epistemology, individual verifications do not prove a theory; and individual falsifications will not refute it in this case because we cannot expect anything more than statistical regularities in the field of social processes. It is, furthermore, an additional theoretical assumption to say that technical development really is a social process; so we can see that a mere empirical investigation would not be worth anything without sufficient theoretical background. As long as no fully elaborated interdisciplinary research program on technics exists (Ropohl, 1979, 1981; Lenk and Ropohl, 1979), it obviously is a task for philosophy of technics to formulate the right questions for theoretical investigation.

First of all it has to be asked how in principle to conceive of technical development. (This includes the question how to designate the phenomenon adequately: as development, change, progress, or whatever.) In other words, it is necessary to find a suitable description of the phenomenon to be investigated. Secondly, what "control" might mean when applied to technical development has to be defined. Next it has to be made clear who, or which entity, might be the controlling instance. And last, it has to be discussed: According to which objectives, goals, and values should technical development be controlled? – supposing, of course, that it really proves to be controllable.

But before proposing a descriptive model in order to discuss these questions, we had better deal first with technological determinism; for, if this theory were correct, it would not make much sense to ask about controllability any longer. Moreover, it might turn out that technological determinism actually starts from a correct observation but gives it the wrong interpretation.

3. TECHNOLOGICAL DETERMINISM AND ITS VARIANTS

There are several theoretical conceptions which, more or less, maintain the autonomy of technical development. Regardless of their partial differences, here they will all be subsumed under the concept of technological determinism; for the hard core of all these conceptions is the assertion that technical development does not depend on external factors but determines and dominates the mental and social situation of men as the driving force of social change.

Occasionally, technological determinism is linked with the work of Karl Marx; in several statements he stresses very strongly the impact of productive forces – technics being one of them – on production relations and on the ideological superstructure. But, to be truthful, these points have to be understood as a polemic exaggeration against the idealistic conception of history. For, after all, Marx as a positive dialectician assumes a reciprocal interrelationship between technics and the social system, with neither of them predominant.

On the contrary, it was non-Marxist philosophy of technics that narrowed the view to technological determinism, ascribing total omnipotence to technics – an approach subsequently criticized by Marxist philosophers over and over again (e.g., Anonymous, 1973; Bohring, 1976; Wollgast and Banse, 1979). Nevertheless, this onesided, pessimistic, and fatalistic view of technics has grown into a powerful intellectual tradition in Western thought in this century; even the currently fashionable ecologism has roots in this tradition, to a certain extent. It may be sufficient to cite some names – like Oswald Spengler, F. G. Jünger, Lewis Mumford, or Jacques Ellul – to stand for the whole movement. In the end, it was the German sociologist Helmut Schelsky who, basing his views on Ellul, condensed the perspective into its most impressive form, sketching a model of the technical state as an anonymous technocracy: "We have to give up the idea that the technical-scientific self-creation of man and his new world would follow a universal blueprint, which we were able to manipulate or even to think about. As it is a matter of reconstruction of man himself, there is no human thinking which could anticipate this process in planning and understanding its performance"; again, "There is no kind of human knowledge which could deduce the world and the human being to come, in advance" (Schelsky, 1965, p. 450). If this were true, technopolitics would indeed have no chance!

Of course, a lot of convincing criticisms of these ideas have been put forward (e.g., Lenk, 1973). But perhaps it will be more fruitful to adopt the proposal of Ullrich (1979) not to condemn technological determinism outright but to give it a more subtle interpretation. Obviously not all of the arguments associated with this view can be absolutely misleading, since so many scholars and so much of public opinion tend to give credit to this understanding of technical development. In fact, the prevailing experience for everyday man is the impression that technical innovations come about without his having been asked or engaged. Even if he may be engaged in some sense – for instance, as an employee, or a customer of some particular producer – he usually feels helpless in face of the actual process. Even if it is admitted that technical development originates in human decisions and actions, the individual contribution is so insignificant that man loses the feeling of being the originator of the process. And Friedrich Rapp, by no means a suspicious mystifier of problems, notes (1978, pp. 153 ff.) the "paradox of technical actions" – namely, that they are rationally planned and performed in detail, but as a whole, as those actions and their consequences detach themselves from their personal originators, assume the character of an alien, autonomous force.

To denote this autonomization of a process which, although originating in human action, has run out of human control, Ullrich (1979, pp. 176 ff.) suggests that we use the Marxian concept of reification ("Verdinglichung"). It is worth emphasizing that Ullrich, differing from orthodox Marxists, does not stick to the economic aspect of capitalist production relations, but widens the concept of reification, making of it a general sociological category which also applies to bureaucracy or to technics, no matter whether capitalism is involved or not. But, unfortunately, Ullrich does not take the final step; he fails to recognize that reification, in this broad sense, applies to every process of socialization and is found especially in the cases of social institutions and social norms. It remains very important, of course, to analyze very thoroughly the phenomenon of reification to find an approach to master this problem either theoretically or practically. An incidental hint of Ullrich (1979, p. 183) indicates the direction the investigation will have to take, as he calls reification a "systems concept"; this, indeed, is the point to take up.

4. A SYSTEMS-THEORETICAL SOLUTION OF THE PROBLEM

Obviously, it is impossible to elaborate a complete systems theory of technics in these few pages. Basing the summary on a more detailed investigation (Ropohl, 1979), we can only sketch some main outlines indispensable for the considerations to follow (Section 5).

To start with, technical development will here be understood as the constant increase and improvement of artificial goods. (Evidently, this is more than mere change, but we hesitate to call it "progress" right off.) These goods, objects, or products, as artificial, result from ideas, decisions and actions of men. Therefore, we conceive the originator of technical development as an "action system" (which does not mean a system of actions, but a system which acts). Figure 1 shows the basic model of an action system,

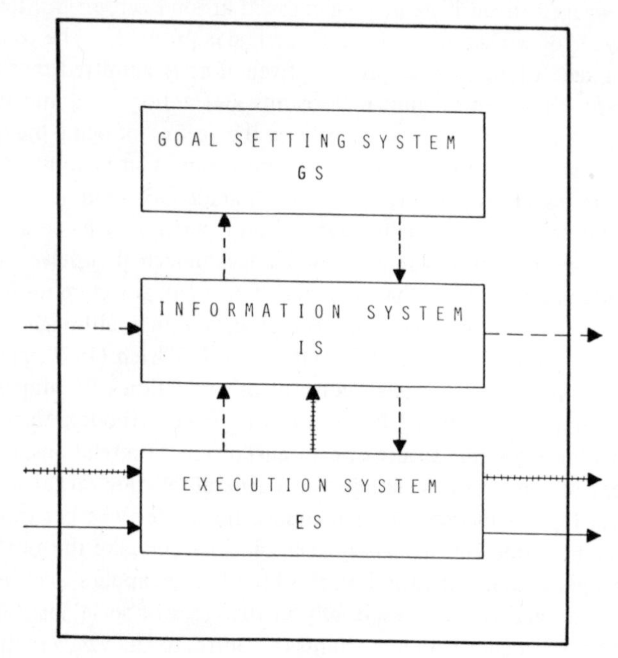

Fig. 1. Rough Structure of a Total Action System (—— = Matter, ++++++ = Energy, - - - = Information).

consisting of three subsystems: for goal-setting, information processing, and execution. We may apply this model to an individual person, like Robinson Crusoe; then the goal-setting system is the motivation center which generates and defines the person's needs and wants; the information processing system finds different ways of solving a given problem, choses one way, and plans the implementation; this last is carried out by the execution system, the body acting on the physical environment.

However, it is the main defect of traditional philosophy of technology to consider technics as a Robinson Crusoe affair, to neglect the social concerns of technical development. To correct this mistake, we assume several levels of action systems, thus forming a hierarchy (in the formal sense). Figure 2

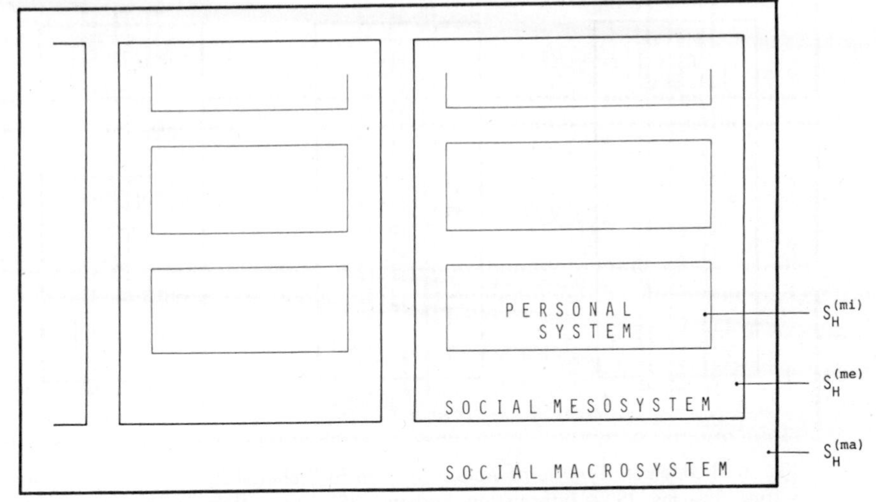

Fig. 2. Hierarchy of Human Action Systems

shows a hierarchy of three levels: the micro-level of individuals, the meso-level of organizations (industrial corporations, administrative offices, families or households), and the macro-level of the national society. For each of these levels, we can explain the action-system model analogously; some intricate interpretation problems concerning the interplay between the different levels (for instance, the participation of organization members in the goal-setting system) have to be omitted here. Moreover, we ought to introduce a fourth level, of world society, for the purpose of this investigation; but up to now we can hardly identify an effective action unit at that level – although we have to keep this idea in mind because it will turn out to be crucial for our central problem.

Now we are able to discuss the overall model of technical development as illustrated in Figure 3. We distinguish three types of action systems on the meso-level: (i) the production of technological knowledge; (ii) the production of technical goods; and (iii) the application of technical goods. Of course, this model is simplified very much; really, the knowledge producing systems add up to hundreds, the systems producing goods to tens of thousands, and the consuming systems to millions – and, furthermore, the interplay between all these action systems must be taken into account. Moreover, individual

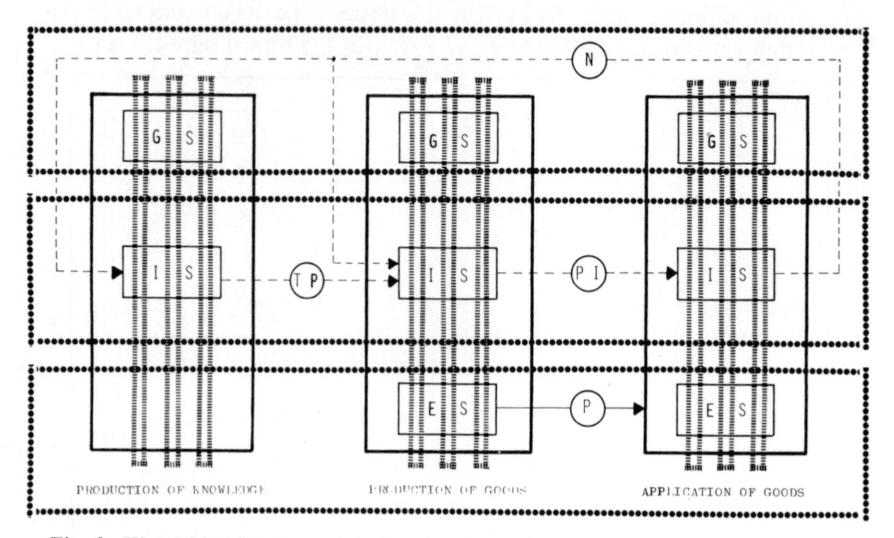

Fig. 3. Hierarchical Texture of Action Systems in Technical Development GS = Goal-setting Systems, IS = Information Systems, ES = Execution Systems, N = Needs, TP = Technical Potentialities, PI = Product Informations, P = Products, ····· Macro-level, —— Meso-level, ıııı Micro-level.

persons in these meso-systems contribute to the goal-setting, information processing, and execution subsystems in many different ways. And, finally, society as a whole has an influence; for instance, by governmental acts setting certain goals or prohibiting certain kinds of actions within the meso-systems or by the personal systems.

Nevertheless, the scheme gives a first impression of the setup or structure of the overall sociotechnical system involved in technical development. Furthermore, the model also indicates some elements of the working structure of the development process. On the one hand, the knowledge-producing systems supply new technical potentialities, in terms of scientific findings and technical inventions which are adopted by the producers of goods and which, when successfully developed, become products distributed to users – no matter whether the users originally had asked for these innovations or not. Technological determinism pretends that this mechanism, the so-called "supply-push," is the only one, assuming that new technical potentialities turn into innovations almost automatically and that innovations create new needs in the application systems automatically as well. On the other hand, the application systems may express their needs and wants to the producing systems and thus stimulate inventions and innovations to fulfill their original needs. This conception, of "demand-pull," is preferred by liberal economists. But actually both conceptions represent less than half the truth. For, first, supply-push and demand-pull overlap and are related in a reciprocal way. Second, both conceptions neglect the crucial role of the goal-setting systems of the producers of goods, who are not so much interested in fulfilling needs as in making profits. And third, both conceptions fail to recognize the possible influences of either individuals or the state.

To complete this section, we must make clear that the processes analyzed here take place when a single technical solution is developed. But the expression "technical development" is ambiguous and principles mentioned also apply to technical development as a whole. This "phylogenetic" development turns out to be a systemic aggregation of all the individual "ontogenetic" developments.

Now we can answer the question as to the way in which the so-called reification of technical development can be conceived of. Indeed, reification refers to the systemic quality of technical development as perceived from the individual's point of view. The system is more than the sum of its parts, the relations between the parts establishing the character of the system; and, obviously, an individual element, exclusively and by itself, does not control the relations since relations by definition depend on other elements. This is true as well for passing from the micro- to the meso-level or from the meso- to the macro-level. The performance of a design team is more than the sum of the contributions of all the individual engineers; even if one engineer has stimulated a certain innovation, the final outcome usually will differ from his initial intentions because teamwork introduces additional objectives, information, and experiences. And, of course, the state of the art in a certain industrial branch is a result of the interplay between competitors rather than the performance of a single corporation. And finally we must recall international competition, as mentioned earlier; it is not up to a single nation to decide whether mankind will be blessed with supersonic transportation systems.

So reification is the result of the highly refined division of labor in modern societies and its manifold patterns of cooperation and competition.

5. SOME TENTATIVE CONCLUSIONS

After we have introduced this model of technical development, we are in a position to give some tentative answers to the questions raised in the second section.

To repeat, we are not going to call increases in and improvements of technical products a mere change; the transition from the electronic valve to the transistor, for instance, introduces a new quality, and the word "change" would not underscore the qualitative nature of that transition. On the other hand, it is not self-evident, at first glance, whether any transition of that kind should be considered an unrestrained contribution to mankind's well-being; the number of silly or even dangerous innovations is too large to subsume all of them under the phrase "technical progress." So we must keep to the notion of development – although we must not associate with it the idea of organismic evolution because there seem to be fundamental differences between man-made and natural processes. There have been some attempts to consider technical development in terms of evolution theory, but this is a vague analogy rather than an adequate explanation. The essence of technical development is man's purposeful activity, and this is the point from which the explanation, as well as the control, of technical development must begin.

It seems appropriate above all to treat the "ontogenetic" developments because they are the elements of technical development as a whole. So we must make clear what it really means to control an individual technical development. Obviously we must distinguish two principal kinds of control, positive and negative. To exert positive control is to initiate a certain development process, to promote it, and to enhance its progress; negative control, on the other hand, is expected to prevent a certain development process, to decelerate it, or to stop it. Similar to the "falsification-verification problem" in epistemology, there is a lack of symmetry between positive and negative control. If positive control succeeds, the success, in principle, is irreversible. An invention, once made, cannot be forgotten; an innovation, once introduced, remains an innovation even if it will not spread widely; and the diffusion of a certain product follows the so-called "logistic law of market penetration," interrupted if at all only by a superior innovation in the same field. Negative control, however, cannot succeed once and for all. Leaving aside the fact that it can by no means be clearly stated whether negative control has really prevented an invention (because nothing at all can be said about it unless an invention has already been made), even the successful suppression of a certain invention or innovation does not guarantee that it will not be repeated elsewhere beyond the range of that specific control activity. Again we must remind ourselves of the supersonic transportation system case. If only for this reason, it seems evident that negative control depends on centralization, whereas positive control may work very well within the decentralized structures of a market economy. (Incidentally, this is likely the reason why innovation policy is much more fashionable in Western countries than technology assessment.)

Both positive and negative control are expected to work on individual technical developments. As recognized in the last section, every development process turns out to be a sequence of individual, organizational, and social actions, each consisting of goal-setting, information processing (and above all decision making), and execution activities. So to control technical developments obviously means to influence those actions: to prevent one action or bring about another, to modify one action or substitute another.

The next step in our investigation is to analyze which action systems and which kinds of action may be subjected to effective control. For this purpose the scheme of Figure 3 may serve as a systemizing aid: We have to check the action systems of the micro-level, of the meso-level, and of the macro-level; the three types of action systems – production of knowledge, production of goods, application of goods – correspond to the three stages of a technical development, invention, innovation and diffusion. And, moreover, we must discuss the goal-setting, the information processing, and the execution components of those actions with regard to their respective levels in the hierarchy. This, of course, is a complete research program and cannot be dealt with in these few pages. We can only pick out some obvious features.

Invention, a positively new arrangement of technological and technical knowledge, usually takes place at the level of the personal system. Although some internal and external conditions of a person's mental activity may be manipulated, invention itself is more or less fortuitous and cannot be controlled in the proper sense of the word. Neither inventing nor failing to make an invention can be forced upon a person – even if a particular sort of social climate, influencing the individual goal–setting and information processing system, may work in favor of or against inventiveness. Moreover, inventions, in principle, are use-oriented, and there is no inventor who could not imagine a useful purpose for his ideas; which means control could succeed in suggesting certain categories and fields of utility or rejecting others. This indicates, no doubt, a field of opportunity for professional ethics.

Diffusion, on the other hand, depends on the consumer's readiness to accept or reject new technical objects. Again, individual ethics may play a certain role, but up to now we can find hardly a case in which consumers, by rejection, have prevented an unpleasant development. And positive control is as well unlikely because consumers can only choose among the goods supplied and usually find no way to communicate their actual needs and wants to inventors.

So the decisive stage of the development process proves to be that of innovation, that is, the technically and economically successful realization of an invention. In the ideal case, an innovation is the result of an optimal match, with regard to economic efficiency, between technical potentialities and human needs. On the micro-level, the ideal type of the classical entrepreneur would perform this way if he regarded profits only as a secondary condition of his productive activity. But things have now changed completely; collective management has been substituted for the individual entrepreneur. Hence the innovation process nowadays can be discussed only at the meso-level, and the formal goal of making profits has come to prevail over the material goal of satisfying needs. So the goal-setting and decision-making structures of industrial organizations are very promising starting points for the control of technical development. And technopolitics is on the right track when it tries to impose social goals on the activities of these industrial meso-systems. The fact that formal economic goals like profit and growth tend to prevail has been an important reason for such autonomization of technical development as there has been. Above all, this explains the failure of negative control, because the only brake on innovation policy within the business corporation is the lack of profit expectations; the proverb, "Rubbish that sells isn't rubbish at all," well known in British business circles, summarizes the whole matter; Thus we are likely to be overstocked with such rubbish as television games, chess computers, electric tooth brushes, electric knives, digital wrist watches, egg boilers, cable television, and other innovations with dubious claims to meet true needs.

Obviously, it is not up to an individual corporation to save society from tosh like that; if one firm renounces such "innovations," another will certainly produce them. Technology assessment, in the sense of negative control, will not work without centralization at the macro-level. This leads us, in our considerations, to the possibility of societal and governmental control. As long as the nature of private enterprise is not modified (for instance, if decision rights were socialized, while still allowing benefits to flow to the owners), the chances of political control are very limited; without such changes public control of industrial organizations is limited to subsidies and fiscal incentives, on the one hand, or, on the other, to the classical measures of legislation and administrative directives. Only in those fields where the state itself sponsors, manages, or applies a certain development, is it free to determine the direction of that development. No doubt the sphere of public influence has grown, but a great many development processes on the meso-level are still outside public control. But whereas in this case political institutions, in principle at least, do exist, and effective control is only a question of constitutional modifications, on the international level there do not even exist the indispensable institutions. So international competition is always an excuse for leaving technical developments uncontrolled at the national level even if negative control would be highly desirable. These considerations already indicate that we have to conceive of a multitude of agents and agencies as responsible for the control of technical developments. When we argue in favor of centralized decisions, at least with regard to negative control, we remain aware of the danger of bureaucracy; we would therefore recommend that any control which is feasible at the micro or meso-level be left to individuals and private organizations. This way, indeed, everybody could share the responsibility, although, of course, no single person can change things completely.

So technological determinism is correct insofar as technical development is characterized by a certain degree of reification. In fact, the individual person, although himself engaged in the process, cannot control development on his own. And if anyone resists this simple fact, he fails to recognize (either in the descriptive or in the normative sense) the systemic quality of that large-scale socialization of which technical development is a part. Consequently, criticism directed against reification is usually connected with philosophical individualism or social romanticism. (This may even be true of the young Marx!) However, technological determinism is wrong when it denies the controllability of technical development because of this reification. Actually, reification can be mastered if it is countered at an adequate level in the systems hierarchy. The development of nuclear power, for instance, may have shown attributes of reification in recent decades; and the individual citizen was, indeed, helpless against it. But since citizens' movements against nuclear power stations have been established all over the Western world, nuclear energy is no longer a superhuman fate. It depends on democratic decisions, which, in several cases, have been able at least to postpone the start-up of operations. So reification, in the sense defined, is an inevitable characteristic of our highly developed sociotechnical system, but it need not degenerate into the autonomization of technical development. Of course, there remain numerous problems to be solved, whether in interdisciplinary research on technics or in political practice; and certain hopes in the sphere of international politics seem somewhat utopian at the moment. But there is no philosophical reason to rule out the fundamental controllability of technical development.

If there still exists a problem for philosophy, it is the value problem. Control, by definition, implies definite objectives, and these objectives are ultimately rooted in certain general conceptions of the quality of life. When essential features of sociotechnical control must be centralized, we need a minimal consensus about the main features of the quality of life; that means that we need an obligatory value system acceptable to the large majority. So the philosophical task is to identify and to substantiate these indispensable values, to participate in a kind of value management within the limits of democratic discourse. This discussion is well on the way already, and we look forward to a feasible conception of "the good life," one that is broad enough to guarantee a variety of lifestyles but meaningful enough to enable a decisive control of technical development. This will close the gap between public policies and private ethics.

Johann Wolfgang Goethe University, Frankfurt

NOTE

[1] A terminological note: In English the term "technology" means both technical objects and the science of those objects. I adopt this idiom only for standard phrases; as a rule, I keep to the older German tradition which makes a distinction between "technics" as the array of objects and "technology" as the science of that array.

REFERENCES

Anonymous. 1973. *Man, Science, Technology*. Moscow and Prague.

Becker, W. 1978. "Technischer Fortschritt und das Freiheitsverständnis der Liberalen Demokratie," preprint for the XVIth World Congress of Philosophy. Düsseldorf, pp. 83–85.

Bohring, G. 1976. *Technik im Kampf der Weltanschauungen*. Berlin.

Lenk, H. (ed.). 1973. *Technokratie als Ideologie*. Stuttgart.

Lenk, H. and Ropohl, G. 1979. "Toward an Interdisciplinary and Pragmatic Philosophy of Technology," in P. Durbin (ed.), *Research in Philosophy & Technology*, vol. 2. Greenwich, Conn., pp. 15–52.

Rapp, F. 1978. *Analytische Technikphilosophie*. Freiburg and Munich.

Ropohl, G. 1979. *Eine Systemtheorie der Technik*. Munich and Vienna.

Ropohl, G. (ed.). 1981. *Interdisziplinäre Technikforschung*. Berlin.

Schelsky, H. 1965. "Der Mensch in der wissenschaftlichen Zivilisation," in H. Schelsky, *Auf der Suche nach der Wirklichkeit*. Düsseldorf and Cologne, pp. 439–480.

Schumpeter, J. 1942. *Capitalism, Socialism, and Democracy*. New York.

Ullrich, O. 1979. *Technik und Herrschaft*. Frankfurt.

Wollgast, S. and Banse, G. 1979. *Philosophie und Technik*. Berlin.

LANGDON WINNER

TECHNĒ AND POLITEIA: THE TECHNICAL CONSTITUTION OF SOCIETY

At the beginning of Western political theory a powerful analogy links the practice of technology to that of politics. In his *Republic*, *Laws*, *Statesman*, and other discourses, Plato asserts that statecraft is a *technē*, one of the practical arts. Much like architecture, the building of ships and other commonly recognized arts and crafts, politics is a field of practice that has its own distinctive knowledge, its own special skills. One purpose of Plato's argument was to discredit those who believed that the affairs of public life could be left to mere amateurs. But beyond that, it is clear that he thought the art of politics could be useful in the same way as any other *technē*, that is it could produce finished works of lasting value. The works he had in mind were good constitutions, supremely well-crafted products of political architecture. *Politeia*, the title of the *Republic* in Greek, means the constitution of a polis, the proper order of human relationships within a city-state. The dialogue describes and justifies what Plato holds to be the institutional arrangements appropriate to the best *politeia*. He returns to this theme in the *Laws*, a discussion of the "second best" constitution, comparing his work to that of a well-established craft. "The shipwright, you know, begins his work by laying down the keel of the vessel and indicating her outlines, and I feel myself to be doing the same thing in my attempt to present you with outlines of human lives. . . . I am really laying the keels of the vessels by due consideration of the question by what means or manner of life we shall make our voyage over the sea of time to the best purpose."[1]

In Plato's interpretation the analogy between technology and politics works in one direction only; *technē* serves as a model for politics, not the other way around. Having employed this notion for his own purposes, he took a further step, seeking to segregate the substance of politics from any other form of *technē* whatsoever. Thus, in the *Laws* he excludes craftsmen from positions of citizenship (because they already have an art that requires their full attention) and forbids citizens to engage in any material craft (because citizenship makes its own full demands upon them) – a judgment that reflects his supreme respect for the dedication *technē* involves and his suspicion of material pursuits in general. Evidently, it did not occur to Plato or to anyone else for a very long time that the analogy could at some

Paul T. Durbin and Friedrich Rapp (eds.), Philosophy and Technology, 97–111.
Copyright © 1983 by D. Reidel Publishing Company.

point qualify in reverse, that *technē* itself might become a *politeia*, that technical forms of life might in themselves give powerful and authoritative shape to human affairs. If that ever did occur, what would the response of political theory be?

The one-sided comparison of technical and political creativity appears again in modern political thought. Writing in the *Social Contract*, Jean-Jacques Rousseau employs a mechanical metaphor to illuminate the art of constitution making. "A prince," he says, "has only to follow a model which the lawgiver provides. The lawgiver is the engineer who invents the machine; the prince is merely the mechanic who sets it up and operates it."[2] At another point in the book, Rousseau compares the work of the lawgiver to that of an architect. With a frustrated ambition reminiscent of Plato's, Rousseau offered himself as a political engineer or architect of exactly this kind, writing treatises on the constitutions of Corsica and Poland in the hope that his ideas might influence the founding of new states.

An opportunity of exactly that kind later became available to the founders of modern nation states, among them the leaders of the American Revolution. From the earliest rumblings of rebellion in the seventeenth century to the adoption of the Federal Constitution in 1787, the nation was alive with disputes about the application of political principles to the design of public institutions. Taking what they found useful from previous history and existing theories, thinkers like Madison, Hamilton, Adams, and Jefferson tried to devise a "science of politics," a science specifically aimed at providing knowledge for a collective act of architectonic skill. Thus, in *The Federalist Papers*, to take one example, we find a sustained discussion of how one moves from abstract political notions such as power, liberty and public good to their tangible manifestations in the divisions, functions, powers, relationships, and limits of the Constitution. "The science of politics, . . ." Hamilton explains in *Federalist* No. 9, "like most other sciences, has received great improvement. The efficacy of various principles is now well understood, which were either not known at all, or imperfectly known to the ancients. The regular distribution of power into distinct departments; the introduction of legislative balances and checks; the institution of courts composed of judges holding their offices during good behavior; the representation of the people in the legislature by deputies of their own election: these are wholly new discoveries, or have made their principal progress towards perfection in modern times." Metaphors from eighteenth century science and mechanical invention – e.g., "the ENLARGEMENT of the ORBIT within which such systems are to revolve"[3] and references to the idea of checks and balances –

pervade *The Federalist Papers* and indicate the extent to which its writers saw the founding as the creation of an ingenious political/mechanical device.

But even as the eighteenth century was reviving the ancient analogy between technology and politics, even as philosopher statesmen were restoring the ancient art of constitution-making, another extremely powerful mode of institutionalization was taking shape in America and Europe. The industrial revolution with its distinctive ways of arranging people, machines and materials for production very soon began to compete with strictly political institutions for power, authority and the loyalties of men and women. *Technē* and *politeia*, long thought to be entirely separate phenomena within the category of human works, were drawing closer together.

Writing in 1781 in his *Notes on Virginia*, Thomas Jefferson noted the new force abroad in the world and commented upon its probable meaning for public life. The system of manufactures emerging at the time would, he argued, be incompatible with the life of a stable, virtuous republic. Manufacturing would create a thoroughly dependent rather than a self-sufficient populace. "Dependence," he warned, "begets subservience and venality, suffocates the germ of virtue, and prepares fit tools for the designs of ambition." In his view the industrial mode of production threatened "the manners and spirit of a people which preserve a republic in vigor. A degeneracy in these is a canker which soon eats to the heart of its laws and constitution."[4] For that reason he advised, in this book at least, that Americans agree to leave the workshops in Europe.

There are signs that a desire to shape industrial development to accord with the principles of the republican political tradition continued to interest some Americans well into the 1830s. Attempts to include elements of republican community in the building of the factory town in Lowell, Massachusetts show this impulse at work.[5] But these efforts were neither prominent within economic patterns then taking shape nor successful in their own right. In the 1840s and decades since, the notion that industrial development might be shaped or limited by republican virtues – virtues of frugality, self-restraint, respect for the public good, and the like – dropped out of common discourse, echoed in the woeful lamentations of Henry David Thoreau, Henry Adams, Lewis Mumford, Paul Goodman, and a host of others now flippantly dismissed as "romantics" and "pastoralists."

In fact, the republican tradition of political thought had long since made its peace with the primary carrier of technical change, entrepreneurial capitalism. Moral and political thinkers from Machiavelli to Montesquieu and Adam Smith had argued that the pursuit of economic advantage is a civilizing, moderating influence in society, the very basis of stable government. Rather than engage the fierce passion for glory that often leads to conflict, it is better, so the argument goes, to convince people to pursue their self-interest, an interest that inclines them toward rational behavior.[6] The framers of the American constitution were, by and large, convinced of the wisdom of this formula. They expected that Americans would act in a self-interested manner, employing whatever instruments they needed to generate wealth. The competition of these interests in society would, they believed, provide a check upon the concentration of power in the hands of any one faction. Thus, in one important sense, republicanism and capitalism were fully reconciled at the time of the founding.

By the middle of the nineteenth century this point of view had been strongly augmented by another idea, one that to this day informs the self-image of Americans – a notion that equates abundance and freedom. The country was rich in land and resources; people liberated from the social hierarchies and status definitions of traditional societies were given the opportunity to exploit that material bounty in whatever ways they could muster. In this context, new technologies were seen as an undeniable blessing because they enabled the treasures to be extracted more quickly, because they vastly increased the product of labor. Factories, railroads, steamboats, telegraphs, and the like were greeted as the very essence of democratic freedom for the ways they rendered, as one mid-nineteenth century writer explained, "the conveniences and elegancies of life accessible to the many instead of the few."[7]

American society encouraged people to be self-determining, to pursue their own economic goals. That policy would work, it was commonly believed, only if there was a surplus that guaranteed enough to go around. Class conflict, the scourge of democracy in the ancient world, could be avoided in the United States because the inequalities present in society would not matter very much. Material abundance would make it possible for everybody to have enough to be perfectly happy. Eventually, Americans took this notion to be a generally applicable theory: economic enterprise driven by the engine of technical improvement was the very essence of human freedom. Franklin D. Roosevelt reportedly remarked that if he could put one American book in the hands of every Russian, it would be the Sears, Roebuck catalogue.

In this way of looking at things, the form of the technology you adopt does not matter. If you have cornucopia in your grasp, you do not worry about its shape. Insofar as it is a powerful thing, more power to it. Anything that history, literature, philosophy, or long-standing traditions might have to suggest about the prudence one ought to employ in the shaping of new institutions can be thrown in the trash bin. Describing the industrial revolution in Britain, historian Karl Polanyi drew an accurate picture of this attitude. "Fired by an emotional faith in spontaneity, the common-sense attitude toward change was discarded in favor of a mystical readiness to accept the social consequences of economic improvement, whatever they might be. The elementary truths of political science and statecraft were first discarded, then forgotten. It should need no elaboration that a process of undirected change, the pace of which is deemed too fast, should be slowed down, if possible, so as to safeguard the welfare of the community. Such household truths of traditional statesmanship, often merely reflecting the teachings of a social philosophy inherited from the ancients, were in the nineteenth century erased from the thoughts of the educated by the corrosive of a crude utilitarianism combined with an uncritical reliance on the alleged self-healing virtues of unconscious growth."[8] Indeed, by the late nineteenth century, an impressive array of scientific discoveries, technical inventions and industrial innovations seemed to make the mastery of nature an accomplished fact rather than an idle dream. Many took this as a sign that all ancient wisdom had simply been rendered obsolete. As one chronicler of the new technology wrote in *Scientific American*: "The speculative philosophy of the past is but a too empty consolation for short-lived, busy man, and, seeing with the eye of science the possibilities of matter, he has touched it with the divine breath of thought and made a new world."[9]

Today we can examine the world of interconnected systems in manufacturing, communications, transportation, and the like that have arisen over the past two centuries and appreciate how they form *de facto* a constitution of a sociotechnical order. This way of arranging people and things, of course, did not develop as the result of the application of any particular plan or political theory. It grew gradually and in separate increments, invention by invention, industry by industry, engineering project by engineering project, system by system. From a contemporary vantage point, nevertheless, one can notice some of its characteristics and begin to see how they embody answers to age-old political questions about membership, power, authority, order, freedom, justice, and the conditions of government. Several of the characteristics that matter in this way of seeing, characteristics that would have interested Plato or Rousseau or Madison, are, very briefly, the following.

First is the ability of technologies of transportation and communication to facilitate control over events from a single center or small number of centers. Largely unchecked by effective countervailing influence, there has been an extraordinary centralization of social control in large business corporations, bureaucracies and, lest we forget, the military.

Second is a tendency for new devices and techniques to increase the most efficient or effective size of organized human associations. Over the past century more and more people have found themselves living and working within technology-based institutions that previous generations would have called gigantic. Justified by impressive economies of scale and (economies or not) always an expression of the power that accrues to very large organizations, this gigantism has become an accustomed feature of the material and social settings of everyday life.

Third is the way in which the rational arrangement of sociotechnical systems has tended to produce its own distinctive forms of hierarchical authority. Legitimized by the felt need to do things in what seems to be the most efficient, productive way, human roles and relationships are structured in rule-guided patterns that involve taking orders and giving orders along an elaborate chain of command. Thus, far from being an expression of democratic freedom, the reality of the workplace tends to be undisguisedly authoritarian.

Fourth is the tendency of large, centralized, hierarchically arranged sociotechnical systems to crowd out and eliminate other varieties of social activity. Hence, industrial production placed craftwork in eclipse; technologies of modern agribusiness made small scale farming all but impossible; high speed transportation crowded out slower means of getting about. It is not merely that useful devices and techniques of earlier periods have been rendered extinct, but also that the patterns of social existence and individual experience based upon them have vanished as living realities.

Fifth are the various ways that large sociotechnical organizations exercise power to control the social and political influences that ostensibly control them. In a process that I have elsewhere described as "reverse adaptation," the human needs, markets and political institutions that might regulate technology-based systems are often subject to manipulation by those very systems.[10] Thus, to take one example, psychologically sophisticated techniques of advertising have become an accustomed way of altering people's ends to suit the structure of available means, a practice that now affects political campaigns no less than campaigns to sell underarm deodorant (with similar results).

There are, of course, other characteristics of today's technological systems that can accurately be read as political phenomena; it is certainly true that there are factors other than technology which figure prominently in the developments I have mentioned. What I want to indicate is that as our society adopts one sociotechnical system after another, it answers *de facto* a number of the most important questions that political philosophers have asked about the proper order of human affairs. Should power be centralized or dispersed? What is the best size for units of organized social activity? What constitutes justifiable authority in human associations? Does a free society depend upon social uniformity or social diversity? What are appropriate structures and processes of public deliberation and decision making? For the past century or longer our answers to such questions have been primarily instrumental ones, expressed in an instrumental language of efficiency and productivity, physically embodied in instrumentalities that often seem to be nothing more than ways of providing goods and services.

For those who have adopted the formula of freedom through abundance, however, the traditional questions have ceased to matter. In fact, it has commonly been assumed that whatever happened to be created in the sphere of material/instrumental culture would certainly be compatible with freedom and democracy. This amounts to a conviction that all technology is inherently liberating, a very peculiar belief indeed.

It is true that on occasion agencies of the modern state have attempted to "regulate" business enterprises and technological applications of various kinds.[11] On balance, however, the extent of that regulation has been modest at best. In the United States absolute monopolies are sometimes outlawed only to be replaced by enormous semi-monopolies no less powerful in their ability to influence the life of society. The history of regulation shows abundant instances in which the rules and procedures that govern production or trade were actually demanded by or later captured by the industries they supposedly regulate. In general, the rule of thumb has been: if a business makes goods and services widely available, at low cost with due regard for public health and safety and with a reasonable return on investment, all is well.

In recent times the question of recognizing possible limits upon the growth of technologies has had something of a revival. Many people are now prepared to entertain the notion that a given technology be limited: (1) if its application threatens public health or safety, (2) if its use threatens to exhaust some vital resource, (3) if it degrades the quality of the environment – air, land, and water, (4) if it threatens natural species and wilderness areas that ought to be preserved, and (5) if its application causes social stresses and strains. Along with the ongoing discussion about ways to sustain economic growth and prosperity, these are about the only matters of "technology assessment" that the general public, decision makers, and academics are prepared to take seriously at present.

While such concerns are valid ones, they do severely restrict the range of moral and political criteria that are permissible in public discussions of technological change. Several years ago I tried to register my discomfort on this score with some colleagues in computer science who were doing a study of the then novel electronic funds transfer systems. They had concluded that such systems contained the potential for redistributing financial power in the world of banking. Electronic money would make possible a shift of power from smaller local banks to large institutions of national and international finance. They asked me to suggest a way of arguing the possible dangers of this development to their audience of scholars and policy makers. In a letter to them I recommended that their research might try to show that under conditions of heavy, continued exposure, electronic funds transfer causes cancer in laboratory animals. Surely, that finding would be cause for concern. My ironic suggestion acknowledged what I take to be the central characteristic of socially acceptable criticism of science and technology in our time. Unless one can demonstrate conclusively that a particular technical practice will generate some physically evident catastrophe – cancer, destruction of the ozone layer, or some other – one might as well shut up.

Even criteria of that kind, however, may soon erode. In a style of analysis now being perfected in the United States and Europe, the outright banning of carcinogenic and mutagenic substances is now considered an overwrought policy. Rather than eliminate from human consumption any substance shown to cause cancer or birth defects in laboratory animals, we are asked to substitute "risk/benefit analysis." In that rapidly developing, highly quantified moral science, people are (in effect) asked to acknowledge cancer and birth defects as among the exhilarating risks – often compared to flying or mountain climbing! – we run in order to live in such a materially abundant society. Like our factories that need a dose of "reindustrialization" to bring them back to life, our nihilism is now being completely retooled, becoming at long last a truly rigorous discipline.

But even before risk/benefit analysis helps dissolve existing public norms of care and prudence – e.g., the Delaney clause of the Food and Drug Act – it is important to notice how very narrow attempts to guide or limit the direction of technological innovation have been up to this point. The development of techniques of technology assessment during the 1960s and 1970s did little to alter the widespread belief that ultimately what matters is economic growth and efficiency. As regards the substance of its many studies, the Office of Technology Assessment established by the U.S. Congress could well change its name to the office of Economic Growth and Efficiency Research. Whatever hope there may have been for contemporary universities to produce critical re-evaluations of science and technology have largely faded. Academic departments and programs in science, technology and society have vastly improved our ability to generate intricate descriptions, explanations and apologies for what scientific and technical institutions do. But such programs are, for the most part, completely sterile as regards any serious questioning of the central practices they study. Criticism of technology is taken to be a sign of bad taste, even bad collegiality, not to mention a threat to S.T.S. groups' sources of funding.

The renewed conversation about technology and society has continued to a point at which an obvious question needs to be asked: Are there no shared ends that matter greatly to us any longer other than the desire to be affluent while avoiding the risk of cancer? It may be that the answer is "No." The prevailing social consensus seems to be that people love the life of high consumption, tremble at the thought that it might end, and are displeased about having to clean up the various messes that modern technology sometimes brings. To argue a moral position convincingly these days requires that one speak to (and not depart from) concerns of this kind. One must engage people's love of material well-being, their fascination with efficiency or their fear of death. The moral sentiments that hold force, that can be played upon in attempts to change social behavior, are those described by Adam Smith and Thomas Hobbes. I do not wish to deny the validity of these sentiments, only to suggest that they comprise an extremely narrow set. We continue to disregard a problem that has been brewing since the earliest days of the industrial revolution – whether our society will find it possible to establish forms and limits for technological change, forms and limits that derive from a positively articulated idea of what society ought to be.

To aid our thinking about such matters, I would suggest we begin with a simple heuristic device. Let us suppose that every political philosophy in a given time implies a technology or set of technologies in a particular configuration for its realization. And let us recognize that every technology of significance to us implies a set of political commitments that one can identify if one looks carefully enough. To state it more directly, what appear to be merely instrumental choices are better seen as choices about the form of the society we continually build, choices about the kinds of people we want to be. Plato's metaphor, especially his reference to the shipwright, is one that a high technology age ought to take literally: we ought to lay out the keels of our vessels with due consideration to what means or manner of life we shall make our voyage over the sea of time to best purpose. The vessels that matter now are such things as communications systems, transit systems, energy supply and distribution systems, household instruments, biomedical technologies, and of course systems of industrial and agricultural production.

In one way or another, issues of this kind are addressed and settled in the course of technological change. Often they appear as subliminal or concealed agendas in discussions that seem to be about efficiency, productivity, abundance, profit, and market conditions. For instance, it is possible to read the dozens of sophisticated energy studies that have been done during the past ten years and interpret them for the social and political structures their analyses and recommendations imply. Will it be nuclear power administered by a benign priesthood of scientists? Will it be coal and oil brought to you by large multinational corporations? Will it be synthetic fuels, directly subsidized and administered by the state? Will it be the soft energy path brought to you by you and your neighbors?

Whatever one's position might be, the prevailing consensus requires that all parties argue their positions solely on grounds of efficiency. Regardless of how a particular energy solution will distribute social power, the case for or against it must be stated as a practical necessity that derives from demonstrable conditions of efficiency. Thus, even those who favor decentralist energy solutions feel compelled to rest their claims solely on the grounds: "This way is most efficient." Prominent among these now is Amory Lovins who writes, "While not under the illusion that facts are separable from values, I have tried to separate my personal preferences from my analytic assumptions and to rely not on modes of discourse that might be viewed as overtly ideological, but rather on classical arguments of economic and engineering efficiency (which are only tacitly ideological)."[12] In his book, *Soft Energy Paths*, Lovins notices "centrism, vulnerability, technocracy, repression, alienation," and other grave problems present in the existing energy sector. He sets out to compare "two energy paths that are distinguished by their antithetical social implications." Then he notes that basing energy decisions on social criteria may appear to involve a "heroic decision," that is "doing something the more expensive way because it is desirable on other more important grounds than internal cost."

But fear not. "Surprisingly," he writes, "a heroic decision does not seem to be necessary in this case, because the energy system that seems socially more attractive is also cheaper and easier."[13] But what if the analysis had shown the contrary? Would Lovins and others who play the energy policy game be prepared to give up the social advantages believed to exist along the soft energy path? Would they feel compelled by the logic of analytic, empirical demonstration to embrace "centrism, vulnerability, technocracy, repression, alienation" and the like? Here Lovins gives up ground that has in modern history again and again been abandoned as lost territory. It raises the question of whether even the best meaning among us are anything more than mere efficiency worshippers.

The same tendency stands out in the arguments of those who seek to defend democratic self-management and small scale technology. More often than not, they feel compelled to show their proposal is more efficient rather than endorse it on directly social or political grounds: That is, indeed, one way of catching people's attention; if you can get away with it, it is certainly the most convincing variety of argument. But victories won in this way are in other important respects great losses. For they affirm in our words, our very methodologies that there are certain human ends that no longer dare speak their names in public. The silence grows deeper by the day.

As long as we respond reflexively to the latest "crisis," energy crisis or some other, as long as we persist in seeing the world as a set of discrete "problems" awaiting ingeniously concocted "solutions," the failure to consider the question of ends will continue. True, the idea that society might try to guide its own sociotechnical development according to preconceived articulations of form and limit has only a tiny constituency, no larger than that which existed among the utopians and social reformers of the early nineteenth century. It is, in fact, a matter of pride for many engaged in the scientific, technical and financial aspects of innovation in material culture that we do not know where we are going until we get there. Nevertheless, it may still be possible to stand outside the stampede long enough to begin regaining our bearings.

My suggestion is that each significant area of technical/functional organization in modern society be seen as a kind of regime, a *regime of instrumentality*, under which we are obliged to live. Thus, there are a number of regimes of mass production, each with a structure that may be interpreted as a technopolitical phenomenon. There are a number of regimes in energy production and distribution, in petroleum, coal, hydroelectric, nuclear power, etc., each with a form that can be seen as a political phenomenon. There is, of course, the regime of network television and that of the automobile. If one were to identify and characterize all of the instrumental/functional sociotechnical systems of our society, all of our regimes of instrumentality and their complex interconnections, one would have a picture of a second constitution, one that stands parallel to and occasionally overlaps the constitution of political society as such. The important task becomes, therefore, not that of studying the "impacts" of technical change – a social scientific approach committed to a passive stance – but of evaluating the material and social infrastructures that specific technologies create for our life's activity. We should try to imagine and seek to build technical infrastructures compatible with freedom, justice, and other crucial political ends.

One set of technical possibilities that I find interesting from this point of view are those involved in the development of photovoltaic systems. Here is a case in which the crucial choices have yet to be made. We can expect to see the events unfold in our lifetime with outcomes that could have many different dimensions. The short version of the story, as I would tell it, goes as follows. If solar cells become feasible to mass produce, if their price in installed systems comes down to a reasonable level (rather than being outrageously expensive when compared to other means of producing electricity), solar electricity can make a contribution to our energy needs. If it should occur that photovoltaic systems do become technically and economically feasible (and many working with such systems find the auspices good), then there will be – at least in principle – a choice about how to structure these systems.

One could, for example, build centralized photovoltaic farms that hook directly into the existing electrical grid like any other form of central power. One could produce fully stand-alone systems placed on the rooftops of homes, schools, factories, and the like. Or one could design and build intermediate sized ensembles, perhaps at a neighborhood level. In one way or another a number of questions will eventually be answered. Will society have photovoltaics in any substantial number at all? How large should such systems be? Who should own them? How should they be managed? Should they be "fully automatic"? Or should the producer/consumer of solar power be involved in activities of load management?

I believe that if one wants technical diversity as part of the infrastructure of freedom in the regime of renewable energy, then a disaggregated, decentralized design of photovoltaic power could be counted a positive good. In saying this, I would not ask anyone to make exorbitant economic sacrifices. But rather than pursue the lemming-like course of choosing only that system which gives the lowest cost kilowatt, perhaps we ought to ask whether or not that lowest cost design simply reproduces the large, centralized, uniform, automatic sociotechnical patterns that contribute to the ways people find themselves to be insignificant, not involved, powerless. To propose an innovation of this kind does not, by the way, presuppose that existing producers and suppliers of electricity are somehow evil. I am speaking of the kinds of patterns in material culture that eventually become forms of social life and individual experience, suggesting that we actively seek to create patterns that manifest (rather than mock) the idea of a good society.[15]

It goes without saying that the people now actually developing photovoltaics have no such questions in mind. Government subsidized research, development and implementation processes in the United States now rest upon the explicit aim of finding that lowest cost kilowatt, and helping it be commercialized in the "private sector."[16] In the meantime, large multinational petroleum companies that have no particular interest in solar power have been buying up companies at work in this field; their motive seems to be a desire to control the configuration of whatever mix of energy sources this society eventually receives and to influence the rates of transition. Rather than simply echo what a growing number of journalists and scholars like to do when confronted with evidence of this kind, namely to look at a particular area of development and decry the role of capitalist corporations yet again, I believe it is important to reclaim the critical sense of possibilities. To keep alive our sense of which alternatives exist, why they matter and why they are not being chosen is the most important role that the study of technology and society can play right now. Otherwise, it seems to me, research and thinking of this kind can become little more than detached description or bitter lament.

In our time *technē* has at last become *politeia*. What appear to be merely useful artifacts are, from another point of view, underlying preconditions of social activity. Our instruments are institutions in the making. In a world already saturated by a myriad of devices and systems (with more and more on the horizon all the time), the modern idea that equates technology, material abundance and freedom can only be a source of blindness. If our society is to recapture the power to determine consciously its own form (rather than flitter wildly about in pursuit of each new mechanical, electronic or biotechnical novelty), each new technology of any significance must be examined with respect to the way it will become a durable infrastructure of human life. Insofar as the possibilities present in that technology allow it, the thing ought to be designed in both its hardware and social components to accord with a deliberately articulated, widely shared notion of a society worthy of our care and loyalty, a notion defined by an open, broadly based process of deliberation. If it is clear that the social contract implicitly created by implementing a particular generic variety of technology is incompatible with the kind of society we would deliberately choose, then that kind of device or system ought to be excluded from society altogether.

A crucial failure of modern political theory has been its inability or unwillingness even to begin this project: critical evaluation of society's technical constitution. The silence of modern liberalism on this issue is matched by an equally obvious neglect in Marxist theory. Both persuasions have enthusiastically sought freedom in sheer material plenitude, welcoming whatever technological means (or monstrosities) seemed to produce abundance fastest. It is, however, a serious mistake to construct one sociotechnical system after another in the blind faith that each will automatically be compatible with a free humanity. Many crucial choices about the form and limits of our regimes of instrumentality must be enforced at the time of the founding, at the genesis of new technology. It is here that our best purposes must be heard.

Through technological creation and in other ways, we make a world for each other to live in. Much more than we have acknowledged in the past, we must own our responsibility for what we are making.

University of California, Santa Cruz

NOTES

[1] Plato, *Laws*, 7.803b, translated by A. E. Taylor, in *The Collected Dialogues of Plato*, edited by Edith Hamilton and Huntington Cairns (Princeton, New Jersey: Princeton University Press, 1961), p. 1374.

[2] Jean-Jacques Rousseau, *The Social Contract*, translated and introduced by Maurice Cranston (Harmondsworth, Middlesex, England: Penguin Books, 1968), p. 84.

[3] Alexander Hamilton, Federalist No. 9, in *The Federalist Papers*, with an introduction by Clinton Rossiter (New York: Mentor Books, 1961), pp. 72–73.

[4] Thomas Jefferson, *Notes on Virginia*, in *The Life and Selected Writings of Thomas Jefferson*, edited by Adrienne Koch and William Peden (New York: Modern Library, 1944), pp. 28–281.

[5] See John Kasson, *Civilizing the Machine: Technology and Republican Values in America, 1776–1900* (New York: Grossman, 1976).

[6] See Albert O. Hirschman, *The Passions and the Interests: Political Arguments for Capitalism before Its Triumph* (Princeton, New Jersey: Princeton University Press, 1977).

[7] Denison Olmsted, "On the Democratic Tendencies of Science," *Barnard's Journal of Education*, 1 (1855–1856), reprinted in *Changing Attitudes toward American Technology*, Thomas Parke Hughes (ed.), (New York: Harper & Row, 1975), p. 148.

[8] Karl Polanyi, *The Great Transformation* (Boston: Beacon Press, 1957), p. 33.

[9] "Beta" (Edward W. Byrn), "The Progress of Invention during the Past Fifty Years,"

Scientific American, 75 (July 25, 1896), reprinted in Hughes (see note 7, above), pp. 158–159.

[10] Langdon Winner, *Autonomous Technology: Technics-out-of-Control as a Theme in Political Thought* (Cambridge, Massachusetts: M. I. T. Press, 1977).

[11] A recent sent of commentaries on the relative efficacy of regulation is contained in *The Politics of Regulation*, James Q. Wilson (ed.) (New York: Basic Books, 1980).

[12] Amory B. Lovins, "Technology Is the Answer! (But What Was the Question?): Energy as a Case Study of Inappropriate Technology," discussion paper for the Symposium on Social Values and Technological Change in an International Context, Racine, Wisconsin, June 1978, p. 1.

[13] Amory B. Lovins, *Soft Energy Paths: Toward a Durable Peace* (Cambridge, Mass.: Ballinger, 1977), pp. 6–7.

[14] See my argument in "Do Artifacts Have Politics?" *Daedalus*, Winter 1980, pp. 121–136.

[15] I suggest a way of addressing such issues in "Technologies as Forms of Life," *Boston Studies in the Philosophy of Science*, forthcoming.

[16] See for example, J. L. Smith, "Photovoltaics," *Science*, June 26, 1981, Vol. 212, No. 4502, pp. 1472–1478.

PART II

TECHNOLOGY ASSESSMENT

STANLEY R. CARPENTER

TECHNOAXIOLOGY: APPROPRIATE NORMS FOR TECHNOLOGY ASSESSMENT

1. INTRODUCTION

The search for an appropriate normative basis for evaluating technology has been conditioned in the last quarter century[1] by a type of policy analysis known as technology assessment (TA). While there is no logical necessity that the philosopher of technology frame his/her own understanding of technological phenomena according to the interpretive model which it provides, TA does aim at a comprehensive picture of the factors which define technological choices and, at the same time, directs our attention to the broader social context that is affected, often unintentionally, when a new technology is introduced, or an existing one is significantly modified. Given its claim to comprehensiveness and its growing use among the major industrialized countries, TA deserves careful and thorough analysis.

It is the claim of the present author that TA is a highly context-dependent activity that carries with it a number of assumptions about the proper role of technology in human culture, the characteristics of "rational" technological actions, and acceptable limits within which technological actors should be held accountable for socially undesirable effects caused by their actions. It is appropriate that philosophers of technology attempt to identify what such assumptions are in as explicit and precise a manner as is possible. Lacking such an identification, technology assessors seem inclined to treat contemporary industrial technology as a fait accompli. The assessment of technology becomes neither a critique of technological means nor ends, but rather a search for strategies for mitigating unwanted "side effects." Additionally, when technologists and assessors are ignorant of the norms of their action, they are more likely than not to rule "out of court" a number of alternative approaches to technology as being unfeasible, impractical, utopian, etc., believing their negative assessments to be objective conclusions of a technical character. In place of an explicit and self-conscious awareness of the norms which inform their activities, contemporary industrial technologists are likely to regard technology *per se* as evaluatively neutral, with ethical questions restricted to human intentions alone. Assessors of technology perpetuate this fact/value bifurcation with the surprising

Paul T. Durbin and Friedrich Rapp (eds.), Philosophy and Technology, 115–136.
Copyright © 1983 *by D. Reidel Publishing Company.*

assertion that TA itself can and should be performed in an objective and neutral manner,

I hope to support the claim that assertions that TA can and should be performed according to a methodology that is evaluatively neutral reflect a striking disregard for the normative assumptions that legitimate technological practices. Furthermore, when these assumptions are codified by terms such as "economic," "feasible," "efficient," etc., and given precise quantitative measures, the normative axioms of economic theory on which they are based are easily lost sight of. Because such terms possess an aura of technical precision, it becomes all too easy to dismiss technological procedures that depart significantly from convention with economic pronouncements that appear to be objective, realistic, and practical.

It is a serious enough matter that persons and firms which create technological changes show little inclination to challenge the received wisdom concerning what is or is not feasible. It is more serious if those concerned with identifying the negative effects of twentieth-century industrial technology, and with formulating realistic alternatives, allow themselves unreflectively to be bound by the same constraints. It would unfortunately appear to be the case that TA methodology constitutes a wholesale and uncritical importation of market economics. The two major components of TA, impact analysis and policy identification, thus tend to be unduly limited from the outset.

It is fair to say that the international technology assessment movement amounts to the institutionalization of doubt that the cure for technologically induced ills is simply more technology of the same type. This alone makes it worthy of examination. When, however, the limited scope of its critique is perceived and is expressly contrasted with more severe condemnations of current practice – for example, those provided by the alternative technology movement – it becomes all the more important to attempt an explanation of these limitations.

With reference to the present forum, the search for an authentic basis on which to frame an evaluation of contemporary technological practices can be aided by a criticism of TA. It can further be argued that some of the current attempts to formulate a philosophy of technology suffer, as does TA, from a failure to adequately question assumptions concerning the norms of practical action. If these common mistakes can be recognized, then it is to be hoped that advancement will have been made toward that goal of authentic technology assessment. To this end the present discussion first provides a descriptive account of the origins of TA and of the more radical program of alternative technology.

2. RECENT ASSESSMENTS OF WESTERN INDUSTRIAL TECHNOLOGY

Within the past quarter century, concerns about undesirable features of modern industrial technology have taken new forms. These challenges have gone beyond the already painfully obvious fact that twentieth-century technology, in concert with evil human intentions, has developed the capacity to obliterate our species. Instead, what is now being questioned are certain systemic properties of industrial technology itself, properties which, despite even good intentions of human actors, lead to unwanted and unanticipated results that are themselves species threatening.

The international technology assessment movement constitutes one form these concerns have taken. TA originated in the U.S.A., due in part to the fact that, given its size, resource wealth, and birth at a time when free-market liberalism was replacing mercantilism, and aversion to technological planning had become particularly ingrained as a feature of its productive systems. Its initial concerns were environmental in character. TA reflected a gradually emerging awareness of the fact that while technologies based on market economics were responsive to short-term consumer demands, certain long-term results were beginning to be recognized that were ultimately life threatening. If, due to its extreme discounting of the future, market economics contained inadequate means for systematically dealing with long-term consequences, then apparently the "invisible hand" that regulated and corrected the enterprise of technology needed additional propping up, over and above the regulatory efforts of government that had already been introduced.

TA was proposed as a new form of policy research that would assist in the isolation of potentially negative impacts that might occur when a new technology was introduced, or an existing one was extended in significantly new ways.[2] TA might scrutinize a particular project, say, a new airport or a pipeline. It might evaluate an entirely new approach to some fundamental economic activity, for example, electronic funds transfer as a replacement for paper transactions in banking or mechanical robots as substitutes for human workers on automobile assembly lines. TA might also examine a problem, say, energy scarcity, and the range of alternatives proposed as solutions. Lest TA be perceived as a threat to traditional patterns of industrial technology, its early proponents were quick to point out that its mission would also include a more systematic identification of the *positive* payoffs of technological innovation and recommendations as to how they might more effectively be realized.

By 1972, eighty-six offices of the U.S. federal government were claiming that their operations included the performance ot TAs.[3] The National Environmental Protection Act of 1969 and the Technology Assessment Acts of 1972 mandated TA and environmental impact analysis for technological projects receiving federal financing. Internationally, TA established a foothold in Canada, Japan, and Western Europe, with England, France, and the Federal Republic of Germany leading the way.[4]

TA should be viewed as a mild form of disenchantment with twentieth-century technology as it has been practiced by the industrial nations of the world. Yet, as I hope to show below, the particular form which this disenchantment has taken – an interdisciplinary team approach to social impact analysis and alternative policy identification – falls short of being comprehensive in scope.

A second form which concerns about present-day technological practices have taken may be characterized as the movement for alternative technology. Based more on intuition than coherent political theory, on small group initiatives rather than societal mobilization, and too often displaying a "hardware fetish," alternative technology – or, as it is variously called, "appropriate," "soft," "careful," "frugal," "participatory technology"[5] – represents a more radical challenge to contemporary practice than TA is disposed to tolerate. Examples of AT include alternative energy devices, agricultural practices and tools, transportation vehicles, and building designs, in all of which the emphasis is upon hardware. Other instances, however, represent attempts to transform the organizational arrangements whereby technology is developed, controlled, and delivered. These include co-operative organizations for medicine, marketing, farming, food delivery, financial credit, communications, fishing, death, and as institutional barriers begin to fall, insurance and banking. In reaction to the anonymity of life in the large urban centers, a re-emergence of emphasis on neighborhood identity is occurring, often tapping the latent artisan skills possessed by various members of the community through so-called "sweat equity" exchanges of services.

A variety of organizations have formed around the AT theme, some of which, in addition to engaging in systematic research, have attempted to buttress their approach with coherent statements of principle.[6] The number of newsletters, magazines, and journals spawned by the movement continues to grow.[7]

From the point of view of AT, the enterprise of technology should embody two fundamental design norms: sustainability and democratic patterns of organization. The concept of sustainability as a design constraint leads to selection of only those practices which can be maintained into the indefinite future. Since some current industrial practices are now recognized as interfering with the regenerative capacities of the earth's life-sustaining processes themselves – processes which provide air, water, fertile land and relatively stable climates – such practices will need to undergo important modification if sustainability is to be achieved.[8] An appreciation of the finite, and in some cases severely limited, stock of fossil fuels and other materials, calls for an economic philosophy which treats these scarce materials as capital rather than intrinsically valueless raw material inputs.

Several of the norms of AT follow from the assumption of sustainability. The artifacts of the future should be made of renewable materials, that is, materials that are grown, not created from finite material stocks.[9] From this perspective, wearing clothes made from petrochemicals or simply burning up the remaining stock of petroleum for the heat it provides is both inappropriate, and, in the not too distant future, avoidable.

Additionally, there is an emphasis on conservation, on limiting the flow of materials from manufacture to consumption.[10] This restraint would be unnecessary were one-hundred-percent recovery possible. Because it is not, even the ideal of sustainability can never be fully realized. AT is sometimes called "soft" technology to emphasize the desire to have a soft or unobtrusive impact on the ecological systems of the natural world.[11]

Sustainability has a higher probability of being realized if diverse ways of achieving the same practical end are incorporated. Emphasis on diversity reflects certain evolutionary principles of survival which suggest that species survival is enhanced when multiple strategies are available for the resolution of any given threat.[12]

AT also contains an emphasis on adapting-to, rather than merely overriding natural patterns. Admittedly, this distinction involves matters of degree. But, as will be discussed below, it is possible to devise technologies that violate natural laws and not get caught. Such short-term successes, however, involve long-term penalties, a situation that a more appropriate design from the beginning could have avoided.[13]

A second determinant of AT practice is the goal of democratic management of technological enterprises themselves. An emphasis upon anti-hierarchical forms of management and control seems unrealistic in the extreme, given the dominant position of contemporary multinational corporations, with their rigid, many-layered management a virtual necessity for effective operation. For this reason, questions of scale assume central importance with AT. An emphasis on the decentralization of productive facilities into small, relatively autonomous units appears to be the only way that democratic self-management could even begin to become a reality.[14]

Proponents of AT argue that technology can be made more democratic in an additional way. When technological tools and products are intelligible to the user a new form of power results. The user is no longer at the mercy of a mysterious, alien object, but instead can adapt, repair, and thus preserve it. In this light, the producer of flimsy, disposable objects becomes both irresponsible and politically suspect.[15]

Finally, AT holds that in fashioning a technology, the character of work itself must be included as a design constraint, rather than mere afterthought. Schumacher has proposed that every job be required to meet three desiderata: (1) a means of attaining a becoming existence; (2) an opportunity to enhance human skill capacities; and (3) the chance to overcome ego-centeredness through joint participation in common tasks.[16] By these criteria, creation of an unrelievedly boring workplace is clearly immoral. Both Schumacher and Illich distinguish between moral and immoral apparatuses, with the distinction turning on whether the pace of production is under human or machine control.

The preceding brief recapitulation provokes the following observation. Haunting the efforts of a philosopher of technology to examine and describe underlying assumptions of AT is a feeling of unreality concerning the enterprise *in toto*. Since AT aims at nothing less than a restructuring of industrial practices in the most highly developed countries of the world, and since those practices are so thoroughly embedded in social and political philosophies which define what it is reasonable and desirable to achieve with technology, the impractical character of such a venture seems all too obvious. As Winner has noted "... the alternative technologists and new community builders of our time face a world of material accomplishments and social adaptations of astounding completeness."[17]

It is therefore important that an effort be made to understand why the proposals of the alternative technologists carry so little weight. In particular, I wish to examine this question from the perspective of TA. Why are the practitioners of this methodology virtually silent about AT? Is AT nothing more than an ill-formed ideological movement, a kind of radical chic for generally well-educated dropouts from the integrated, capital-intensive society? Perhaps so.[18] If, however, it were possible to develop systematic and coherent variants upon historical patterns of industrial organization, along the lines, say, of Robert Owen's community experiments of New

Lanark, New Harmony, and Villages of Cooperation,[19] or again, as envisaged by the anti-Marxist anarchists such as Kropotkin,[20] would these capture the sustained attention of contemporary practitioners of TA?

The answer is, probably not, and the justification would very likely take one of two forms. The first, a sophisticated technological determinism, would regard the existing order as a natural working out of forces in the material world, an explanation amounting to a special case of evolutionary theories of biology in general. Rapp identifies Moscovici and Lem with such a position.[21] Complex variants of technological determinism must be ascribed to Gehlen and Ribeiro,[22] as well as to Ellul, and certainly Marx.

Among the practitioners of TA it is more likely, however, that contemporary industrial practice is viewed not as the culmination of inexorable social and physical processes, but rather as an example of rational and prudent systems of allocating scarce resources. AT would be dismissed because it lacks feasibility. But judgments as to what is or is not feasible are strongly context-dependent. It is, therefore, unlikely that the best intentions of concerned assessors of technology will lead to balanced appraisals, unless this context, its assumptions, definitions and value commitments are made explicit. Furthermore, it seems probable that it must be philosophers of technology who inject such considerations into the TA process itself by, among other things, becoming involved in the actual performance of TAs. The role of the armchair philosopher of technology seems to be a highly suspect one.

3. THE EVALUATIVE NEUTRALITY OF TA

A bedrock assumption of TA appears to be a belief that the entire enterprise can assume a stance of detached neutrality. Particularly in the case of impact analysis is this true. Impact analysis, typically performed by technically trained persons – usually economists, scientists, and engineers – assumes that the identification of impacts is fundamentally an exercise in scientific prediction. Presented with a particular instance of technology, the assessors are asked to predict what effects its introduction may be expected to produce. Obviously, the most easily identified results will be those intended by the inventors. TA, however, extends its predictions beyond this point to other effects which lie outside the main focus of inventive intentions. In particular, assessors endeavor to identify effects that may not be welcomed (though unintended results that may turn out to be fortuitous should also be sought). Typically, such indirect results are called "higher-order" effects, though this distinction seems somewhat arbitrary.[23] In general, higher-order effects appear to be connected to the original technological innovation by circuitous causal paths. For example, Lynn White Jr. observes that the medieval discovery of the chimney principle made it possible in Northern Europe to replace the manor house, with its great room and common central fire, with multi-story houses, with fireplaces in each room. As a consequence of the new heating technology, privacy was greatly enhanced, disparate social classes became increasingly separated into "upstairs" and "downstairs" worlds, and the institution of feudalism was strengthened.[24]

Frequently, impact identification in TA is structured according to categories: economic, legal, environmental, institutional, social, political, and technological (other technologies than the one under assessment).[25] To the extent possible, with a major limitation being the level at which the TA is funded, each category of impacts is examined by a person technically trained in that area.

The entire enterprise can take on a descriptive, evaluatively neutral character. Consider the following

> Technology Assessment is the process of taking a purposeful look at the consequences of technological change. It includes the primary cost benefit analysis of short-term localized market place economics, but particularly goes beyond these to identify affected parties and unanticipated impacts in as broad and long-range fashion as is possible. It is ***neutral*** and *objective*, seeking to enrich the information for management decision. Technology assessment is a tool for the renewal of our basic decisionmaking institutions – the democratic political process and the free market economy.[26]

Again, a report prepared by the prestigious National Academy of Engineering proposes that TAs can be effectively utilized by parliaments provided those actually performing the TAs are insulated from the overtly political milieu of the parliaments themselves. A small full-time staff is proposed as a buffer which would be "... placed in a position to have direct relationships with Congress as well as with the performers of the assessments, so that its [the assessment] results are produced in an environment *free from political influence or predetermined bias*."[27]

In both of these cases the assumption is clear; if the proper precautions are taken, and the technical experts are allowed to do their analyses unhindered, an objective, evaluatively neutral assessment can be produced. It is hard to avoid the conclusion that such claims amount to a blatant and naive positivism. One is asked to assume that impacts identified by this objective and neutral process will, as it were, burst into view on their own. One has only to apply the proper scientific model, turn the crank, and out come the impacts. What is missing, of course, is evidence that these practitioners appreciate the fact that every application of technical knowledge has its own selective focus. The impacts that one will predict are, in a fundamental sense, constrained by what one is looking for. Since one cannot look for everything, he must choose according to an assessment of importance. But judgments about importance are themselves conditioned by what are taken to be the appropriate scope and limits of technological action itself.

One effort to correct this misconceived notion of impact identification has been termed "adversary TA."[28] Following the example of adversarial legal process, a TA is presumed to be unavoidably biased. Identifiable groups among those whom the technology will affect are invited to perform their own impact identifications and to compile their preferred policy recommendations. A critical public debate is then held in which all interests are represented and disagreements about technical issues are separated from political choices. Finally, informed consensus can lead to democratically based decisions. The process has been criticized as being cumbersome, wasteful of effort, and politically naive in assuming technical and political questions are clearly separable.[29] In addition, concerns have been voiced that the overtly negative tone of the assessments produced by opponents of some envisaged technology may in fact cause undue alarm with the general public. Scientific and technological advance may be retarded. That this would indeed be too high a price to pay is presumed to be obvious. To assume, however, that every showdown between public sentiment and proposed scientific and technological advance must eventually yield to an inbuilt technological imperative is to reflect a technocratic contempt for democratic forms of decision-making themselves.

TA attempts to maintain an evaluatively neutral stance with its second main component – policy analysis:

> Policy analysis relates the impact assessment to the concerns of the society. It compares options for implementing technological developments and for dealing with their desirable and undesirable consequences. The policy making sectors that can deal with the options must be identified. The policy options available to these sectors are then laid out and analyzed. Probable consequences are studied and presented. Explicit policy recommendations may or may not be appropriate, depending on the preferences of the study's sponsors, users, and performers.[30]

Using the impact analysis already performed, the assessment team next pinpoints those predictions that are problematic. The team then attempts to devise a variety of alternative strategies that might lessen such impacts. For example, new institutional controls may be proposed as one option, or better ways suggested for monitoring the technology. Public education may be recommended to remove unjustified fears about the technology. In some cases the option of modifying or retarding introduction of the new technology may be spelled out. As a possible, though rarely exercised, policy option, an outright ban on further work on the technology may be proposed. The mere possibility that such might be the outcome of performing a TA led early opponents to brand it "technology harrassment" or "technology arrestment."[31]

The stance of political neutrality that TA attempts to adopt is a central feature of this methodology. The concluding section of a TA should contain a menu of alternative policy options. Choice and implementation should be left to the political process. In practice this usually means that actual selection from the menu is reserved for legislators functioning at the state (provincial) or federal level, since these individuals are assumed to be politically accountable.

The identification of policy options is even more value laden in its fundamental approach than is impact analysis. It would be a waste of time – always a scarce commodity for the assessment team – to list those options considered to be unrealistic, uneconomical, or possessing small likelihood of adoption. Yet, the apparent reasonableness of this approach amounts to tacit acceptance of the ground rules established by industrial technology itself. TA is therefore deprived of an evaluative stance that could call in question, in fundamental ways, the norms of prevailing practice. In the worst cases, then, instead of standing in critical judgment on contemporary technology, as a genuine assessment ought to do, TA becomes a legitimating influence for present practices and policies and serves as a positive obstacle to the consideration of alternative approaches to technology.

4. UNCRITICIZED ASSUMPTIONS OF TA

If TA is to achieve a genuine critical perspective, it will need to become self-reflexive about the norms that legitimate existing practice. It will need, for example, to realize that a dichotomy of impact categories into economic and social categories reflects an uncritical acceptance of a political philosophy and ethical theory – classical liberalism and utilitarianism – that is particularly inimical to the sustainable technological systems. Because this realization, in the main, is absent from the perspectives of technically-trained economists, engineers, and scientists, the formulations of analysis in quantified disciplinary terms acquires a false objectivity.

To elaborate on this point, a brief recounting of operating assumptions is in order. The system of thought underlying the industrial practices of the past three centuries is generally accepted as having been articulated and systematized by a line of thinkers from Locke to Bentham.[32] Reflective of the natural order or the material universe which had achieved comprehensive theoretical elaboration in the great achievement of Newton, this new wave of thinking about the social world sought to ground human actions on an equally firm system of laws. New possibilities for free human action were envisaged for a system of private ownership, on the basis of which individual skills, ambitions, and material resources, in combination with patterns of contractual arrangements, would contribute to an overall process that was maximally efficient from the perspective of production, and was also intrinsically self-regulating. The inducements for human actors that would make realization possible for such a system consisted in protection of the right of unlimited appropriation. Indeed, Locke held this right to be inalienable, not one that was bestowed by the society.[33]

The right of unlimited appropriation received ethical grounding in the theory of utllitarianism, according to which man was pictured as primarily a creature of desires, unlimited in extent, both innate and acquired in origin and alternatively sensual or rational in character. Macpherson observes,

> But if the postulate of man as infinite appropriator was too stark, there was another that appeared more moral and would serve as well. This was the postulate that man was essentially an unlimited desirer of utilities. . . . Man is essentially an infinite consumer.[34]

However, while the postulate of man as infinite consumer does not necessarily make him an infinite appropriator, only a simple additional minor premise is needed to convert him into that. The premise required is merely that land and capital must be privately owned to be productive (a premise which Locke, for instance, explicitly made). Then to realize his essence as consumer, man must be an appropriator of land and capital. Man the infinite consumer becomes man the infinite appropriator.[35]

The Locke to Bentham theory of society was questioned in the nineteenth century in the form of a democratic challenge to the principles of utility. Socialists, anarchists, and even neo-utilitarians such as John Stuart Mill sought to demonstrate that, at best, the combination of market economics and utilitarian ethics produced a system that was maximally efficient in production. In no sense did it follow, nor did actual practice even begin to bear it out, that such a system resulted in efficient, that is to say equitable, distribution. From the agonizing reappraisals of Mill to the twentieth-century welfare economists, however, attempts to augment the stark theories of market economics with greater concerns for equity among those not operating within totally planned economies have left the right of unlimited private appropriation essentially intact.

With few exceptions, the postulates of infinite human desires and the implied need for an ever expanding production system to provide for their satiation, have been addressed by the axiom of growth.[36] According to this principle, and even assuming a stabilization in the human population, a successful economic system that can both meet the ever-increasing stock of human desires and, at the same time, offer hope toward improving the lot of the poorest members of the society, must unceasingly consume more of the earth's raw materials and produce greater quantities of products. The goal of sustainability for technological systems is thus seen to be in fundamental conflict with contemporary industrial practice and theory – a conclusion which, if true, helps to explain the air of unreality surrounding proposals for alternative technology.

TA should be viewed as operating within the framework of economic theory just described. In fact, the centrality of cost-benefit analysis to TA methodology would seem to position it in the tradition from Mill to the contemporary welfare economists, especially in its identification of negative social impacts and in its belief that offsetting policies in the form of taxes, regulation, and other forms of governmental response, can provide sufficient correctives to technological abuse. Like welfare economics, TA represents an acquiescence to an economic system which defines what are, and what are not, legitimate factors to be included in the costs of production. Invariably, it takes as its starting point a particular envisaged technology, the feasibility of which has already been determined by private developers, utilizing accounting procedures which consider certain economic factors such as cost, marketability, existing government regulations, and which relegate a host of others to the limbo of "externalities."

The concept of externalities of social costs has received extensive treatment by welfare economists. Intrinsic to the principle of unlimited acquisition, in concert with rights of private ownership, is the fact that only those items are included in the computation of cost of production for which the entrepreneur has had to pay a price. At the same time, the production process itself will produce effects that transcend his domain of responsibility and cost. Neo-classical economist A. C. Pigou expresses it in terms of "marginal social product" and "marginal private net product." By the latter he means "that part of the total net product of physical things or objective services due to the marginal increment of resources in any given use or place which accrues in the first instance – i.e., prior to sale – to the person responsible for investing resources there." On the other hand, "marginal social product" includes the "total net product of physical things or objective services due to the marginal increment of resources in any given use or place, no matter to whom any part of this product may accrue."[37] Because the "marginal social product" can, and frequently does, include costs that must be borne by parties other than the producer, it tends to be lower than the "marginal private net product."

One of the most comprehensive studies of this phenomenon is provided by Kapp.[38] With careful data to buttress his case, he documents the social costs that society must pay for its commitment to private enterprise. Costs are identified, and quantified where possible, in the areas of impairment of human health, unemployment, air and water pollution, depletion and destruction of animal resources, depletion of energy resources, soil depletion, erosion, and deforestation. Additional systemic costs are identified with distribution and transportation. The existence and indeed marked growth of monopoly capitalism is shown to exacerbate social costs, and to reduce production efficiency. It is also noteworthy that Kapp shows no fondness for planned economies either, but to the contrary identifies the social costs which public utilities, public enterprises, and government agencies themselves generate. In an introduction to the reissuance of his book first published twenty-one years prior, Kapp holds that there is now even less basis for approval of the cost-accounting procedures of the modern industrial society. New forms of social costs are now recognized. Environmental contamination from radioactivity, heat, and pesticides is an acknowledged fact. Definitions of environmental pollution have had to be broadened to include noise and aesthetic intrusions. Chemical waste disposal is now claimed to be the number one threat to human life (excepting nuclear war). The regenerative capacities of the world's oceans themselves are threatened.[39]

5. IMPLICATIONS FOR THE PHILOSOPHY OF TECHNOLOGY

My remarks have been directed to a substantive criticism of TA.[40] It has been argued that an assessment of current technological practices and accounting procedures that accepts a division of issues into internal costs, which are computed with sophisticated economic tools, and external costs which are outside its concern, is flawed in principle. If this claim is correct, a partial explanation will have been given as to why proposals for alternative technologies, which challenge this bifurcation, receive such short shrift.

The failure of TA to adequately appreciate its self-limitations – its acceptance of the assertion that the effects of technology can be divided into "first-order" effects, involving direct costs, and "higher-order" effects which have social costs – can be instructive to the philosopher of technology. The payoff in terms of theoretical rigor that is gained when internal costs are isolated and analyzed is undeniable. Indeed, the application of cost-benefit analysis to the externalities is itself an example of flattery by imitation. The hope is that the same objective, quantified rigor that characterizes market economics can be applied to the mitigation of social costs. Yet this objectivity is illusory, and it should be part of the program of philosophy of technology to demonstrate this point. Such has, however, not always been the case.

Kotarbinski, for example, has sought to formulate a science of effective practical action – praxiology – based on the assertion that it is possible sharply to distingiush praxiological concerns from matters of ethics, aesthetics, and politics. An extended analysis of his careful arguments would seem to be in order.[41] Can presumed praxiological concepts such as "instrumentalization," "specialization," "planning" be given technical definition outside a particular economic theory? Early in his work Kotarbinski cities Frederick Taylor's scientific management as an example of a praxiological approach to business. Yet others have persuasively argued that Taylor's "scientific" findings amounted to rationalizations of management interests, and that Taylor's efficient "one best way" presupposed a theory of appropriate work that was highly controversial for its normative assumptions.[42]

When production expenses are limited to internal costs, the concept of effectiveness, or the more familiar designation "efficiency," can take on an air of objectivity. As an electrical engineering student I was taught to compute it by straightforward formulas. The ratio of output divided by input, multiplied by 100, gives the efficiency. Considerations of social cost expose just how value-laden and arbitrary the process of calculation actually is. Kapp observes that early procedures for petroleum recovery made it rational, once a strike had been successful, to drill as many wells as possible and extract the oil as soon as one could. Given the peculiar properties of oil to migrate to a neighbor's field – in contrast, for example, to coal – one had to act as fast as possible. By the same token, an adjacent neighbor was acting with maximal efficiency if he too sunk many wells in as short a time as was possible. After all, some of the original strike was possibly coming from his land. In the process, natural gas was often blown off in the air.[43]

There is, of course, an inexorable logic at work in an example such as this. Given the promise of private ownership of a valuable commodity that defies precise containment, the process is understandable. Yet, the social costs that have been generated are very much in evidence today. Hardin has termed the mechanism of market economics that forces the prudent developer of any commonly held resource to act as quickly and efficiently as possible, the "tragedy of the commons."[44] Whether that common resource be air, water, flora, fauna or mineral, the private developer who holds back is acting inefficiently and will lose out to the competition. If it is possible under these circumstances to present a definition of "efficiency" in terms that are free of ethics, aesthetics and politics, that case has yet to be made.

If the assessment of technology is an appropriate task for philosophers – and I take the existence of the present forum as an assertion that it is – then terms such as "efficiency" will require sustained analysis. Fundamental inconsistencies between the constraints of a particular production philosophy and those that seem appropriate in terms of long-term species survival will need to be exposed. A preoccupation by philosophers of technology with fine-grained examination of the ways our contemporary high cultures actually perform technology, in the form of analysis of technical rules and maxims, discussions of the role of science in technology – in short, technoepistemology – is incomplete, at best.

More seriously, such a preoccupation can lead to errors of the following kind. The so-called "marriage of science and technology" that characterizes late-nineteenth and twentieth-century industrial practices has been used to explain the success of the enterprise. A philosophical analysis of the concept of "possibility" reveals logically distinct types of constraint on practical action.[45] Successful technologies are explained by their conformities to maxims of action that are first, self-consistent, that is, not contradictory, and secondly, consistent with the laws of nature, a requirement that applied science allows one to realize. Finally, at a less basic level, technological acts are claimed to be consistent with the maxims of good economic practice, in accordance with which, designations of "economic," "efficient," "feasible," "economic," etc., broadly define what is possible. If one then assumes, as this hierarchy of levels of possibility would suggest, that technologies that are, in fact, feasible must of necessity have violated no laws of nature, then existing technological practices take on an objectivity grounded in the objectivity of science itself.

Such an assumption I now believe is in error, resting on a mistaken notion of the ways science is utilized by technology. The role of science in technology is contingent, piecemeal, and subject to the highly selective focus of technologists themselves. It is distinctly possible to fashion technologies that violate scientific laws and to achieve success in so doing. If, for example, the violation is one which will take decades or centuries for its realization, and practical intentions are governed by economic considerations of a short-term nature, the human actors can simply choose to ignore the violation. Technologies which deplete the earth's mineral reserves or threaten to upset the temperature-regulating mechanisms of the planet must certainly fall into this category. Actions which violate long-term laws of nature thus possess their own rational basis, and internal coherence. Accepted practice dictates which of nature's laws one can violate and not get caught, and which others one is prudent to attend to. By isolating technoepistemology from technoaxiology one runs the risk of fashioning an analysis possessing "pseudo-objectivity."

6. CONCLUSION

The purpose of this discussion has been to examine a recently developed tool for the identification of social policy entitled Technology Assessment (TA). The very existence and widespread practice of TA bears witness to growing concerns that technological practices which originated with the modern industrial era may produce unintended and often unwanted results. Because TA claims to provide a rational way of dealing with such results, it is appropriate that philosophers of technology examine this claim. Is it, in fact, rational? Is the way TA characterizes technology sufficiently comprehensive? Is it accurate?

Our response to such questions has been to argue that substantive problems exist with the general approach. The main point of the argument has been a demonstration of how completely TA chooses to frame its assessments within the social and political framework of market economics and utilitarian ethics. By failing to consider criticisms of common practice that would be implied by alternative frames of reference, TA both limits its range of impact analyses, and also rules out apriori certain policy options that are worthy of consideration. Operating from within this perspective, radically different approaches, such as are discussed by proponents of Alternative Technology (AT), appear so unrealistic as not to merit consideration by the assessors.

In a similar manner, the enterprise of philosophy of technology perpetuates an unduly narrow picture of technology if it too restricts itself to an analysis of common practice of the modern industrial society, including, for example, attempting explications of terms such as "efficiency," "feasibility," "economic," etc.

It has not been my intention to suggest that either TA or analytic philosophies of contemporary technology are fallacious. My assertion of substantive problems with TA implies that it is insufficiently comprehensive. Similarly, as I believe my colleague Kristin Shrader-Frechette has recently demonstrated,[46] a philosopher can perform an important clarifying function within the limits of TA methodology, both by clarifying conceptual fuzziness of concepts central to TA practice – e.g., the interpretation of the Pareto Principle in cost-benefit analyses – and also by forcing the performances of actual TAs to measure social costs more consistently and accurately and to achieve greater equity in their distribution. To do this literally requires involvement of the philosopher as an active team member.

In a sense, however, the broader scope for a comprehensive program of technology assessment which I have been calling for amounts to a subcase of the internalist/externalist arguments that have proliferated within the philosophy of science, especially since the midpoint of the present century.[47] The crux of that disagreement appears to be the issue of whether one is likely to obtain an adequate understanding of science by reconstructing logical schemas of statements of scientific law, observation, prediction and explanation, or whether a true picture of science is more likely to be obtained when one considers the process of scientific change itself, patterns of theory modification and replacement, and the procedures, standards of proof, and central concepts that define and set one research program and community off from another.

The analytic philosophy of technology, and in a particularly applied way, TA methodology, follows an internalist program, one which takes modern high technology, its practices and underlying picture of society as a given, and seeks to render explicit the concepts and norms that are operative. On the basis of the limitations inherent in this program, which it has been the objective of this paper to identify, a program of technology assessment should also be pursued that considers alternative economic philosophies, definitions of good work, and alternative technological artifacts and procedures as appropriate candidates for detailed analysis and criticism.

Georgia Institute of Technology

NOTES

[1] For a description of the origins of the TA movement, see Porter, A., Rossini, F., Carpenter, S., and Roper, A. *A Guidebook for Technology Assessment and Impact Analysis* (New York: North Holland, 1980), esp. Chapter 3.

[2] Daddario, E. (1968) "Subcommittee Report: Science, Research and Development, Committee on Science and Astronautics," U. S. House, 90th Congress, First Session. Illustrative examples of TAs: Jamaica Bay Environmental Study Group (1971) *Jamaica Bay and Kennedy Airport*, Washington, DC: National Academy of Sciences and National Academy of Engineering; U. S. Congress, Office of Technology Assessment (1978) *A Technology Assessment of Coal Slurry Pipelines*, Washington, DC: U. S. Government Printing Office; Little, Arthur D. Inc. (1975) *The Consequences of Electronic Funds Transfer*, Cambridge, MA: Prepared for National Science Foundation under contract NSF – C884; BDM Corporation (1977) *Study of Alternative Strategies and Methods for Conserving Energy*, McLean, VA: Prepared for NSF under contract NSF – C885; U. S. Congress, Office of Technology Assessment (1978) *Application of Solar Technology to Today's Energy Needs*, Washington, DC: U. S. Government Printing Office.

[3] Coates, V. T. *Technology and Public Policy: The Processes of Technology Assessment in the Federal Government* (Washington, DC: George Washington, 1972), University Program of Policy Studies in Science and Technology, vol. 1, p. 5.

[4] Vlacos, E. "Transnational Interest in Technology Assessment," paper presented at International Society of Technology Assessment Conference (Ann Arbor, Michigan 13 October, 1976).

[5] An extended list of the characteristics of alternative technology is provided in Dickson, D. *The Politics of Alternative Technology* (New York: Universe Books, 1975), pp. 103, 104. It should be noted that while "intermediate technology" is sometimes included as an example of alternative technology, its actual application amounts to a special case. Intermediate technology is proposed for Third World countries as a more satisfactory program of development than is represented by conventional attempts to replicate technological practices of the major industrial countries within the developing world. The context-specific concerns of intermediate technology are not addressed by the present discussion. For a treatment of these issues, cf. Schumacher, E. *Small is Beautiful: Economics as if People Mattered* (New York: Harper Torchbooks, 1973).

[6] A representative listing of this rapidly expanding movement would include: New Alchemy Institute, Woods Hole, Massachusetts and Prince Edward Island, Canada; Farallones Institute, Point Reyes Sta., California; RAIN, Portland, Oregon; Brace Institute, Montreal; Institute for Local Self-Reliance, Washington, DC; Intermediate Technology Development Group (founded by E. F. Schumacher), London; Domestic Tech, Denver; De Kleine Aarde, Netherlands; Lindisfarne Association, Long Island; U. S. National Center for Appropriate Technology, Butte, Montana. It must be admitted, however, that the list of casualties is probably as long as that of the survivors.

[7] Cf. *Mother Earth News, Foxfire Books, Journal of the New Alchemists, Co-Evolution Quarterly, Rain, Science for the People, Workforce, Shelter, Self-Reliance, Alternative Sources of Energy, The Ecologist, Undercurrents, Resurgence, Appropriate Technology, Alternatives*, and an even greater number of newsletters.

[8] Cf. MacDonald, G. "The Modification of the Planet Earth by Man," *Technology Review*

71, 1 (October–November 1969), pp. 27–35, for an early expression of this concern.
[9] A. Lovins notes that soft technologies "rely on renewable energy flows that are always there whether we use them or not, such as sun and wind and vegetation: on energy income, not on depletable energy capital": "Energy Strategy: The Road Not Taken," *Foreign Affairs* 55, 1 (October 1976), pp. 77, 78. Lovins develops his program of soft energy paths more fully in *Soft Energy Paths: Toward a Durable Peace* (Cambridge, MA: Ballinger, 1977).
[10] There is a greater tendency these days on the part of national leaders and policy makers to pay lip service to the norm of conservation than has heretofore been the case. Proposed cuts in conservation research programs, along with decisions to proceed at full speed with new energy production schemes expose such pronouncements as rhetoric. A serious and analytical demonstration of the benefits of energy conservation is provided in Stobaugh, R., and Yergin, D. *Energy Future: Report of the Energy Project at the Harvard Business School* (New York: Random House, 1979), pp. 136–182; see also BDM Corporation, *op. cit.*, note 2.
[11] R. Dubos reminds us that man always changes the environment of which he is a part. The ideal is not to leave the planet untouched – a clear impossibility – but rather to interact with nature in creative and harmonious ways that continue to enhance present and future existence: "St. Francis and St. Benedict," *Psychology Today* (May, 1973), pp. 54–60.
[12] *Ibid.*
[13] The design of houses provides an illustration of the difference between technologies that adapt to the environment and those that simply override the inclemencies of the weather. A growing number of homes are being built according to principles of "passive solar architecture." By taking into account average yearly solar insolation, prevailing wind patterns, and careful siting, it is possible to design buildings that can be heated and cooled to comfortable levels without dependence on large quantities of energy supplied by non-renewable energy sources. Existing practices, on the other hand, borne of an era of artificially depressed fossil fuel prices, simply override the environment, by creating living spaces, the comfort of which is mechanically contrived. A definitive book on the principles of passive solar design is provided by E. Mazria. *The Passive Solar Energy Book* (Emmaus, PA: Rodale Books, 1979).
[14] The emphasis on decentralization of social organization for work and community places AT in a direct line of descent from the philosophical anarchists of the 19th and 20th century. Early quarrels which the anarchists had with the Marxists, who argued for centralization of production, are repeated in contemporary writings. From the perspective of alternative technology, neither the market economies of the West nor the planned economies of the Eastern bloc represent sustainable modes of technological action. An example of the older anarchist position is provided by P. Kropotkin *Mutual Aid* (Boston: Extending Horizon Books, 1914) and *Fields, Factories and Workshops* (1899), republished (New York: Harper Torchbooks, 1974). For an example of modern-day anarchist argument, cf. M. Bookchin *Post-Scarcity Anarchism* (San Francisco: Ramparts Press, 1971).
[15] Illich, I. *Tools for Conviviality* (New York: Harper & Row, 1973). For a phenomenological analysis of instruments, cf. D. Ihde *Technics and Praxis* (Boston: Reidel, 1979), pp. 28–50.

[16] Schumacher *op. cit.*, note 5, p. 51.

[17] Winner, L. "The Political Philosophy of Alternative Technology: Historical Roots and Present Prospects," in Lovekin, D. and Verene, D. (eds.), *Essays in Humanity and Technology* (Dixon, IL: Sauk Valley College, 1977), p. 131.

[18] Such was the conclusion of J. Coates, one of the founders of TA methodology, in remarks before the annual meeting of the American Association for the Advancement of Science in a symposium on AT, Denver, 22 February, 1977.

[19] Owen, Robert, *The Book of the New Moral Order* (London: 1836; German trans., 1840, *Das Buch der Neuen Moralischen Welt*, Nordhausen. Additions to the original: Parts II, III, 1842; IV–VII, 1844, London). For a recent analysis of the Owenite movement, cf. J. Harrison *Quest for the New Moral World* (New York: Scribner's, 1969).

[20] Kropotkin, *op. cit.*, note 14.

[21] Rapp, F. *Analytische Technikphilosophie* (Freiburg: Alber, English translation of this work by S. Carpenter and T. Langenbruch as *Analytical Philosophy of Technology* (Boston: Reidel, 1978), p. 146.

[22] *Ibid.*, p. 138.

[23] Cf. my discussion in (1977) "Philosophical Issues in Technology Assessment," *Philosophy of Science* **44**, 4 (December), pp. 574–93, esp. p. 588.

[24] L. White, Jr. (1974) "Technology Assessment from the Stance of a Medieval Historian," *American Historical Review* **79**, 1 (February), p. 9. White's thesis concerning TA, supported with characteristic richness of detail and beauty of style, is: ". . . if it is not to be misleading (TA) must be based as much, if not more, on careful discussion of the imponderables in a total situation as upon the measurable elements." Whereas White has his doubts about TA as practiced, P. Drucker has none. Noting that the U.S. Congress had recently set up an Office of Technology Assessment, he concludes, "This, one can say with certainty, is going to be a fiasco. Not only is 'technology assessment' of this kind impossible but also it is likely to lead to encouragement of the wrong technologies and to discouragement of the technologies we need. The future impact of new technology is almost always beyond anyone's imagination": (1973) "New Technology: Predicting Its Impact," *New York Times* (8 April).

[25] Porter, A. *et al., op. cit.,* note 1, pp. 157, 158.

[26] Carpenter, R. "The Scope and Limits of Technology Assessment," in *Technology Assessment*, R. Kasper (ed.) (New York: Praeger, 1972), emphasis added.

[27] U.S. National Academy of Engineering (1969) *A Study of Technology Assessment*, Committee on Public Engineering Policy (COPEP), prepared for U.S. House Committee on Science and Astronautics, Washington, DC, pp. 13, 14, emphasis added.

[28] Green, H. (1970) "The Adversary Process in Technology Assessment," *Technology and Society*, 5, pp. 163–67; see also U.S. Senate (1972) "Technology Assessment for the Congress," Staff Study of the Subcommittee on Computer Services of the Committee on Rules and Administration, Washington, DC: U.S. Government Printing Office (1 November), p. 15.

[29] Wynne, B. (1975) "The Rhetoric of Consensus Politics: A Critical Review of Technology Assessment," *Research Policy* **4**, 12–14.

[30] Porter *et al., op. cit.,* note 1, p. 60.

[31] U.S. Senate, *op. cit.*, note 28, p. 22.

[32] Cf. Macpherson, C. (1967) "Democratic Theory: Ontology and Technology," in

Political Theory and Social Change, D. Spitz (ed.) (New York: Atherton Press), pp. 203–20; see also Ferkiss, V. *The Future of Technological Civilization* (New York: Braziller, 1974), esp. pp. 3–61.

[33] "Thus Locke's social contract differed from that of Hobbes in that it created not rights but a mechanism for protecting the rights which already existed in the state of nature," V. Ferkiss, *ibid.*, p. 25. This claim acquired the status of a founding principle of the United States by its inclusion in the preamble to the U.S. Declaration of Independence.

[34] Macpherson, *op. cit.*, note 32, pp. 209, 216.

[35] It should be noted that the system of classical economic thought that formed around these assumptions was not quite as brazen in its original conception as it has become. Adam Smith, for example, held that the basis for market economics, in addition to the principle of free competition, must include government establishment of public institutions and public works that would never be "profitable" in the market sense, as well as the ethical ideal that competitors restrain their activities with "sympathy" and "moral sentiments": A. Smith *The Wealth of Nations* (New York: Modern Library, 1937 edition), p. 681.

[36] Among attempts to formulate coherent alternatives to growth economics, cf. Georgescu-Roegen, N. *The Entropy Law and the Economic Process* (Cambridge, MA: Harvard Univ. Press, 1971); Daly, H. (ed.) *Toward a Steady-State Economy* (San Francisco: Freeman, 1972). Among the challenges to the growth axiom, cf. E. Mishan (1971) "On Making the Future Safe for Mankind," *The Public Interest* (Summer), 33–61 and *Technology and Growth* (New York: Praeger, 1969). Additionally, cf. E. Goldsmith *et al.* (1972) "A Blueprint for Survival," *The Ecologist* 2; and the Club of Rome sponsored studies including D. Meadows *et al. The Limits to Growth* (New York: Universe Books, 1972); Mesarovic, M. and Pestel, E. *Mankind at the Turning Point* (New York: Dutton, 1974).

[37] Pigou, A. *The Economics of Welfare* (London: Macmillan, 1932), 4th ed., pp. 134–135.

[38] Kapp, K. *The Social Costs of Private Enterprise*, revised edition (New York: Schochen, 1971).

[39] *Ibid.*, p. xxv.

[40] For methodological criticisms of TA, cf. Porter *et al.*, note 1, 460–67 and my (1978) "The Problematic Character of Technology Assessment," *Proceedings: 16. World Congress of Philosophy*, Düsseldorf (27 Aug. – 2 Sept.) pp. 151–54. Cf. also Weinstein, J. (1981) "Is Social Impact Analysis Good Social Science?" Draft of remarks for presentation at 44th annual meeting, Southern Sociological Society, Louisville, KY (April 8–11).

[41] Kotarbinski, T. *Praxiology: An Introduction to the Science of Efficient Action*, translated by O. Wojtasiewicz (London: Pergamon, 1965).

[42] Guest, R. "The Rationalization of Management," in *Technology in Western Civilization*, vol. II, M. Kranzberg and C. Pursell, Jr., (eds.) (New York: Oxford University Press, 1967), p. 59.

[43] Kapp, *op. cit.*, note 38, pp. 108–111.

[44] Hardin, G. (1968) "The Tragedy of the Commons," *Science* **162** (13 December), 1243–48. Hardin elaborates on this thesis in *Exploring New Ethics for Survival* (Baltimore, MD: Penguin Books, 1972).

[45] Cf. my own discussion along these lines in (1977) "Philosophical Issues in Technology Assessment," *Philosophy of Science* **44**, 4 (December), 579–83; see also H. Skolimowski (1970) "Problems of Truth in Technology," *Ingenor* **8**, pp. 41–46.
[46] Shrader-Frechette, K. (1980) "Technology Assessment as Applied Philosophy of Science," *Science, Technology, & Human Values* **33** (Fall), pp. 33–50.
[47] Cf. Toulmin, S. (1977) "From Form to Function: Philosophy and History of Science in the 1950s and Now," *Daedalus* **106**, 3 (Summer), pp. 143–162.

HANS SACHSSE

COMMENT: WHAT IS ALTERNATIVE TECHNOLOGY? A REPLY TO PROFESSOR STANLEY CARPENTER

Modern man is dissatisfied with the technology we have and would replace it with another. Three very different reasons are given for this. I will divide my treatment of them under three headings, listing the reason in each case together with a reply.

(1) The *harms* associated with technological progress can be serious – though in principle they can be avoided. To prevent these harms, what are needed are certain provisions for the protection of man and the environment.

Laws and regulations of this sort have been around since the beginning of modern technology: for example, child labor laws, regulation of working hours, job safety rules for technical workers, and many other rules of the same sort. As technological development has progressed, unfavorable secondary consequences have taken on greater weight to such an extent that today, to channel technological development and prevent dangers to society, a great deal of research as well as legislation is required. Taking all this into account, it is clear that all forms of technological activity require ethical, limit-setting control; the end does not justify the means.

(2) There are serious dangers associated with the *misuse* of technology. Anyone who controls technological means and methods has power, and power can always be misused – for exploitation and repression, for violence and war, for deception and manipulation.

But this is a common human problem, not something peculiar to technology; not a problem of technological progress, but of the aims of technological activity. But the aims of technology are non-technological, and as with all human aims are both good and bad. Setting suitable goals for the management of technology is not a problem for technology but for the social and political system.

(3) The third type of reason concerns the very *structure* of technology. That is that the power conferred by technology is so great that man, psychologically, is no longer capable of trying to resist its misuse. Furthermore, because of the long-drawn-out production process, the extreme division of labor, and the impenetrable character of modern science, the whole system has lost its intelligibility, leading to alienation, even to an escape from the control of experts. Back of this often stands a critique of completely

Paul T. Durbin and Friedrich Rapp (eds.), Philosophy and Technology, 137–139.
Copyright © 1983 *by D. Reidel Publishing Company.*

dispassionate, instrumental thought, claiming that the result, the consequence of this development, is that life becomes inhuman. This leads to a demand for an "alternative technology" according to "human scale." The characteristics of alternative technology are claimed to be these: small, intelligible, with as little division of labor as possible.

This last objection in fact does get to the essential feature of technology. But it can be countered that, in the strong sense, there is no such thing as alternative technology. In every case, technology is a means to make some goal more easily attainable. If there are different ways of proceeding to attain these goals, whatever has the smaller cost-to-yield ratio is better. It is important, thus, to weigh cost against yield in a critical fashion, as well as to consider social aspects and future consequences. This is the measure, and (together with ethical and legal limits) there is no other way to evaluate technological methods. There are surely alternatives to technological progress since we are overwhelmed with choices among methods and must always seek out the best way in each situation; however, there is, to technology as a whole, no alternative which would have us follow a measure other than the one mentioned.

The optimum size of a business depends, in each situation, on the givens: costs for capital, labor, raw materials, as well as the infrastructure of distribution and sales systems and the community in which the business must be integrated. Depending on the differences among projects especially in different countries, one can come to different results. That is how it goes with the weighing of goods. Although most often one large enterprise is more profitable and less damaging to the environment than many small ones, at other times this also leads to greater risks. Hence that principle known to every engineer by heart: as massive as necessary, as light as possible. A fixed "human scale" there cannot be. Man, who travels the sea in storms, hunts wild game, and turns the soil into plowed fields, Sophocles rightly calls "monstrous." Every era stands before new possibilities which present themselves to it, before a new weighing of goods, of risk and renunciation – where the renunciation involves not only material but spiritual goods, such as freedom and the expansion of knowledge.

A similar weighing takes place with regard to the division of labor. We make use of it for the advantages that specialists working together make possible, but given the problems specialization brings here also the same thing makes sense: as much as necessary and as little as possible. Anyone who would not put up with this essential feature of technology is not opting for an "alternative technology"; rather is he necessarily, however willingly, giving up, to a greater or lesser extent, the possibilities that come with technology. In fact and therefore, whoever demands an "alternative technology" is in reality demanding *less* technology. There is no question that the fundamental problems of human existence are increased and intensified by technology. We can thus understand it if people no longer feel equal to the task and cut themselves off, seeking a "simple" life. This sort of adjustment should be respected as an authentic way of life. To be sure, those concerned must then also be ready to give up the corresponding advantages that come with modern technology.

It would take us too far afield to list these advantages here. It might suffice to suggest only that modern technology has doubled life expectancy, has freed mankind from stress, need, and labor, and has, by way of today's forty-hour week (counting ten hours for eating, sleeping, and personal hygiene), provided for twice as much leisure time as work time.

Technology also supports spiritual as well as material existence. Its only meaning is to set free, to create the conditions for freedom, to put at our disposal a general conscious sharing of life and inner enrichment by way of comprehensive information that conquers space and time. To be sure, we know that we have not yet made much of a beginning with the use of this free time, and one sign is that we do not yet understand what technology offers us. Free time should be used in a more meaningful way for that in which professional specialization comes up short: namely, general culture. It is a mistake to believe that, because of purely thing-oriented, instrumental thought and narrow concentration on detail, our humanity must necessarily become stunted. Insofar as technology sets man free, it creates room for general culture and provides the possibility of pursuing universal interests. And the right environment for technology also demands this general culture in order to overcome "cultural lag," that moral residue that has arisen from the onesided favoring of practical knowledge.

Thus, in order to integrate technology in our life in a meaningful way, we do not need an "alternative technology" – the different character of which, in any case, has no clear definition and involves no concrete, well-worked-out proposals. (One start toward this has been provided by O. Renn's book, *Die sanfte Revolution – Zukunft ohne Zwang*? [The gentle revolution: future without constraints?], Essen, 1980.) Instead, it would be more beneficial to develop technology optimally, using restraint and impartiality, so that it aids, on a global scale, in the pursuit of freedom from hunger and misery and creates the freedom for everyone consciously to take part in and contribute to public life.

University of Mainz

FRIEDRICH RAPP

THE PROSPECTS FOR TECHNOLOGY ASSESSMENT

This paper is designed to draw attention to specific problems of Technology Assessment (TA) and to indicate the range of results that can reasonably be expected from this emergent procedure.[1] In doing this only the *formal* features are discussed. Their consideration is necessary and at the same time sufficient to elucidate the epistemological status of TA. The complementary *material* problems – which concern, for example, the "true needs" and the ideal of the good life, and also our responsibility towards future generations or towards nature – are not discussed here. Needless to say, the interest of TA is aroused by these concrete problems. In what follows they are referred to in an overall way, but their (controversial) details are not discussed, the underlying premise being that a metatheoretical analysis, as performed here, can contribute *indirectly* to a better understanding of these problems. By its very nature such an analysis has a double function. It demarcates a sort of safe ground – i.e., a sphere within which theoretically justified conditional assessments can be expected – and at the same time it excludes the problematic and unsafe area of dogmatic prescriptions.

TA is not a completely *new* phenomenon. In the past, technological innovations were not put to use at random but rather with regard to their expected effects. Under the circumstances given and relying on culture-dependent criteria man has always aimed at what he regarded as the highest possible efficiency in his actions. Only in this way is he able to use his potentials economically and at least partly reconcile the constant disparity between the always limited resources at his disposal and the potentially unlimited needs in all spheres of human activity. In any situation of choice, i.e., when at least two different courses of action are possible, a decision must be made. If this decision is not to be the result of mere intuition but deliberately selected and theoretically considered, and hence a rational one, then, at least in part and implicitly, some sort of assessment of the relative values of the alternatives given has to take place. This is to say that an order of priority for the effects desired and for the means available for attaining them will be taken into account.

The point is that hitherto TA was mostly performed implicitly and on an intuitive basis rather than explicitly and by means of scientific methodology.

Paul T. Durbin and Friedrich Rapp (eds.), Philosophy and Technology, 141–150.
Copyright © 1983 by D. Reidel Publishing Company.

This is only natural, since in the past the effects of technological innovations were, roughly speaking, on a comparatively limited scale, and did not too far exceed the immediately desired results, which were usually considered only in terms of engineering efficiency and economic profit. In contrast to this, many people nowadays feel that the far-reaching side effects of (large-scale) technological innovations on the social and the physical environment are reaching a critical threshold, so that deliberate assessment of them becomes imperative.

The character of TA as conducted hitherto indicates a basic problem, that the (desired) status of TA within the *process of decision-making* is far from clear. As technological innovations do not come about by themselves but are intentionally brought about by man, their genesis must in principle be open to a decision-theoretical analysis. But in fact the complex process of social choice which results in concrete decisions about technological innovations involves various stages (research and development, construction of the first prototype, broad diffusion), different levels (managerial and/or planning decisions, marketing, advertising, consumer behavior), and many institutional frameworks (economics, government, public opinion). Within scientific disciplines such as decision theory, the theory of innovations, economics, sociology, and political science, specific aspects of this process are investigated. It is no coincidence that in all of these disciplines only a conceptually isolated segment of the whole process of technological innovation is taken into account.

In fact there is no overall approach that would be capable of assigning to all of these disciplines a systematic place within a comprehensive and all-embracing account, let alone a well-elaborated supertheory. Thus it is not surprising that in the TA programmes performed up to now the desired interdisciplinary collaboration of scholars from different disciplines did not result in the coherent synthesis ideally aimed at. It would indeed be unfair to demand that in specific ad hoc programmes of TA a degree of synthesis should be attained to which no parallel in the whole range of scholarship in the social sciences exists. One could even go so far as to claim that due to the manifold and mutually interfering intentions and decisions and the complex network of interacting processes involved in bringing about technological innovations, it is unavoidable that different explanations will appear, just as a photograph of a house can be taken from different points of view. This being the case, the idea of a perfectly stylized scheme in which TA holds an unambiguous and unchallenged place must be given up. The vague scientific and methodological status of TA almost necessarily extends to the concrete role TA is to play within the process of technological innovation. Evidently the efficiency of TA is bound to stay within the limits allowed by the general theoretical and institutional framework into which it is integrated. Thus, for example, it is not clear whether TA is to take place during (before or after) the earliest stage – namely that of research and development – or before final implementation. Similar considerations apply to the means by which the assessment arrived at is actually put into practice. The sanctions that could in principle be considered here include broad education of the public, legal restraints, fiscal restrictions or incentives, and downright state interference.

For the purpose of an analytical distinction, three different types of TA can be distinguished approximately at present:

(1) TA can be exploited to *justify political aims*. As (allegedly) scientific arguments tend to increase public support, politicians and other decision-makers are interested in gaining support for their claims through appropriate TA reports. Needless to say, this type of TA is not really of scientific, theoretical significance, although it is of practical importance.

(2) One may also use TA as a means of (political) decision-making by pointing out alternative ways of attaining pre-set goals, predicting the consequences to be expected from certain modes of action, and assessing the relative values of the foreseeable consequences. In this case the TA statements are of a descriptive and hypothetical nature. No single solution is put forward which the decision-maker ought to accept. On the contrary, different options are discussed and compared, so the that decision-maker can arrive at a well-considered choice of his own, based *descriptively* on the maximum information possible and *normatively* on his own responsibility.

(3) In a more far-reaching version TA is regarded as having a prescriptive and categorical character; it is no longer treated as an instrumental means but rather as an authoritative procedure, which can design the very goals that have to be attained and specify the desirable state of affairs. In this case the *only appropriate and "true"* way is suggested. The philosophically relevant point is that this prescriptive type of TA is just a disguised revival of the traditional idea of philosopher-kings.[2] In this new, modern version it is no longer the philosopher, but the person or the group performing the TA that knows about the "true" needs of people, the ideal of the good life, and the best state of affairs to be attained for the common weal.

All three types of TA, which tend to merge in practical use, raise specific philosophical problems: In the first case, i.e., pretext TA, the problem is the methodological one of deciding what *scientific criteria* make it possible to distinguish would-be TA from genuine TA. Generally speaking, though not in every detailed case, the established standards of scholarship should be sufficient to exclude mere tricks. These standards demand the formulation of explicit and consistent models for the theoretical premises and the highest possible degree of corroboration of the statements about the observable state of affairs, and also a clear distinction between the prescriptive and the descriptive statements[3] that enter into the process of TA.

Concerning (2), i.e., TA as a means of decision-making, a host of questions arise. The idea of assessing the effects of technological innovations implies that due to the causal laws of the physical world and the social realm certain effects will (objectively) obtain and that we are (subjectively) able to *predict* these effects with sufficient certainty. As is well known, in fact our capacity for prediction is severely restricted. Even in the physical world only the results of well-defined and/or controlled variables allow reliable predictions. About earthquakes, hurricanes, and even the weather next week, only rough forecasts are possible. Concerning the impact of complex technological systems on the ecological environment (and its capacity for self-regeneration) similar uncertainties obtain. The predictability of the social and cultural consequences of the scientific and technological innovations that will be possible in the future is even more restricted. As a result, the forecasts implied in assessing technological innovations cannot achieve more than a greater or lesser degree of probability. It is this feature that opens the way for the experts to put forward different predictions, and in fact all of them may be compatible with the state of the art of the discipline concerned.

The range of reasonably justified expectations concerning social and cultural consequences often turns out to be extremely broad, whereas predictions about natural phenomena, such as the use of resources, or the pollution of the environment, are far more reliable. It may be worth mentioning at this point that the notorious problem of induction unavoidably comes into play, since the task attempted is to make statements about the future state of affairs on the basis of empirical evidence. As even in the limited field of economics (or the even more limited one of econometrics) only highly conditional and in practice rather unreliable predictions are possible, one need not wonder about the problems TA is faced with here. This becomes especially evident in retrospect. When the initial steps towards building an automobile or an airplane were made, nobody could predict the diverse and ramified, direct and indirect, social and cultural results that actually followed from these technological innovations.

Furthermore, one must take account of the fact that the results to be expected do not exist in isolation but are elements of a *larger system*, which in the last analysis comprises the whole earth. Concerning the problems of ecology, resources, and energy, this is evident. But due to the complex interrelations between technology, economics, politics, social structure, and the sphere of culture, the context considered ought, strictly speaking, to be as broad as possible in all the fields. In view of world-wide economic competition all attempts at control of technology on a merely national level face severe difficulties. From the point of view of perfect TA, anything less than comprehensive can only be considered as a makeshift.

Similar arguments apply to the *technological items* considered. In our times technological systems and procedures are based on the interaction of highly complex and mutually dependent subsystems on different levels. Even an apparently minor and unimportant innovation – e.g., in electronics or in synthetic products – will usually penetrate into other fields, where it may give rise, or at least contribute, to further, far-reaching technological changes. For this reason the consistent and complete control of technology would imply that along with conspicuous large-scale innovations the possibly far-reaching results of the (at first sight) smaller innovations should also be assessed. In order to attain the highest degree of *theoretical* consistency in the assessment of technology and optimum *practical* results in controlling it, one may even, in the last analysis, be induced to go beyond merely considering the future and aim to include the present (and hence indirectly the past as well) in the sphere of assessment.

The situation is aggravated by the fact that all predictions are based on usually implicit *ceteris paribus* clauses which allow only for the hitherto foreseeable course of events and exclude unexpected occurrences such as economic changes and shifts in consumer behavior. But in fact nobody can be sure to what degree scientific discoveries and technological innovations of the future will alter trends that have prevailed up to now, so that completely new situations might arise. The very same type of argument applies to the preferences we ascribe to future generations. Nobody can be sure that our successors will adhere to the same hierarchy of values as our society.

But even if all the methodological and epistemological issues involved in predicting the factual state of affairs and the normative patterns that will obtain in the future were settled in a satisfactory way, this would not yet touch the problem of *assessing* these states. To do this, criteria are needed that allow desirable states to be distinguished from undesirable ones. For the purpose of a rough, overall assessment one may rely on a vague, intuitive notion of desirability which corresponds to the supposed average opinion.

But TA is purposely designed to go beyond this intuitive and overall notion and to give an explicit and detailed assessment. In doing this, the *diverging preferences* of different individuals and of various groups in society can no longer be ignored. This is perfectly natural, since individuals usually have different characters and particular biographical experiences; furthermore, they belong to specific social groups, are shaped by their professional lives, have different political opinions, etc. In all cases in which people are free to express their opinion, such divergences are clearly observable. They existed in the past and in the present, and – we must assume – will also persist in the future. Hence, even if we were to possess perfect knowledge about the preferences of coming generations, this would not solve our problem. In doing TA one cannot avoid presupposing a hypothetical resolution of normative differences, since any process of TA will by definition result in some one evaluation of the available alternatives in terms of a clearly specified value pattern.

It is the pluralism of value systems that must lead us to reject the prescriptive, categorical type (3) of TA. From a logical point of view a single recommendation instead of a list of alternative suggestions is possible only if the divergence of preferences is settled in one way or another. As soon as diverging reference scales are taken into consideration, one can never arrive at a single solution; one is forced to point out different alternatives and leave the choice to the individual or collective decision-makers – which in fact amounts to giving up the categorical type (3) of TA in favor of the hypothetical type (2). This general argument depends neither on the content of the preference scales considered nor on the nature of the normative categories taken into account. It holds good for whatever ultimate prescriptive elements are taken into account, be they diverging values (as treated in moral philosophy), diverging needs (dealt with in social anthropology), or diverging interests (discussed in political theory). At first sight the question of whether explicitly different alternatives are listed, none of them being presented as the "right" one, or one single solution is recommended as the final result of the TA process, may seem to be merely methodological. But on closer inspection the type of presentation used turns out to be indicative of the philosophical position (tacitly) presupposed in this context.

What arises here is a crucial problem of social and political philosophy, that of how to shape *practically* the divergent and even conflicting aims of the members of society into a well-defined political intention and how to ascertain *theoretically* the *true* needs of the common weal – which may perhaps be different from the ostensible ones. The overall final results that emerge with respect to the practical problem are visible in the obvious course of events; whatever happens constitutes the outcome of the "decision" actually resulting from the complex, mutually interacting (and interfering) intentions and actions of the individual and political and economic institutions involved. History goes its way, as it were, even without theoretical justification.

But philosophical reflection cannot be satisfied merely by pointing to what actually happens. Its task consists in conceptual analysis that leads to theoretical understanding and guidance for practical action. Here two simplified, but still highly revealing, philosophical models are pertinent: the Pragmatic vs. the Platonic approach to settling political affairs. The continuing philosophical dispute between the adherents of pragmatic solutions and piecemeal engineering and the followers of theoretically-based ultimate ideas is only too well known.[4]

Evidently neither of these views can claim to give an exhaustive answer, since (1) even in well-considered *pragmatic solutions* rational deliberation about the pro's and con's of the different solutions will obtain, and (2) due to the finiteness of all human affairs and the constraints of the concrete situation given even the most subtle *philosophical inquiry about ultimate goals* can be effective only if it finally comes to an end, irrespective of whether all the theoretical issues have been settled or not. In speculative terms, two points, in particular, are involved here: firstly, the *manipulated and alienated* vs. the *"true" and non-alienated* needs (and, in the last analysis, the ideal of the good life and of the *summum bonum*); and secondly, the *free*, spontaneous decision of the *individual will* favored by the Pragmatists as opposed to the *binding*, universal *rational insight* common to all men, emphasized by the Platonists. TA cannot solve these problems in its own terms; it has to refer to the philosophical discussion relevant here.

In view of these controversies, two arguments can be adduced in favor of the pluralistic approach and against the unified solutions suggested by the partisans of the Platonic philosopher-kings. The first is a theoretical one. Up to now in philosophical inquiry it has not been possible to establish an authoritative and undisputed theoretical understanding of the true needs and of the ultimate aims of human conduct. In view of the contingent, culturally and historically changing conditions of individual and social life, one may even doubt whether it is reasonable to expect in this area undisputed and strictly universal ideas. As a result, no safe and definite point of reference with respect to the content of ethical norms exists. This, of course, does not preclude adherence to the ideal of a conclusive truth; this in fact forms the usually tacit presupposition of all philosophical discussions. But this search for truth concerns the intention and thus constitutes a "regulative idea" in the Kantian sense and does not refer to an actual given entity.

The second argument is based on practical consequences. Since value pluralism *exists*, one could make an orthodox, codified idea of the good obligatory for everybody only by imposing it on the social group or nation in question. But this would deprive man of his freedom and invalidate the basic aim of TA, which consists in securing "the good life" for all members of society. Thus the idea of achieving control of technology at the cost of imposing complete control on society would in fact be self-destroying.

Of course this is only a highly simplified model, but it is nonetheless revealing. Fortunately in reality there are ways of finding a solution that lies somewhere between the extremes of pure laissez faire and perfect control. (Reports of commissions, science and technology "courts," the system of checks and balances, etc.) All such endeavors are designed to shape divergent value patterns into a single course of action. And it is precisely in this context that TA has a perfectly legitimate and methodologically indispensable function.

At this juncture an unavoidable, inherent tension within the very concept of TA becomes evident. TA is designed to moderate or, if possible, eliminate those negative effects of modern technology which are directed against the autonomy, self-determination, and spontaneity of man – effects which impede an authentic and fulfilled way of life. In order to achieve this goal TA necessarily involves technocratic and utopian elements: the notion of scientifically-based, rationally justified, and efficiently implemented solutions, and the utopian vision of a perfect state of human affairs free from uncertainties, evils, and defects. Now if we turned into an epigram the concept of categorical TA (3), we would say it combines the ideal of rational control with that of an earthly paradise, yielding a technologically implemented utopia. Clearly, arguments in favor of this conception are as legitimate and respectable philosophically as are arguments that can be adduced against it. But since the concrete physical environment and the social and institutional structure of modern industrial societies are shaped throughout by technology and scientific methods, any measure aimed at controlling technology will necessarily remain within the given framework – and hence itself be of a technical nature. (This becomes most evident with the idea of control, which is in fact the key concept of modern technological thinking.) Seen from this perspective, only a general rejection of modern technology, as in certain versions of Alternative Technology, seems to be a methodologically consistent counter-approach. But even in this case one cannot simply deny or escape from the given concrete technology-shaped situation; one cannot renounce carefully considered (and hence scientific) and effective (and hence technological) means in order to change things in the direction of the humanistic ends aimed at!

This does not imply that we cannot counteract an apparently almighty and ever-present "technological progress" exclusively following its own internal logic without human interference. After all, technology only exists as deliberately produced by man; hence man is in principle free to change the future course of technological development.[5] But it should be kept in mind that technology is an inherent part of our material culture and hence by derivation of our social and individual culture as well – so that unqualified and complete renunciation of technology would amount to renouncing the whole way of life that has emerged in the course of history and is thus part of our heritage. In fact, at any time in history the human condition was far from perfect; it is not by chance that Thomas More named the fictitious island where ideal conditions of life existed Utopia, i.e., nowhere.

To sum up, three points have been raised against perfect and omnipotent TA:

– the complex and many-faceted decision-processes obtaining,
– the limits to reliable predictions,
– the plurality of value systems.

Since TA is confronted with these problems, it cannot serve as a panacea for all the deficiencies of modern technology.

This does not, of course, imply that TA should be given up. TA is not perfect, but it is indispensable. To discuss the potential of a new approach, to specify the problems involved, and to point out limits not only precludes unjustified expectations but at the same time clears the ground for fruitful work. In order to arrive at a balanced account, the critical evaluation put forward above needs to be complemented by pointing out the *positive aspects* of TA – a task not undertaken here. (These positive aspects include the need for TA, and the fact that concrete results can reasonably be expected from it.) Within the limitations indicated, TA is indeed the *only* means of making the problems concerned explicit and open to rational discussion on an interdisciplinary level; it can do this by taking adventage of all the relevant information, including scientific knowledge and philosophical understanding. The broader discussion thus induced is urgently needed in order to arrive at a well-founded judgment about the factual and moral problems of ever-increasing technological change. To the extent that the risks of the future are foreseeable, we should face them with open eyes and with a broad consensus of opinion, based on clear-cut and sober-minded TA.

Technical University, Berlin

NOTES

[1] The recent work, *A Guidebook for Technology Assessment and Impact Analysis* by A. L. Porter, F. A. Rossini, S. R. Carpenter, and A. T. Roper (New York/Oxford, 1981) gives an authoritative and comprehensive account of the state of the art.
[2] This is the conception put forward by Plato in his *Republic*, Book V (457 c–e).
[3] A broad selection of articles about the problem of value judgments is compiled in *Werturteilsstreit* ed. by H. Albert and E. Topitsch (Darmstadt, 1971). One of the results is that, on the object-theoretical and the metatheoretical level of social research, value judgments will necessarily be present in the attitudes of the men being studied and in the selection of the problems for investigation by the research worker respectively. But they must be made explicit and distinguished from statements about factual matters.
[4] Relevant are, for example, K. R. Popper, *The Poverty of Historicism* (London, 1957), and J. Habermas, *Toward a Rational Society* (Boston, 1971).
[5] Cf. pp. 132–143 of the author's *Analytical Philosophy of Technology* (Boston Studies in the Philosophy of Science, vol. 63) (Dordrecht, 1981).

KRISTIN SHRADER-FRECHETTE

TECHNOLOGY ASSESSMENT AND THE PROBLEM OF QUANTIFICATION

1. INTRODUCTION

Until this century, there was usually great harmony between the predominant *ethics* of the culture and the underlying assumptions of its writers on *economics*. Adam Smith mirrored the atmosphere of eighteenth-century egoism, for example, while Bentham epitomized the climate of nineteenth-century utilitarian reform. In recent times, however, there has been little contact between ethics and economics.[1] Perhaps this is because economics has in part been assimilated to "positive science," in the Comtean sense, while philosophers have made only tortoise progress in clarifying important societal problems of values.

The disjunction between ethics and economics has created especially grave difficulties in the area of technology assessments. Technologists typically have based their analyses primarily on engineering-cost estimates,[2] which they erroneously describe as "value-neutral," "objective," or "atheoretical."[3] In reality, however, nearly all technology assessments have been dominated by key econometric models whose ethical assumptions have been ignored, in large part because they were not recognized. As a consequence, unexamined technicalism has helped to sanction methodologically implicit challenges to justice in legislation, regulation, and public policy regarding technology.[4]

One way to deflate "unexamined technicalism" is to *examine* important *technical* assumptions central to accepted economic methodology. Basic to the core concepts of Pareto Optimization and compensating variation employed in technology assessment, one such presupposition governs the way technological costs and benefits are calculated. Economists generally adhere to the methodological assumption that no quantitative values ought to be placed on qualitative, nonmarket, or "subjective" costs and benefits, such as the aesthetic impacts of technology.[5] There are a number of reasons for their typically avoiding what has been called "economic Philistinism," i.e., "operating on the premise that any meaningful benefits . . . can be expressed in dollars, and cents,"[6] or at least quantified. Most obviously, they argue that "the worst economic fetishism" is "money fetishism," because all costs and benefits simply cannot be put either in dollar terms or in quantitative terms.[7]

Paul T. Durbin and Friedrich Rapp (eds.), Philosophy and Technology, 151–164.
Copyright © 1983 *by D. Reidel Publishing Company.*

In this brief paper I wish to challenge the assumption that, when one quantifies all market parameters in order to perform a cost-benefit analysis, he ought not to attempt to quantify nonmarket variables. I will argue that, although no such quantification ever approaches being ontologically adequate, although "price" never equals "value," and although fallacies of aggregation can occur as a result of complete quantification, it is practically necessary to quantify all relevant assessment parameters whenever one compares market costs and benefits as a basis for public policy decisions. Even though I basically support analytical, rather than intuitive, modes of risk assessment and technology-related evaluation,[8] I will argue neither in favor of analytical methods generally nor in favor of monetary quantification specifically. Analytical assessment techniques, especially those utilizing common monetary parameters, are susceptible to many complicated methodological problems which I cannot evaluate here. My argument, instead, is a limited, conditional and strategic one. *If* market parameters are quantified and compared, *then* it is reasonable to attempt to quantify everything, rather than to leave nonmarket items out of consideration in aggregating costs and benefits. While avoiding (in this essay) the question of the *a priori* desirability of cost-benefit analysis and quantification, my thesis is merely that, *if* cost-benefit analysis is used, *then* any quantification ought to include all relevant parameters.

This modest conditional thesis is more significant than might appear, since federal courts in the U.S. have recently rejected proposed environmental rules and technological regulations, e.g., on benzene exposure, on the grounds that no cost-benefit data were presented to justify them.[9] Moreover, nearly all U.S. regulatory agencies (with the exception only of the Occupational Safety and Health Administration, OSHA), including the Environmental Protection Agency, now base their assessments in part on cost-benefit analysis and risk quantification.[10] This means that the question of whether nonmarket parameters ought to be quantified in cost-benefit analysis is one which potentially touches nearly every technology assessment or regulatory agency rulemaking decision.

2. ARGUMENTS AGAINST QUANTIFICATION

The clear presupposition, of most of those who wish to avoid quantification of subjective or qualitative values, is that only the market system provides an objective means (or at least a system more objective than any of those available) for quantifying certain costs and benefits, and that whatever has no market price has no objective value.[11] There are strong reasons, however, for believing that market *price* does not provide a wholly objective procedure for determining *value*.

Oscar Wilde accused the cynic of knowing the price of everything and the value of nothing. This suggests, of course, that "value" includes something "price" does not. Hobson believed that price was a function of the intensity of human wants, but not necessarily a function of the intensity of correct or desirable human wants. Moreover the value of goods is not causally determined by economic exchange any more than the amount of water in a vessel is causally determined by one's measuring it; all this suggests that using "price" as a measure of "value" gives price too great an ontological status.[12]

Apart from whether Hobson, Anderson, Boulding, and others are philosophically correct, however, there are some clear reasons, economically speaking, why market prices diverge from authentic values. Several of these are (a) the distorting effects of monopoly;[13] (b) the failure to compute effects of externalities;[14] (c) speculative instabilities in the market;[15] and (d) the absence of monetary-term values for natural resources and for "free goods" (e.g., air) or "public goods" (e.g., national defense), even though these items obviously have great utility to those who use them.[16]

Admittedly many theorists recognize the existence of normative bias in the market mechanism, and nevertheless resist attempts to quantify nonmarket costs and benefits. According to them, such resistance helps to avoid further bias and arbitrariness. Kenneth Boulding, for example, warns of the danger of using quantitatively measurable economic indices, especially in terms of money.[17] He believes that people are likely, when they see such measures, to forget all the assumptions and *caveats* built into their employment, and instead to use them uncritically in deliberating about complex problems of public policy-making.

A final important reason why most welfare economists appear to resist quantifying nonmarket, subjective, or qualitative parameters is that they have other procedures open to them, such as the trial-designer model. In this model, there are no cost-benefit calculations. Instead, "the designer just produces more or less detailed particular designs for the client to look at, all of which he certifies as at least feasible and attaches perhaps a rough costing to them." This procedure is like buying shoes, except that the options are not tried on.[18]

In response to all these arguments against employing the methodological principle that even nonmarket or qualitative costs and benefits ought to be quantified whenever market parameters are, what is the "minority view" of welfare economists? Why do a few of them argue in favor of quantitative assignments?

3. ARGUMENTS IN FAVOR OF QUANTIFICATION

Philosophically speaking, they often claim that the valuation process central to welfare economics always implies reducing a heterogeneous aggregate to a common uniform measure, such as money; otherwise, they maintain, it is impossible to compare or aggregate costs and benefits, a procedure necessary for accurate assessment.[19] In this regard, consider the example of planning for radioactive pollution control according to the $1000 per man-rem criterion of the NRC. Although I would be the first to argue that the monetary value assigned to each life lost (because of radiation emissions) is seriously inadequate, nevertheless the procedure of fixing such an amount, as a basis for pollution control expenditures, appears to be necessary. Given a society in which a zero-risk pollution standard is not recognized, this seems to be the only way to provide government and industry with a clear criterion for required action. Hence the real question appears to be, not whether some quantitative value ought to be used, but how its assignment can be made more consistent, rational, equitable, and democratic through informed decision-making involving the public.[20]

Despite all the obvious problems with attempting to quantify qualitative or subjective costs and benefits, the major argument in its favor is that the alternatives to quantification are even less desirable. As one student of economic methodology expressed it: it is "often politically convenient to avoid quantitative statements, since it is then easier to assert that the policy was successful."[21] In the absence of publicly justified quantifications of costs and benefits, it is more likely that purely political, undemocratic, misleading, or secretive methods of policymaking will be employed.[22] For instance, consider one famous analysis of nuclear technology, WASH-1400. Here authors writing under Nuclear Regulatory Commission contracts argue that atomic energy is cost-effective, since the per-year, per-reactor probability of a core melt is only 1 in 17 thousand. Because explicit numerical parameters were used both for this probability and for the costs and benefits of nuclear-generated electricity, it has been much easier for concerned citizens to investigate the issue according to established principles of scientific methodology.

Without the presence of quantitative values affixed to specific parameters, controversies over nuclear technology would be reduced to vague generalities or to debates over the credentials of various spokespersons. Moreover, in the face of some costs and benefits which are quantified and some which are not, the tendency of most persons is to ignore those in the latter group, often because they do not know how to take account of them.[23] Without quantification of nonmarket costs and benefits, a modern-day, double-barreled Gresham's Law tends to operate: "monetary information tends to drive out of circulation quantitative information of greater significance, and quantitative information of any kind tends to retard the circulation of qualitative information."[24] If this is true, then failure to quantify nonmarket parameters serves the interests of those who wish either to consider only market costs and benefits, or to limit the scope, and therefore the validity, of cost-benefit analysis.

What many of the pro-quantification arguments come down to is the trite observation that "the devil you know is better than the devil you don't know." The main reason for quantifying all parameters is that the methodological assumptions about their values are more obvious, easier to "get at," and therefore easier to criticize than are the presuppositions behind qualitative, subjective, or intuitive means of formulating costs and benefits. Therefore one might reasonably reject the principle that no quantitative values ought to be placed on nonmarket aspects of technological costs and benefits.

4. TECHNOLOGY ASSESSMENTS AND THE FAILURE TO QUANTIFY

What happens in specific cases when all parameters are not quantified? As was pointed out earlier, one of the likely results is that nonquantified costs and benefits will be ignored in the analysis. Consideration of representative technological studies illustrates both that many assessors have in fact failed to consider whatever parameters were not quantified, and that their omissions have led to a number of undesirable results. Some of the consequences include: (1) a failure to determine the actual cost effectiveness of various technological programs and policies; (2) a tendency to neglect consideration of social and political solutions to pressing environmental and technical problems; (3) an apparent willingness to draw specific conclusions unwarranted by the data; (4) a reluctance to investigate the social costs of technology; and (5) a tendency to exhibit pro-industry, pro-technology, pro-status-quo bias in the analyses.

Consider first a recent OTA assessment of whether to develop the flows of the Uinta and Whiterocks Rivers in central Utah. Surprisingly the calculations included as the *only costs* of the project, those for construction of a dam, reservoir, and canal. The hazards of pollution and congestion, likely to result from industrialization, were not considered. Moreover the authors of the report assumed that undeveloped land had no recreational benefits, and they ignored the negative social, political, and environmental effects of development. Instead they concluded that the aggregate benefits of development obviously far outweighed the costs.[25]

Likewise the authors of recent OTA studies on pest management, for example, ignored all the "qualitative" social, medical, and environmental hazards of using chemicals. Instead they included only three easily quantifiable parameters (cost of pesticides, and value of crops with and without employment of herbicides and insecticides). As a consequence, they concluded that use of chemical pest control was not only cost-effective, but more cost-effective than biological means of crop management. (Costs of biological control were not calculated.)[26]

A similar, erroneous methodology was followed in the OTA assessment of oil tanker technology. The study contains *no calculations* for the costs of oil spills (including damage to recreational beaches, to the commercial fishing industry, and to aesthetic quality).[27] In fact, the economic consequences of accidents were ignored completely; the only costs calculated, those regarding safety improvement, were for various types of ship design, navigational aids, and alternative control systems for the vessels.[28] Since the report ignored the social and economic parameters that were difficult to calculate, it is not surprising that social solutions (e.g., restriction of travel in some enclosed waters) and economic answers (e.g., increases in liability as a deterrent to operation of some ships) to the oil spill problems were also ignored. Instead only technological solutions (e.g., improved navigation systems), whose costs were easily quantifiable, were considered in the analysis. This suggests that application of the accepted methodological principle, of not calculating qualitative costs, not only may lead to erroneous aggregations (of costs and benefits) which fail to support the alleged conclusion of the assessment, but also that, because their costs and benefits are difficult to quantify, social and political solutions to the relevant problem are ignored.

Certain nontechnological solutions were assuredly omitted from consideration, for example, in a recent cost-benefit study of railroad safety done by the OTA. The authors of the report specifically denied that one type of social-political solution, increased regulation of the railroads, was cost-effective.[29] The denial is puzzling, however, because regulation intuitively appears to be at least a possible solution to the number one cause of increased accidents: deferred maintenance, especially on tracks. Regulation, in the form of mandatory use of the "Hazardous Information Emergency Response" form (used by Canadian railroads), also seems to be at least a possible solution to the problem that 65% of all U.S. railway cars loaded with hazardous materials are involved annually in accidental releases of these substances.[30]

Because the assessors omitted cost-effectiveness studies of programs whose costs and benefits are difficult to quantify – regulatory, inspection, and track-maintenance programs – there were no hard data to prove that regulation was/was not cost-effective. The authors even admitted that railroad policy, especially as regards safety, was based on market-determined risks and benefits and not on inclusion of other assessment parameters, such as social cost.[31] They also explicitly admitted that the reason why the cost-effectiveness of certain programs was not calculated was that they followed the methodological principle under consideration, and therefore eschewed parameters and goals which were not "measurable."[32] Such items were not included, they said, because of "data gathering difficulties."[33] This leads one to suspect that, in the absence of cost-benefit "proof" of means to reduce risks or to lower social costs, the industry or technology is assumed innocent until proven guilty, and is left free to pursue business as usual.

Similar pro-technology bias is clearly evident in a recent OTA study of coal slurry technology. Since only easily quantifiable market "costs" were considered (e.g., pumping water for use in the pipelines),[34] while more qualitative costs were *not* priced (e.g., any use of water where it is a scarce natural resource, as in the western U.S.),[35] the authors of the assessment were readily able to draw a conclusion in favor of the technology. They claimed: "Slurry pipelines can, according to this analysis, transport coal more economically than can other modes [of transport]."[36] The same sort of unwarranted conclusion was drawn in another study, an OTA assessment of the technology for transport of liquefied natural gas (LNG). The authors concluded that existing U.S. Coast Guard standards were cost-effective, i.e., adequate to avoid catastrophe as a result of LNG ship failure.[37] This conclusion was reached in spite of (and perhaps on account of?) the fact that cost-benefit calculations were not done for a major LNG spill. They were not done, presumably, because researchers claimed they would be too subjective; instead they said that decisions should be based on "nonquantitative approaches."[38] As a result, the assessors implicitly judged that the benefits outweighed the costs. They sanctioned the status quo, a continuation of the use of the technology as currently employed,[39] even though LNG transport accidents in the past have killed hundreds of persons and caused billions of dollars in damage.[40] In the light of such potential hazards, one wonders how "nonquantitative approaches" are sufficient to justify a "business as usual" conclusion.

A particularly interesting case of pro-technology bias, apparently resulting from the failure to quantify all costs and benefits, is evidenced in a recent assessment of the computed tomography (CT) scanner. Although the assessors considered factors such as purchase price of the machines, operating costs, and profits to the hospital as a result of their use, they excluded the nonmarket parameters necessary for determining the true costs and benefits of a scan. They noted, however, that high profits were made on CT scans,[41] that head scans tended to be used unnecessarily,[42] and that a typical CT scan exposed the patient to thirty rads of radiation,[43] but they failed to compute the allegedly subjective costs of increased cancers and genetic injuries occasioned by such exposure.[44] Despite these ommissions in the cost-benefit data, the assessment team concluded that computed tomography scans are "a relatively safe and painless procedure."[45] Obviously, however, their safety is a function of radiation risk as measured against the benefit of more accurate radiological diagnosis. Since the assessment team mentioned criteria for situations in which the CT scan was likely to be negative (suggesting, therefore, cases in which its use was contraindicated because of radiation risk),[46] the cost-effectiveness of use of the scan in different situations easily could have been assessed. Because the risk of cancer and genetic damage was not quantified, however, the risk of radiation was ignored, and the scanners (although moderately dangerous) seem likely to continue to be overused in certain situations.[47]

Failure to quantify nonmarket costs has also generated a bias in favor of the status quo in various studies of solar energy, nuclear power, and coal-generated electricity.[48] Since solar power can be shown to be more, or less, cost-effective than conventional energy sources, depending on whether allegedly subjective parameters (e.g., health risks) are, or are not, quantified and included, ignoring nonmarket costs has biased assessments in favor of conventional energy technologies.[49] According to the OTA, though onsite solar devices *could supply* "over 40 per cent of U.S. energy demand by the mid 1980's,"[50] whether they *will do so*, in fact, will depend upon factors such as the public's perception of their costs. Hence their real costs ought to be calculated, so far as possible, in technology assessments of comparative energy sources.

Just as future policy regarding solar power may be skewed erroneously by methodological presuppositions governing determination of its relative costs, so also future policy regarding the automobile is likely to be controlled in part by methodological presuppositions employed in calculating its cost-effectiveness. Consider, for example, the likely consequences of British, versus American, assessment procedures. Whereas British government studies of auto technology always quantify and include the nonmarket costs of accidents in their calculations, U.S. government analyses of the same technology have not done so.[51] The British *Transport Policy* based its calculations on the costs of pain and suffering, for instance, which are said to be approximately 50% as great as total property damages for the accident and 600% higher than total medical expenses.[52] This suggests that if the British figures are good estimates of the real costs of accidents, and if cost-benefit analysis is followed, then assessors of automobile technology might have to revise the essentially pro-auto policy embraced in the most recent U.S. studies.[53]

One of the major reasons for the pro-auto conclusions of the U.S. reports appears to be that their authors have employed the methodological principle of not including allegedly subjective or nonmarket parameters in their aggregations of costs and benefits. In the most extensive and up-to-date U.S. government assessment, for example, the authors include only the individual's yearly *market* cost (for depreciation, insurance, gas, etc.) of owning and operating a car.[54] The calculations exclude prices for items such as air pollution, noise, risk of death and injury, and resource depletion.[55] Because these latter factors were ignored, and since no quantitative figure was assigned to the alleged "inalienable right" of "personal mobility" via auto travel, the assessors concluded that the benefits of private transport outweighed both the benefits of mass transit and the total costs of employing the automobile.[56]

5. CONCLUSION

Earlier I pointed out that many economists opposed quantification of nonmarket parameters, on the grounds that such pricing was an example of "money fetishism" and "economic Philistinism." If failure to quantify allegedly qualitative costs results in their being excluded from technology assessments, as I believe I have illustrated, with the consequence that human risk, pain, suffering, and death are thereby ignored, then nonquantification also results in a very practical sort of "economic Philistinism," and perhaps one more serious than that arising from imperfect attempts at quantification.

University of California, Santa Barbara

NOTES

[1] A. L. Macfie, "Welfare in Economic Theory," *The Philosophical Quarterly* **3**, No. 10 (January 1953): 59.

[2] Sergio Koreisha and Robert Stobaugh, in "Appendix: Limits to Models," in Robert Stobaugh and Daniel Yergin (eds.), *Energy Future: Report of the Energy Project at the Harvard Business School* (New York: Random House, 1979), p. 234; hereafter cited as: "Appendix," in *EF*.

[3] See, for example, E. C. Pasour, "Benevolence and the Market," *Modern Age* **24**, No. 2 (Spring 1980): 168, and Milton Friedmann, "Value Judgments in Economics," in Sidney Hook (ed.), *Human Values and Economic Policy* (New York: New York University Press, 1967), pp. 85–88; hereafter cited as: Hook, *HV and EP*.

Technology assessments are widely held to be unbiased, outside the realm of policy or value judgments, nonpartisan, and objective (Congress, OTA, *Annual Report to the Congress for 1977* (Washington, D.C.: US Government Printing Office, 1977), p. 4; hereafter cited as: Congress, OTA, *AR 1977*.

[4] Koreisha and Stobaugh, "Appendix," in Stobaugh and Yergin, *EF*, p. 11.

[5] For confirmation of the fact that welfare economists generally do not quantify non-market costs and benefits, see R. M. Hare, "Contrasting Methods of Environmental Planning," in K. E. Goodpaster and K. M. Sayre (eds.), *Ethics and the Problems of the 21st Century* (Notre Dame: University of Notre Dame Press, 1979), pp. 64–68 (hereafter cited as: Goodpaster and Sayre, *Ethics*); E. J. Mishan, *Welfare Economics* (New York: Random House, 1969), p. 86; hereafter cited as: *WE*. See also M. W. Jones-Lee, *The Value of Life: An Economic Analysis* (Chicago: University of Chicago Press, 1976), pp. 21–28 (hereafter cited as: *Value*); and L. H. Mayo, "The Management of Technology Assessment," in R. G. Kasper (ed.), *Technology Assessment: Understanding the Social Consequences of Technological Applications* (New York: Praeger, 1972), p. 78 (hereafter cited as: Kasper, *TA*).

Employment of this principle of nonquantification raises an interesting epistemological issue. (1) Can welfare economists be said *not* to be using the notions of Pareto Optimum and "compensating variation," since they do not include *all* cost-benefit parameters in their calculations? (2) Or, on the other hand, may they be said to employ modified versions of these two concepts, since they are not practically usable as defined in economic theory? Whether either (1) or (2), or neither, is the case will not substantially affect the discussion in this section. Although most economists would probably agree with (2), the point of examining the methodological principle here (regarding nonquantification of some parameters) is to assess its desirability and not to determine its status as Pareto-based or not.

[6] B. M. Gross, "Preface," in R. A. Bauer (ed.), *Social Indicators* (Cambridge: MIT Press, 1966), p. xiii; hereafter cited as: Bauer, *SI*.

[7] Nicholas Georgescu-Roegen, *Energy and Economic Myths: Institutional and Analytical Economic Essays* (New York: Pergamon Press, 1976), p. 56; hereafter cited as: *EEM*.

[8] Although I will not argue this broader point here, my claim is that rational assessment procedures require an unambiguous means of comparing policy alternatives. Unless various options can be expressed in terms of a "common denominator" (e.g., ordering, preference-ranking, rating, quantification), one not necessarily based on a "numerical" or monetary system, then there is little assurance that technology assessment will be as

rational or as objective as might be possible. For a discussion of analytical, versus intuitive, modes of risk assessment, see C. Starr and C. Whipple, "Risks of Risk Decisions," *Science* **208**, No. 4448 (June 6, 1980): 1114–1119; hereafter cited as Starr and Whipple, Risks.

[9] L. J. Carter, "Dispute over Cancer Risk Quantification," *Science* **203**, No. 4387 (March 30, 1979): 1324–1325.

[10] Starr and Whipple, Risks, p. 1118.

[11] L. H. Mayo, "The Management of Technology Assessment," in Kasper, *TA*, p. 78.

[12] J. A. Hobson, *Confessions of an Economic Heretic* (Sussex, England: Harvester Press, 1976), pp. 39–40; hereafter cited as: *Confessions*. See also B. M. Anderson, *Social Value: A Study in Economic Theory Critical and Constructive* (New York: A. M. Kelley, 1966), pp. 26, 31, 162; hereafter cited as: *Social Value*. See K. E. Boulding, "The Basis of Value Judgments in Economics," in Hook, *HV and EP*, pp. 67–69; hereafter cited as: "Basis." See also Anderson, *Social Value*, p. 24.

[13] See K. E. Boulding, "Basis," in Hook, *HV and EP*, pp. 67–68. Oskar Morgenstern, *On the Accuracy of Economic Observations* (Princeton, N.J.: Princeton University Press, 1963), p. 19 (hereafter cited as: *Accuracy*), ties the "errors of economic statistics," such as price, in part to the fact of the prevalence of monopolies. In an economy characterized by monopoly, he says, statistics regarding price are not trustworthy because of "secret rebates granted to different customers." Moreover, he claims, "sales prices constitute some of the most closely guarded secrets in many businesses." For both these reasons it is likely not only that price ≠ value, but also that actual price ≠ official market price.

[14] R. C. Dorf, *Technology, Society, and Man* (San Francisco: Boyd and Fraser, 1974), pp. 223–40 (hereafter cited as: *TSM*), and H. R. Bowen, Chairman, National Commission on Technology, Automation, and Economic Progress, *Applying Technology to Unmet Needs* (Washington, D.C.: US Government Printing Office, 1966), pp. v–138; hereafter cited as: *Applying Technology*. See also K. E. Boulding, "Basis," in Hook, *HV and EP*, pp. 67–68, and E. J. Mishan, *Cost-Benefit Analysis* (New York: Praeger, 1976), pp. 393–94; hereafter cited as: *Cost-Benefit*. Externalities (also known as "spillovers", "diseconomies", or "disamenities") are social benefits or costs (e.g., the cost of factory pollution to homeowners nearby) which are not taken account of either in the cost of the goods produced (e.g., by the factory) or by the factory owner. They are "external" to cost-benefit calculation, and hence do not enter the calculation of the market price. For this reason, says Mishan, *The Costs of Economic Growth* (New York: Praeger, 1967), p. 53; hereafter cited as: *CEG*, "one can no longer take it for granted that the market price of a good is an index of its marginal price to society." Another way of making this same point (Mishan, *CEG*, p. 57) is to say that diseconomies cause social marginal costs of some goods to exceed their corresponding private marginal costs; this means that the social *value* of some goods is significantly less than the (private) market *price*.

[15] See K. E. Boulding, "Basis," in Hook, *HV and EP*, pp. 67–68, and E. F. Schumacher, *Small is Beautiful: Economics as if People Mattered* (New York: Harper, 1973), pp. 38–49: hereafter cited as: *Small*.

[16] There are no monetary-term values for natural resources because the "cost" of using natural resources is measured in terms of low entropy and is subject to the limitations imposed by natural laws (e.g., the finite nature of nonrenewable resources). For

this reason, viz., the theoretical and physical limit to accessible resources, the price mechanism is unable to offset any shortages of land, energy, or materials.

[17] Quoted by B. M. Gross, "Preface," in Bauer, *SI*, p. xviii. See also Bauer, "Detection and Anticipation of Impact: the Nature of the Task," in Bauer, *SI*, pp. 36–48; R. M. Hare, "Contrasting Methods of Environmental Planning," in Goodpaster and Sayre, *Ethics*, pp. 64, 65, and D. E. Kash, Director of the Science and Public Policy Program, University of Oklahoma, in Congress of the US, *Technology Assessment Activities in the Industrial, Academic, and Governmental Communities.* Hearings before the Technology Assessment Board of the Office of Technology Assessment, 94th Congress, Second Session, June 8–10, 14, 1976 (Washington, D.C.: US Government Printing Office, 1976), p. 198. Hereafter cited as: Congress, *TA in IAG*.

[18] R. M. Hare, "Contrasting Methods of Environmental Planning," in Goodpaster and Sayre, *Ethics*, pp. 64, 68, 70.

[19] K. E. Boulding, "The Basis of Value Judgments in Economics," in Hook, *HV and EP*, p. 64. See also Anderson, *Social Value*, p. 13.

[20] This example is taken from K. S. Shrader-Frechette, *Nuclear Power and Public Policy* (Dordrecht: D. Reidel, 1980), pp. 115–116; hereafter cited as: *Nuclear Power*.

[21] Morgenstern, *Accuracy*, p. 125.

[22] J. Primack and F. von Hippel, *Advice and Dissent: Scientists in the Political Arena* (New York: Basic Books, 1974), p. 33.

[23] Mishan, *CEG*, p. xx. This same point is also made by John Davoll, "Systematic Distortion in Planning and Assessment," in D. F. Burkhardt and W. H. Ittelson (eds.), *Environmental Assessment of Socioeconomic Systems* (New York: Plenum, 1978), p. 12; hereafter cited as: Burkhart and Ittelson, *EA*. See also R. A. Tamplin and J. W. Gofman, *"Population Control" Through Nuclear Pollution* (Chicago: Nelson-Hall, 1970), p. 82; and R. A. Bauer, "Detection and Anticipation of Impact: the Nature of the Task," in Bauer, *SI*, p. 35.

[24] B. M. Gross, "The State of the Nation: Social Systems Accounting," in Bauer, *SI*, p. 222. See also p. 260, where Gross discusses the "selectivity-comprehensiveness paradox." There is a tension, perhaps resulting from the prevalence of the application of Gresham's Law, between choosing to measure quantitatively only a few parameters ("selectivity") and deciding to attempt to measure quantitatively a more comprehensive list of items ("comprehensiveness"). When one opts for the former, he gets an *exact*, but *irrelevant* indicator. When he chooses a more comprehensive list of values to quantify, he obtains a more *relevant* (i.e., applicable, realistic, or usable) indicator, but a much less exact one, since he necessarily encounters more difficulty in quantifying qualitative or subjective factors. Hence, even if Gresham's Law is avoided via quantification, one still faces considerable difficulty in the form of this paradox.

[25] Congress, *TA in IAG*, pp. 248–50.

[26] Congress, OTA, *Pest Management Strategies*, vol. 2 (Washington, D.C.: US Government Printing Office, 1979), pp. 48–51, 68–81; hereafter cited as: *Pest M.S.*

[27] Congress, OTA, *Oil Transportation by Tankers: An Analysis of Marine Pollution and Safety Measures* (Washington, D.C.: US Government Printing Office, 1975), pp. 26–37, 173; hereafter cited as: *Oil Tankers*.

[28] Congress, OTA, *Oil Tankers*, pp. 38–71.

[29] Congress, OTA, *An Evaluation of Railroad Safety* (Washington, D.C.: US Government Printing Office, 1978), p. xi; hereafter cited as: *RR*.

30 Information concerning these problems was taken from Congress, OTA, *RR*, pp. 14, 141–161, and Congress, OTA, *Railroad Safety – US-Canadian Comparison* (Washington, D.C.: US Government Printing Office, 1979), pp. vii–xi; hereafter cited as: *RR-US-C.*

31 Congress, OTA, *RR*, p. 37.

32 Congress, OTA, *RR*, p. 160.

33 Congress, OTA, *RR*, pp. x–xi, 125.

34 Congress of the US, Office of Technology Assessment, *A Technology Assessment of Coal Slurry Pipelines* (Washington, D.C.: US Government Printing Office, 1978), p. 84; hereafter cited as: Congress, OTA, *Coal Slurry*.

35 Congress, OTA, *Coal Slurry*, pp. 84, 99. For discussion of the problem of "pricing" natural resources, see M. A. Lutz and K. Lux, *The Challenge of Humanistic Economics* (London: Benjamin Cummings, 1979), pp. 297–308, esp. 305–308; hereafter cited as: *Challenge.*

36 Congress, OTA, *Coal Slurry*, p. 15.

37 Congress, OTA, *Transportation of Liquefied Natural Gas* (Washington, D.C.: US Government Printing Office, 1977), p. 42; hereafter cited as: *LNG.*

38 Congress, OTA, *LNG*, p. 62. See also pp. 63, 66.

39 See note 37.

40 Congress, OTA, *LNG*, p. 8.

41 See Congress, OTA, *Policy Implications of the Computed Tomography (CT) Scanner* (Washington, D.C.: US Government Printing Office, 1978), pp. iii, 9, 105; hereafter cited as: *Scanner.*

42 See Congress, OTA, *Scanner*, pp. 6, 67, 71, 105; this conclusion is based on the facts that most head scans are done merely because of headaches and, in the absence of other abnormalities, are almost always negative. Even with other symptoms, up to 90% of all head scans are negative.

43 Congress, OTA, *Scanner*, p. 38. This dose is 177 times greater than the average annual dose of radiation to which a person is exposed.

44 In the 30-year period following a CT-scan exposure to 30 rads of ionizing radiation, for example, 2 of every 100 persons exposed will contract cancer simply because of this one CT scan. This calculation is based on the standard *BEIR* report of the NAS, used by the US government to yield dose-response statistics. If there is a latency period (for cancer) of 30 to 40 years, and if 0002 cancers per year are induced by exposure to one rad of radiation, then 2 of the 100 persons exposed to 30 rads will contract the disease. Calculational data from government dose-response studies and from the *BEIR* report may be found in K. S. Shrader-Frechette, *Nuclear Power*, p. 26; see also p. 115 for calculations regarding genetic deaths of offspring when parents are exposed to radiation; genetic deaths are higher, by factor of 10, than induced cancers when equal moments of radiation exposure occur. Cost-estimate data for cancers and genetic deaths may be computed on the basis of the discussion throughout Jones-Lee, *Value.*

45 Congress, OTA, *Scanner*, p. 105.

46 Congress, OTA, *Scanner*, pp. 8, 71.

47 See note 44.

48 In WASH-1224, for example, in which costs and benefits of nuclear-versus-coal-generated electricity were compared, the authors concluded that the former was a more cost-effective means of producing power. They did not include the cost of radioactive waste storage in their calculations, however. If this one allegedly "nonmarket cost,"

waste storage (as given on the basis of US government expenditures), had been included in the computations, then the opposite conclusion would follow. That is, coal-generated power could be shown, by an even wider margin, to be cheaper than nuclear electricity. (For a complete analysis of this example, see K. S. Shrader-Frechette, *Nuclear Power*, pp. 49–68).

49 When nonmarket parameters are included in the computations for the cost-effectiveness of various energy technologies, solar power and conservation (for the first 10 million barrels per day of oil equivalent) are cheaper than all other conventional energy sources; when nonmarket parameters are excluded, on the other hand, then solar and conservation are said to be more expensive than conventional power sources. (Stobaugh and Yergin, "Conclusion," in Stobaugh and Yergin, *EF*, pp. 216–33, esp. p. 227; see also Congress, OTA, *Application of Solar Technology to Today's Energy Needs*, Vol. 1 (Washington, D.C.: US Government Printing Office, 1975), pp. 3, 12, 21; hereafter cited as: *Solar I*.

50 Congress, OTA, *Solar I*, p. 3.

51 For British statistics, see M. R. McDowell and D. F. Cooper, "Control Methodology of the U.K. Road Traffic System," in Burkhardt and Ittelson, *EA*, pp. 279–98. For US data, see Congress, OTA, *Technology Assessment of Changes in the Future Use and Characteristics of the Automobile Transportation System*, 2 vols. (Washington, D.C.: US Government Printing Office, 1979), vol. 1, pp. 16, 21, 25–31; hereafter cited as: *Auto I or Auto II*.

52 See note 51.

53 See Congress, OTA, *Auto I*, p. 25.

54 Congress, OTA, *Auto I*, p. 31.

55 Congress, *Auto II*, p. 251; see also pp. 75–294.

56 Congress, OTA, *Auto I*, p. 25.

WALTHER CH. ZIMMERLI

FORECAST, VALUE, AND THE RECENT PHENOMENON OF NON-ACCEPTANCE: THE LIMITS OF A PHILOSOPHY OF TECHNOLOGY ASSESSMENT

INTRODUCTION

It is not for an engineer to engage in the philosophy of technology; just as little is it for the philosopher of technology to take over or even criticize the work of an engineer. If we think through this apparent triviality a little, it turns out to have many interesting consequences. It follows, for instance, that technology assessment and, more particularly, technological forecasting are not philosophical tasks. And I do indeed think that this is correct.

But what, then, is the task of the philosopher of technology? At first it seems in principle very little. He has the same range of tasks that he has in all his other areas of activity. At the object level but as not oriented toward empirical research, the philosopher provides speculative constructions (possible-world metaphysics) and in certain (rare) cases apart from that, material ethics; on the meta-level he provides us with general methodological considerations, and formal or analytical ethics. The contributions at the object level are made largely with constructive intentions, at the meta-level largely with critical intentions. The same holds, *mutatis mutandis*, for philosophers in the realm of the philosophy of technology – and in the case of my present subject, in the realm of philosophical concern and activity with respect to technology assessment (TA). Since TA is an area or aspect of technological *evaluation*, its relation to the realm of *values* is immediately evident, as is the fact of its material and/or formal value-ethical tasks and problems: The philosophy of TA has to indicate meta-ethically the structures according to which relations to values in the realm of TA (might) come to be, and, in accord with the rather critical intentions involved in meta-philosophical investigations, that means that the limits of the legitimate relations to values have to be indicated, and at best one of the alternative possible value systems has to be justified. Since TA has to do with the *consequences* of technology as well, a methodological problem area intrudes itself immediately, namely that of investigating the bare possibilities, that is, the limits of the forecasting of consequences and side effects of the application of (new or already implemented) technologies. If we look more carefully, it becomes clear that a "metaphysical" problem area is thus uncovered – namely that of a

Paul T. Durbin and Friedrich Rapp (eds.), Philosophy and Technology, 165–184.
Copyright © 1983 *by D. Reidel Publishing Company.*

substantive philosophy of history, since the combination of technological forecasting and considerations of desirability can only be undertaken against the background of a theoretical conception of the course of human history.

In what follows, the specific problematic nature of TA is, in a first step (1) embedded historically and is examined with respect to its reducibility to "risk analysis." In this way the conditions are created which enable a further step to be taken (2) in which the central "problem of forecasting" is analyzed, at first generally, and subsequently with reference to TA; this leads into ramifications within the philosophy of history, initially (3) by way of a critical discussion of an evolution-theoretic outline. A sketch is then given (4) of a conception of the end of the "modern age of enlightenment." Some meta-ethical (5) and a few formal value-ethical thoughts (6) follow.

1. TECHNOLOGY ASSESSMENT OR "RISK ANALYSIS"?

Technical innovation has unquestionably always provoked criticism and rejection. But it is characteristic of our contemporary situation that the criticism which occurs, the contemporary form of non-acceptance in connection with science and technology, can no longer be simply out-maneuvered, ignored, or, because of alleged higher technological insight (or something similar), be otherwise passed over. In this context it is significant, not only that a change in the political system as well as in the basic idea of democracy has led – at least in the Western industrial nations – to non-acceptance becoming a political factor, but also, in addition, that a certain change in fundamental attitudes, in consciousness (see below), must be presupposed.

At the risk of repeating what is already known, I shall quickly recount the relatively short history of TA (cf. Paschen, Gresser, and Conrad, 1978, esp. pp. 81 ff.). After a broad discussion, centered around the U.S. Congress, concerning the negative side effects of the applications of technology had taken place in the late 1960s, the Technology Assessment Act was passed in 1972 and in 1973 the OTA (Office of Technology Assessment) was founded. Since then, the office has operated with a considerable expenditure of money and manpower, its returns being variously judged. It is designed to provide information concerning the consequences of technology for members of Congress. Apart from the OTA there are a large number of agencies which commission TA studies, both in government and private industry. Besides its own researchers, public, semi-private, and completely private research institutions are commissioned; and in rare cases even universities play a role. It might be of interest to note here that only about 10% of all the studies carried out represent extensive TA studies in the narrow sense – i.e., which to some extent attempt to do justice to the full range of parameters. For these 10% it is characteristic "that they were designed to influence political decision processes, were staffed by multi-disciplinary teams, and were set up with at least $100,000" (Paschen *et al.*, 1978, p. 85). This might serve as an indication of the size of such studies.

Lively discussions concerning TA have taken place in the Netherlands, Sweden, Great Britain, and France, but especially in West Germany, in parliament and government generally. The high point was reached between 1973 and 1975, culminating in the still valid report entitled *Bundesbericht Forschung VI* (1979). The main question treated concerns the possible institutionalization of TA at the federal level. TA is of course *de facto* carried out by various research establishments supported by the Federal Republic and its constituent states – for example by the Atomic Energy Research Center (*Kernforschungszentrum*) in Karlsruhe, the Fraunhofer Society, the Battelle Institute, or the Centre for Scientific Research (*Wissenschaftszentrum*) in Berlin. In characterizing the activities of TA two things are of interest. In the first place it is pragmatically important to ask constantly (as has indeed been done), whether the commissioned studies in question were just cover-ups in order to legitimize a point of view or whether they were studies with potentially open outcomes. (This depends upon both the situation and the commissioning body.) In the second place – and this is decidedly more important for our subject – it is to be borne in mind that in the early days of TA, extremely exaggerated expectations were held out for this new instrument of decision making. Nothing less than a total factor-sensitive picture of the future was supposed to be distilled out in the prognostications; TA was supposed to be futurology. Discussion concerning technological forecasting was correspondingly intensive, too (cf. Mitroff and Turoff, 1973, for example).

Today only the layman or the politician can harbor such illusions. It is becoming increasingly obvious that TA, as understood purely in the sense of "technological forecasting," is to mean that after TA studies have been provided, the problem of acceptance or rejection has not only not been solved; it has not even been viewed correctly. I can forecast the consequences or side effects of new or already implemented technologies as exactly and as far into the furture as you like, but one thing I will not have thereby accomplished is the clarification of the *desirability* of those consequences and side effects. And even if that were done, it would still not emerge that that which was objectively desirable would be subjectively desired or accepted by those concerned. As we know, both things have become the object of investigation; the clarification of objective desirability goes under the name of "Indices of Social Acceptability" (cf. Meyer-Abich, 1979, 1981), the clarification of the connection between objective desirability and subjective acceptance under the name of "Acceptance Research."

It is well known that in the last few years the discussion of energy technologies and energy systems has aroused quite a high degree of feelings. This discussion has reduced the relationship between desirability and subjective acceptance to the problem of risk – perhaps not least because of the prior existence of the theoretical tool of "risk analysis" (which originated within the mathematics of insurance; it depends, to a large degree, on theorems from game and decision theory; cf. Starr, 1969; Rowe, 1975; Lowrance, 1976; Otway, 1976; Council, 1977; Conrad, 1978). I am personally of the opinion that in spite of the undeniable advances in analytical clarity which have taken place due to considerations based on risk analysis, it represents a narrowing of the problem. Apart from the fact that it will always be an open decision system and quantification which is obtained, all that is observed in the risk-analytic procedure is the ratio of expected utility (benefit) to the concomitant disutility which has to be taken into account (risk) – the so-called "risk-benefit ratio" (cf. Starr *et al.*, 1976). Since most of those concerned with gaining acceptance of new energy systems (or rejecting them) are not aware of what this ratio actually is, risk analysis leads only to a fixing of ideal rational acceptability, never to an understanding of actual acceptance. This means that, using this instrument, only that can be discovered and established which can be demanded or expected of someone from the viewpoint of a rationally acting "*homo ecònomicus*" (cf. Gäfgen, 1968).

There is, however, a large area in which the non-acceptance movement is active which is not determined by calculations of risk. I should like to call this area that of ethics. It may suffice for the time being to adduce some phenomenological evidence. What I have in mind are attitudes towards undifferentiated pictures of the future, anxious expectations which are diffuse and not connected with nameable risks; the problem of responsibility for future generations (inter-generational justice); attitudes of resistance towards opaque centralized bureaucracies; anxiety reactions concerning unsurveyable domains of space-time and their uncalculable negative consequences; emotive attitudes of resistance towards assumed or actual coercion by industrial concerns and state establishments which exercise a monopoly over the supply of energy. It is easy to see that such examples could be extended indefinitely; but it is more important here to note that rational risk analysis is ineffective against such attitudes and dispositions. We would have to formulate and make known the deeply rooted feelings of responsibility and of anxiety which lie at the base of the matter before the measures or correctives which are relevant in making a choice, e.g., between energy systems, could be put into operation.

2. THE PROBLEM OF FORECASTING

Discussion concerning prediction – as carried on within philosophy and in particular within the philosophy of science – is an old one which is yet to be definitively decided. In accordance with our main theme, only a particular selection of problems interests us here, namely those problems connected with TA. I shall begin by listing and analyzing the types of forecasts which can be distinguished in principle; I shall then formulate and discuss the problem of forecasting as it affects TA.

First we can distinguish forecasts in a narrow sense from those in a wide sense. Forecasts in the narrow sense are statements which can be deduced logically from law-like propositions together with initial conditions (cf. Popper, 1935, and Hempel, 1948). Since forecasts in this sense are thought to coincide with explanation, I shall follow Knapp's terminology (1978, p. 15), and call forecasts in the narrow sense "explanatory forecasts." Strictly speaking, explanatory forecasts occur exclusively in connection with the knowledge of specifics subsumed under general law-like propositions (although this may certainly occur in the context of other sorts of knowledge – "background knowledge," for example). Although such forecasts obviously play a role in TA, it is not a central one.

What is important, however, for our case is the following. We can interpret what we have seen in the case of explanatory forecasts in such a way that not every statement may be called a "forecast" but only those statements which are characterized by two factors. Every such statement must be a *statement of expectation*, and the preferred expectations expressed must be justified. In the case of explanatory forecasts, the justification takes place by recourse to knowledge of law-like propositions. When the forecast is correctly deduced from the law-like propositions and the initial conditions, were I to doubt that it really *was* a forecast (something very unlikely), it could only be because I questioned the validity of the knowledge of the law-like propositions.

I come now to forecasts in the wider sense. In fact, this designates something like what a physician would call "clinical prognosis." It is likely that very many things are understood by "forecast" in this wider sense, because, on the one hand, there is no clear etiology, and, on the other, its various guises are similar to one another.

But how can preferences of expectation be justified where no specified knowledge of general law-like propositions is available – that is, where we have such "inexact forecasts"? Of course, the most diverse techniques of justification have been developed, and correspondingly diverse typologies of justification have been elaborated both in the realm of economic and technical planning and in that of political planning systems. These can, however, always be reduced to two basic types, and these in fact are linked with one another. Either a trend analysis is undertaken on which a trend extrapolation is subsequently based, or else experts are polled (pragmatically speaking, this represents an earlier stage). In the first case we speak of "forecasts using the trend procedure" – of "trend forecasts" for short; while in the second case, in virtue of the oracular character of the forecasts, we speak of "forecasts according to the Delphi procedure" – or "Delphi-forecasts" for short (Helmer and Rescher, 1959). With respect to both types, it is important to note that the information gathered to support the justification must be "relevant" information. (There is already a further difficulty with the so-called "inexact" problems of forecasts here: namely, the circularity of implicit definitions.)

In the case of a trend prognosis it is immediately obvious how a trend (e.g., sales of cars in the following month) can be extrapolated from a given body of data (e.g., cars sold per month). Since in this case I have a finite body of data, I can, in principle, formulate any number of trend curves; according to the relevant form of the curve (oscillating, rising, falling), the forecast for the next month will be defined. I thus need some more qualified reasons beyond this in order to prefer a particular mapping of the previous course of events which would yield a preference for a particular extrapolation into the future as opposed to other extrapolations.

Something similar happens with Delphi forecasts. We need relevant information here too – for one, *who* the experts are. Furthermore, with respect to problem and content, we have to produce reasoned assumptions concerning the required degree of expert agreement in a particular case for us to be able to speak of a "forecast" or well founded expectation preference at all.

How do things stand concerning the evaluation of consequences in science and technology? If I see things correctly, at least the following must be demanded of TA: (a) provision of alternative forecasts of side effects with respect to the material or social technologies already or about to be introduced;

(b) evaluation of side effects on the basis of forecasts concerning the development of internalized deontic structures; and (c) an evaluation of the development of internalized deontic structures that provides for optional courses of action (Paschen, Gresser, and Conrad, 1978, p. 19). And here the problem of forecasting shows itself clearly: At least in the area of forecasts of technological side effects, we can in some cases have recourse to explanatory forecasts, but in some cases not – especially when, for example, indications of social acceptability are counted as side effects, in which case value assumptions come back into play. It is completely clear in the case of forecasts concerning changes of internalized deontic structures that it can at best be a matter of weak social-scientific forecasts, whether of the trend or the Delphi variety. The evaluation of the interrelation between forecast side effects and forecast changes of internalized deontic structures can, after all, only be descriptively captured or pragmatically justified or postulated by an apparently irrational procedure. The appeal to qualified or relevant information is certainly insufficient for the justification of optional courses of action.

It is possible to elaborate a chain of scenario-dependent if-then forecasts, extendable at will. But the essence of forecasting – determining the preferred expectation of a *specific* conditional forecast – and also the preferred expectation itself (with respect to prognoses concerning changes of internalized deontic structures) can, in virtue of the high number of combinatorially possible scenario interdependencies absolutely indispensable for operationalization, be justified, for the most part, only in a pragmatic and situation-specific manner.

It is quite clear that a substantive philosophy of history (cf. Danto, 1965) is missing, factually as well as functionally. The same is true, if we do not choose to shy away from such constructions, of a "logic of evolution" which would contain base levels and sequences of progressions – rather like that which Habermas (1976) has attempted to construct, building on Piaget and Kohlberg. So long as there is no model (even a normative one) of the total course of our immediate and distant future, only very little can be regarded as capable of justification and even less as having been justified.

3. THE PHILOSOPHY OF HISTORY AND EVOLUTION

This issue demands a systemic way of looking at things. Obviously we seek a model of the totality of interdependencies which is synchronically as well as diachronically plausible. But certainly we do not require a universal model of history in every single TA study; in most cases we need not consider more than a century, provided that our conceptions are complex enough and we have enough free variables at our disposal.

Now one seductively extensive and apparently non-speculative total concept has been developed recently which has the additional advantage of being able to have recourse to an already elaborated mathematical formalism in its foundations. This is the theory of evolution – enriched in its physical-chemical-biochemical basis and mathematical superstructure in the so-called "synthesis theory" version – as originally formulated by Julian Huxley (1942) and subsequently refined. This theory has been extended explicitly to technology by Hans Sachsse (1978). In fact, claims (of applicability at the universal-historical and social level) and performance (limited to the micro-biological domain) are still far apart; however, what it promises cannot be ignored. It is supposed to provide a theory of strictly law-like random development at all levels (Eigen and Winkler, 1975). With regard to performance: would not this theory make good the lack of a logic of development and provide a genuine possibility of forecasting?

If we assume the theory of evolution, the last and highest system – which, as such, would have no further environment – would be the self-developing system of nature so designed as to keep itself in existence. Various sub-systems have formed within this system, quasi-naturally, which for their part too act to stay in existence, that is, to stay alive. One of these diachronically self-developing sub-systems is the human species. This species is distinguished from other species and types in having developed a specific *Weltbildapparat* which helps enormously in its survival in its specific environment. This has been constantly differentiated and improved, becoming science and technology in our contemporary sense. For their part, these two continue to develop. This whole process of adjustment can be called learning; or in a stricter biochemical idiom, one can even speak of "DNA acquisition."

It is at this point, however, that a fatal difficulty turns up in the situation. It cannot be decided whether we find ourselves in an evolutionary dead end – whether, for example, the preservation of the sub-system called "the human race" has already been made impossible by its own evolution; that is, whether we would in consequence damage our survival chances enormously were we to continue with or even accelerate our evolution – or whether we are in fact in the main trunk of evolution and so may confidently look forward to further or higher developments. This has the consequence for our considerations that the model of evolution says nothing about whether the possibility of our TA exertions could, decisively, be called "self-fulfilling" or "self-destroying." But we expect a substantive background theory to provide us at least with some idea as to where we now stand, so that our forecasts can in some sense be made relevant to the future.

I should like to summarize what I call the "dead-end paradox" of a systemic evolutionary framework as the background for our concrete TA investigations in the following way: That we still live is at best a self-defining sign of the fact that we have been capable of survival until now – not that we still are or that we will remain so in the future. If we find ourselves in the main trunk of evolution – just as much as if we were to find ourselves in a side branch – our previous evolution would have to be cited as the reason. If we are in an evolutionary branch, then the continuation of the previous development will be damaging to our species; if we are in the trunk, then the very same behavior will prove to be beneficial. We can thus see that not only can we not be provided with a helpful theory for the total orientation of our forecasting activity in the realm of TA by evolutionary-systemic attempts; rather, however things turn out, we would be able to explain them with the help of the theory of evolution. This means that at any given time we would be able to deduce contradictory explanatory forecasts on the basis of the theory of evolution. But that would allow us, on grounds of the logic of science, to draw the conclusion that the theory of evolution must be a contradictory and mistaken theory, since "*ex falso quodlibet*."

Of course, however, it cannot be deduced from this that systemic or evolutionary thought is false. The conclusion is relevant only to the widespread theory presently under discussion. What I should like to suggest, rather, is a different systemic-evolutionary method of thought in which two ideas are retained as fundamental:

(1) Man as a part and product of nature, who at the same time is the producer of everything which has been made into culture out of nature, is at the same time the point of self-reflection of the whole system. Nature conceives and reflects upon itself in man; through this reflection nature becomes culture. For this reason, neither self-help, nor self-destruction, nor the revenge of nature against man is to be hoped for. The only form in which nature cures itself is in its "self-enculturation" through man.

(2) Man has in fact an absolutely decentralized position in the cosmos, spatially as well as temporally, the survival or extinction of the human race being neither good nor evil in itself. But for man the continuation of life, and in fact life which is formed and given values by man, is his only goal. Transcendent provision of goals and values is not available, there being no "mission leading beyond the life of man" (Nietzsche). This nihilistic thesis,

however, contains a challenge which originates in the repudiation of a transcendent provision of goals and values: *to provide immanent values and goals oneself.*

These two thoughts to be retained imply for one thing that man must make himself the central point of significance and reference, no matter his total-systemic decentralization and peripheral existence; also, that he must recognize himself to be just that, in the realm of everything which is not human. The hope that the provision of goals and values might, through a part of culture – that is, through nature which has been reformed by man in some way (for example, in the theory of evolution) – be objectively given and scientifically blessed (that is to say, the hope for a higher immanent transcendence by means of scientific predetermination of values) must be jettisoned as illusory.

As we have seen, we nevertheless require some evolutionary total concept in order for our forecasts to be placed against a particular background. When neither transcendent authority nor the institutions of its ideological progeny (in this case the sciences), is in a position to provide us with such a concept (nor could be), there remains only one consequence: *We shall have to do it ourselves.*

4. THE END OF THE MODERN AGE OF ENLIGHTENMENT

My own philosophical position with respect to the philosophy of history, formulated initially extremely tentatively and in retrospect, is that we in the industrialized countries of Western civilization find ourselves at the end of an epoch. I should like to call it the "modern age of enlightenment," without employing the notion of enlightenment in too strict a sense. In another place I called this (with similar qualifications) the "end of the technological-scientific age" (Zimmerli, 1980); I intended this rather precisely as the end of the illusion that science and technology by themselves have the power to provide orienting meaning, values, and goals. This will perhaps become clearer by looking at a wider context of enlightenment. Regarded as one of the mainsprings of modern Europe, the Enlightenment was characterized by the fact that religion, considered as a transcendent provider of meaning and values, was displaced by a secular autonomy of man in his thinking and doing; this was then, as it were, condensed into the aforementioned illusion. It can no longer be denied that there is an increasing consciousness of the fact that the function of science and technology, with respect to the orientation of life and the quality of existence, is missing something, that these two closely related areas of human instrumental behavior require a prior provision of meaning, value, and goals. This is one indication of the end of this age.

The growth in student and youth revolt and the increasing hostility towards science and technology might be regarded as isolated phenomena, as romanticism, or as a mere generation gap. The sudden explosion in religiosity outside the Church, together with the drop-out and ego-trip scenes, might be called mere coincidences. But I should consider all this to be a little too much coincidence at once, preferring rather to discern a turn therein.

I shall attempt to characterize the content of this turn, calling it (for this purpose) the "reflexive turn" (cf. Zimmerli, 1978, pp. 124 ff.). What I mean by this is that in nearly all socially relevant areas of Western European civilization, forms of a movement the inverse of development are manifest, or at least can begin to be made out. "Limits of growth," "qualitative, not quantitative growth," "soft energy paths," "extension of the welfare state," "development of so-called 'goal' sciences" (cf. Böhme, van den Daele, and Krohn, 1973; Hübner *et al.*, 1976), as well as "participation" and "community initiative," are key phrases which might serve to illustrate this movement.

It would be a mistake to assume that these are purely external symptoms. During the last ten years, as already indicated, massive changes of consciousness have taken place in this respect – although perhaps many have not noticed them. That today we regard demands for energy saving, environmental protection, or active support for the investigation of alternative energy sources as something entirely obvious – something we would have regarded ten years ago as unrealistic day-dreaming – is a manifestation of such a change of consciousness. Last but not least, this change is also to be found (objectively as well as subjectively) in the domain of the natural sciences and technology, or at least in societal and political spheres which are oriented towards the natural sciences and technology.

Although many questions may well be left open concerning the implementation or even the feasibility of the suggestions made in the *Bundesbericht Forschung VI* – as C. O. Bauer's report (1980) clearly shows – scarcely anyone in the know would fail to agree completely that some estimation of the consequences of technology was necessary. It has become clear to all of us who have worked on these questions during the last few years that the state (but not exclusively so) in implementing, expanding, and altering technologies burdens itself with costs which have to be anticipated. The slogan, "Shut your eyes and charge ahead!" is as unconvincing as the saying, "Après nous le deluge!" Everybody is quite clear that it is scarcely possible any more to simply give free reign to the process of technological development – which must in any case always be a development of natural science. If, then, the grand sounding expressions, "the end of the technological-scientific age" or "the end of the modern age of enlightenment," are to have even a speculative plausibility (given that we all know that things cannot continue as they have and no one knows exactly what the future holds), it follows that there is general agreement that – unlike in the time of the modern age of enlightenment – we ought to make every possible effort to bring into the sharpest possible focus the likely consequences and side effects of the implementation, extensive application, or alteration of technologies, together with the corresponding state of knowledge and research within the natural sciences.

In this light, the retrospective sketch given in this section (which in its talk of the "end" certainly concealed an element of forecasting) does not provide us with a conclusive total picture of human history that would suffice to justify individual forecasts. It does, nevertheless, provide us with certain negative footholds to which we may return in connection with our ethical considerations.

5. VALUES, NORMS, AND THE JUSTIFICATION OF NORMS

The frame of reference within the philosophy of history which we must develop in order to provide relevant information for the justification of TA forecasts would, in the first instance, consist in grasping and formulating the medium and long-term aids which mankind – or that part of mankind with which we are concerned in a particular TA study – should develop for itself. The transition to the problem of values and norms can be seen here quite clearly. The presupposed values are expressed in the aims which the aforementioned frame of reference would contain.

Values can be defined as basic convictions of various sorts which are ordered in an individual dynamic hierarchy and which are relevant for judging and deciding upon courses of action. These convictions may be thought of as being in a hierarchy only latently, in the sense that, in order for the continuance of the personal identity of the person possessing them, some basic although latent values must remain unchanged.

Norms, in contrast, are general prescriptive maxims involving value preferences applied to a given situation for which trans-individual and societal validity is claimed – precisely in virtue of their general prescriptive character (Zimmerli, 1978). The preference rules (norms) which are to hold in individual value systems in a given situation determine in consequence the goals towards which history is thought to be moving.

In fact, a problem of justification arises here, too – this time as the justification of norms. By "justification of norms" I understand a tracing back of that to be justified to something accepted by those who are entitled to demand the justification of the norm. In order that any such activity designated as "justification" might be carried out with a fair likelihood of success, certain "rules of retrogression" are needed. According to the requirement level of those entitled to demand the justification, these will correspond, by and large, to a more or less well elaborated formal or deontic logic. (In most cases, only the most rudimentary forms of this logic will be needed.) The authority to which a justified norm is, or has to be, traced back, must not, on pain of a mere consensus relativism, depend upon a mere factual consensus, obtained *ad hoc*; otherwise arbitrarily many contradictory norms could be justified according to the composition of the group entitled to demand the justification. The final authority to which appeal can be made must be an authority which has become such by its previous and/or present normative status. I call such norms "basic bridging norms." They make available the basis of justification; at the same time they embody a bridge between the descriptive and the prescriptive. Thus to "justify" in this context means to trace back from that to be justified to a basic bridging norm whose previous validity has been so pragmatically guaranteed that those entitled to demand a justification are already in agreement before they explicitly engage in the activity of justification (Zimmerli, 1981).

Now, as shown earlier, the problematic aspect of forecasting in connection with TA rests on the assumption that a historical change of value systems, and thus of norms, and finally of the justification of norms, is involved. "Change in the justification of norms," according to everything developed here so far, must mean a change in tracing back to basic bridging norms or a change in these norms themselves. For simplicity's sake we might imagine this as follows. The specific pattern of value preferences of each individual – although it represents only one small section and only one hierarchical combination of a possibly finite but nevertheless unsurveyable number of possibilities of such patterns and combinations – unmistakably defines the individual identity of the relevant person in a given situation. A particular range of such patterns and combinations of hierarchies by contrast, defines the identity of higher level systems (families, villages, countries, societies, historical periods, etc.). If one thinks of all the values, as well as those preference relations between values called "norms," as arranged in an ordered system or pattern, the basic bridging norms would have to be those norms which, at a given time, occurred in the value-preference pattern of every individual. Historical change in the justification of norms can occur insofar as at least two of the values which are correlated as value preferences in the basic norm change positions. (All relevant expressions of opinion seem agreed that we are experiencing such a change now.)

6. "DEFENSIVE" ETHICS AND TA

If we recall how the justification of norms was defined previously, the step to the point where the defining characteristics of the change become visible is but a short one. To "justify" means to trace back to a basic bridging norm. Such a tracing back can occur positively or negatively. At the moment at which consequences of an unwanted variety follow from adherence to particular basic bridging norms (as is manifest in the case of non-acceptance), it becomes not only possible but positively in accord with political rhetoric's aim to win consensus (i.e., it becomes nothing short of opportune) to justify particular norms ethically by showing that they stand *in contradiction* to other norms which can be reduced to now-incriminated basic bridging norms.

That new attitude to technology which first became manifest in the recognition of the problem and then subsequently in non-acceptance (something that has become particularly clear in the case of energy systems) can be regarded as a return once again to the original existential ego-groundedness of science and technology. The result of this reflexive turn consists in the insight that the complexity of the situation – partially reduced by science and technology but also partially created anew by it – is a complexity which fails, as it were, to "get through" to the basis, namely, to those concerned.

Here the most complex situations reduce themselves to simple yes-no decisions. Forms turn up in this connection which are certainly familiar to us as the existential pattern of the 1950s – forms of human responsibility for decision making in border-line situations which could only be explained and justified by reference to immediate existential involvement, not by reference to the careful and rational weighing of diverse objectified means-ends relationships.

This has the following consequence for basic ethics in its relation to technology (and *a fortiori* in relation to the question of TA): It must involve a "defensive" ethics (as I have called it), or a "residual" ethics, since it has to show (and also justify this being the case) that its norms stand in contradiction to those basic bridging norms – "aggressive" norms – which are at the center of a progress-oriented technological-scientific ethics of growth. In concrete terms, this means that the new basic ethics will be an ethics of prevention. One foundational norm which follows from this defensive ethics can be formulated in a Kantian way:

> (1) Refrain from any action about which, on the basis of your assessment of consequences, you could not be sure whether you would want its expected consequences or not.

This foundational norm is purely formal; its initial point is to reduce the number of impending decisions or options, without making any preliminary substantive decisions. A further reduction of the number of impending options requiring a decision can be obtained, in the same context, if we formulate a second norm thus:

> (2) Refrain from any action which, on the basis of your assessment of consequences, you can assume will lead to an intensifying of quantitative growth.

I suspect that these defensive maxims and norms already possess a deeply entrenched ethical self-evidence for those concerned with non-acceptance.

Further formal norms which serve to reduce the number of options can be discerned in the debate concerning atomic energy and the environment. What is of significance in this connection can obviously not be cut down to risk analysis proportions. It is rather that, in this context, problems of dimensions and unwanted consequences which transcend individual responsibility (e.g., the problem of final storage of atomic waste) play a role; this in turn presupposes the implicit validity of a principle of "inter-generational responsibility." Moreover, "nature" as a concept which provides its own values, comes to adopt a type of position as the supreme preference, such that there has even been talk of the "responsibility of stewardship" (Jonas, 1979). Two further formal norms result:

> (3) Refrain from any action which is not in conformity with the principle of inter-generational responsibility.

> (4) Refrain from any action which is not in conformity with the principle of responsibility or stewardship with respect to nature.

To discover positive, substantive ethical preference relations beyond these formal ones will most probably be difficult. This could be suspected from the fact that the underlying orientation of non-acceptance is unitary only to the extent that it regards the status quo negatively, presenting itself as an alternative. Nevertheless, something like a basic outline can be discerned in the attitude; and this basic outline must – as a latent societal utopia – give expression to the aims, as well as the value hierarchy, which define the preference relations or norms. Some of the "alternative" ideas can to a large degree be gathered together under the following rubric.

At the individual existential level, a satisfying feeling towards life, enriched by sensuality, is to obtain and to be integrated within a setting of communal life and shared intellectual content. The communal forms of life are to be peace-loving and not centered on achievement. Communal responsibility of a new sort is to be cultivated within these forms – but it will express itself only in the future and among those who are to come. Society, seen politically, need not be centralized in a state; it can be built up of small decentralized cells – something which would entail the disappearance of an increasingly anonymous and alienating bureaucracy. The economic basis for this is to be prepared, not in massive centralized industry, but in decentralized, small, limited technologies and agricultural holdings. Expanded involvement in "soft energy paths" is presupposed here. A tightly organized political system would thus become as superfluous and obstructive as the way society is organized in high or late capitalism.

In outlining this latent basic conception of future society, one thing becomes quite clear which, in the disparate utterances of its various advocates, hides more than it reveals: namely, that the main struggle is no longer against the system of late capitalism. Correspondingly, in virtue of the lack of a clearly defined classical proletariat in Marx's sense, neither is the main struggle against the private ownership of the means of production. According to the system as described, this will also become superfluous and dysfunctional. The orientation towards alternatives – however misty and vague it all appears – is nothing less than a new comprehensive picture of man.

The normative contents of this conception cannot be disregarded. In so far as we proceed on the assumption that the model of values of which I spoke finds expression in this conception, there is some likelihood that, *mutatis mutandis*, it will be adopted in the near future. A future built on post-material values (even on post-materialistic values) will probably reinforce non-acceptance of those energy systems which contradict the aforementioned norms. The sensibility required has been awakened and is being handed down. On the other hand, precisely this argument could be used to justify the suspicion that, with the debate about atomic energy, the battle concerning the level of the acceptance threshold has been fought once and for all – since anything that happens has to seem harmless by comparison with atomic energy.

In a certain sense, both features may turn out to be true; the "alternative" movement is also oriented towards alternative energy systems and is prepared to put up with other things so long as they do not too glaringly contradict the criteria listed.

Whether one is of the opinion that acquiescence in whatever is the case is what is needed or one considers the "alternative" tendency worth fighting against, no theory concerned with conditions of acceptance within the realm of TA can afford to ignore it.

What has just been developed, as an example and by way of residual ethical questions or problems, can once again be related meta-theoretically and generally to the problem of TA. We see there that mid-range forecasts of development must be measured against the contents of our value hierarchy – that is, of our norms as rules of value preferences. In comparison with our problem as initially formulated, an easy simplification now becomes manifest. We may indeed be dealing with changes in internalized deontic structures, but we can neglect them where they diametrically oppose present value systems. Whenever men undertake evaluation, whether with respect to past or to future situations, they always relate the scales of evaluation back to their own scale of evaluation. Thus, at least with respect to mid-range assessments, we may proceed on the assumption that internalized deontic structures will not change fundamentally.

Of course this does not mean that, for example, no changes of internalized deontic structures between generations are to be expected; it only means that even those changes, in so far as we can anticipate them, are evaluated by us now, and they are not only *de facto* judged but must be judged against our prevailing scale of values. That we decide in favor of one and only one alternative future scenario – under the objectifying cloak of "desirability" – already demonstrates that we are subjecting future changes in internalized deontic structures to our evaluation.

On the basis of these clarifications and limitations, it becomes clear that ethics and the philosophy of history, in so far as they are connected with TA, are reduced to the tiny fragment of formal ethics described above. Talk of "the goal of history," of the future towards which we must direct our thought and action, in consequence does not present us, at the level of values, with the feared problems of forecasting; it means nothing more than projecting into the future what we consider desirable and worth keeping of our present value preferences. It is consequently a matter of coming to agree on what we want, of obtaining a common picture of the desired future – and not just of the middle-term future. At this point it becomes clear that not only TA efforts, but also those efforts whose intensification we have been able to observe for some years (which can be summed up under the title of "professional ethics"), get transformed into politics in the sense that they require an attempt to create agreement, to transform non-acceptance. TA studies can also provide casuistic arguments in the form of if-then scenarios selected on the basis of probability. At the same time, that scenario which will be the reality of the future will be co-determined by a great variety of professional ethics efforts and a political value system, hierarchically structured, of medium range, and capable of creating agreement.

Note: I should like to thank Dr. Mark Helme for his help in making the translation from the German.

Technical University, Braunschweig

REFERENCES

Bauer, C. O. 1980. "Technik-Folgeabschätzung – jetzt amtlich?," *Wirtschaft und Wissenschaft* 3/4 4–9.

Berg, M., Chen, K., and Zissiz, S. 1976. "A Value-Oriented Policy Generation Methodology for Technology Assessment," *Technological Forecasting and Social Change* 8 401–420.

Berg, M., Chen, K., and Coates, J. 1976. "Response" (to Professor Skolimowski), *Technological Forecasting and Social Change* 8 426–435.

Böhme, G., van den Daele, W., and Krohn, W. 1973. "Die Finalisierung der Wissenschaft," *Zeitschrift für Soziologie* 2 128–144.

BMFT (Bundesministerium für Forschung und Technologie). 1979. *Bundesbericht Forschung VI*. Bonn.

Conrad, J. 1978. "Zum Stand der Risikoforschung – Kritische Analyse der theoretischen Ansätze im Bereich des Risk Assessment." Bericht für das Bundesministerium des Inneren, BF-R-63. 012–100/3. Frankfurt: Battelle Institute.

Council of Science and Society. 1976. *Superstar Technologies*. London.

Danto, A. C. 1965. *Analytical Philosophy of History*. London and Colchester.

Eigen, M. and Winkler, R. 1975 *Das Spiel: Naturgesetze steuern den Zufall*. Munich.

Gäfgen, G. 1968. *Theorie der wirtschaftlichen Entscheidung*. Tübingen.

Haas, H. (ed.). 1975. *Technikfolgen-Abschätzung*. Munich.

Habermas, J. 1976. *Zur Rekonstruktion des historischen Materialismus*. Frankfurt.

Helmer, O. and Rescher, N. 1959. "On the Epistemology of Inexact Sciences," *Management Science* 5 25–52.

Hempel, C. G. and Oppenheim, P. 1948. "Studies in the Logic of Explanation," *Philosophy of Science* **15** 135–175.

Hübner, K., Lobkowicz, N., Lübbe, H., and Radnitzky, G. (eds.). 1976. *Die politische Herausforderung der Wissenschaft*. Hamburg.

Huxley, J. 1942. *Evolution: The Modern Synthesis*. London.

Johnston, D. F. 1970. "Forecasting Methods in the Social Sciences," *Technological Forecasting and Social Change* **2** 173–187.

Jonas, H. 1979. *Das Prinzip Verantwortung: Versuch einer Ethik für die technologische Zivilisation*. Frankfurt.

Knapp, H. G. 1978. *Logik der Prognose: Semantische Grundlegung technologischer und sozialwissenschaftlicher Vorhersagen*. Freiburg and Munich.

Lowrance, W. 1976. *Of Acceptable Risk*. Los Altos, Calif.

McDonagh, E. 1975. "Wertpräferenzen im Bereich der Technik," *Concilium: Internationale Zeitschrift für Theologie* **11:12** 667–678.

Meyer-Abich, K. M. 1979. "Soziale Verträglichkeit – ein Kriterium zur Beurteilung alternativer Energieversorgungssysteme," *Evangelische Theologie* **39** 38–51.

Meyer-Abich, K. M. 1981. "Zum Problem der Sozialverträglichkeit verschiedener Energieversorgungssysteme," in J. v. Kruedener and K. v. Schubert (eds.), *Technikfolgen und sozialer Wandel: Zur politischen Steuerbarkeit der Technik*. Cologne, pp. 41–56.

Mitroff, J. I. and Turoff, M. 1973. "Technological Forecasting and Assessment: Science and/or Mythology?" *Technological Forecasting and Social Change* **5** 113–134.

OECD (Organization for Economic Co-operation and Development). 1975. *Methodological Guidelines for Social Assessment of Technology*. Paris.

Otway, H. 1976. "Risk Assessment," *Futures* **12** 122–134.

Paschen, H., Gresser, K., and Conrad, F. 1978. *Technology Assessment – Technologiefolgenabschätzung*. Frankfurt and New York.

Popper, K. R. 1971. *Logik der Forschung*. 4 ed., Tübingen. (First published 1935.)

Quick, H. J. 1976. *Organisationsformen der wissenschaftlichen Beratung des Parlaments: Eine Untersuchung zur institutionellen Verankerung einer Technologiebewertungseinrichtung beim Deutschen Bundestag*. Berlin: Schriften zum Öffentlichen Recht, 306.

Reese, J. *et al.* 1979. *Gefahren der informationstechnologischen Entwicklung: Perspektiven der Wirkungsforschung*. Frankfurt: Gesellschaft für Mathematik und Datenverarbeitung, Institut für Planungs- und Entscheidungssysteme.

Ropohl, G. *et al.* 1978. *Masstäbe der Technikbewertung*. Düsseldorf: Verein Deutscher Ingenieure.

Rowe, W. D. 1975. *An Anatomy of Risk*. London.

Sachsse, H. 1978. *Anthropologie der Technik: Ein Beitrag zur Stellung des Menschen in der Welt*. Braunschweig.

Skolimowski, H. 1976a. "Technology Assessment in a Sharp Social Focus," *Technological Forecasting and Social Change* **8** 421–425.

Skolimowski, H. 1976b. "Where the Technicians Fear to Tread: A Rebuttal," *Technological Forecasting and Social Change* **8** 435–438.

Starr, C. 1969. "Social Benefit versus Technological Risk," *Science* **165** 1232–1238.

Starr, C., Rudman, R., and Whipple, C. 1976. "Philosophical Basis for Risk Analysis," *Annual Review of Energy* **1**.

Steinbuch, K. 1971. *Mensch – Technik – Zukunft: Basiswissen für die Probleme von morgen*. Stuttgart.

Vester, F. 1979. "The Biocybernetic Approach to Understand and Plan our Environment – Presentation of a New Sensitivity Model," in H. Buchholz and W. Gmelin (eds.), *Science and Technology and the Future*. Munich, New York, London, and Paris: Proceedings and joint report of World Future Studies conference and DSE preconference held in West Berlin, 4–10 May 1979; part 2; pp. 1420–1425.

Zimmerli, W. C. 1976. "Technokratie und Technophobie – wohin steuert die Menschheit?" in *Technik – oder wissen wir, was wir tun? Philosophie aktuell* 5. Basel and Stuttgart, pp. 146–159.

Zimmerli, W. C. 1978. "Von Kernenergie und menschlicher Verantwortung am Ende des Wachstumsdenkens," in W. C. Zimmerli (ed.), *Kernenergie – wozu? Philosophie aktuell* **13**. Basel and Stuttgart, pp. 122–140.

Zimmerli, W. C. 1980. "Erinnerungen an die Philosophie: Am Ende des wissenschaftlich-technischen Zeitalters," *Schweizer Monatshefte* **60**:7, 575–588.

Zimmerli, W. C. 1981. "Gesellschaftliches System und Wandel ethischer Normenbegründung: Grenzen der systemtheoretischen Betrachtungsweise bei der aktuellen Suche nach einer Ethik der Technik," in J. v. Kruedener and K. v. Schubert (eds.), *Technikfolgen und sozialer Wandel: Zur politischen Steuerbarkeit der Technik*. Cologne, pp. 181–204.

PART III

RESPONSIBILITIES TOWARD NATURE

BERNARD GENDRON

THE VIABILITY OF ENVIRONMENTAL ETHICS

Does nature have any rights against being exploited? The answer given by modern Western philosophy is a resounding "No!" It has been a dominant theme in the ideology of the industrial West that it is always wrong to exploit human beings but never wrong to exploit nature. Indeed, we who have grown up in this milieu have been given the distinct impression that it is highly desirable (even on moral grounds) to exploit nature to the utmost.

This tradition is now clearly being challenged. Many environmentalists (e.g., Aldo Leopold) are trying to get us to accept what is alternatively called a "land ethic" or an "environmental ethic."[1] They are telling us that biotic systems, and some of the animals, plants, and minerals that make them up, should be granted rights against unlimited exploitation. We are now being forced to ask: why should humans monopolize for themselves all moral rights against being exploited? Is this not a case of bald human chauvinism masquerading as high philosophical theory?

I want to consider two very influential objections to the viability of an environmental ethic. The first says that such a stance is philosophically confused. It makes no sense to grant rights to nonconscious beings. The second says that this stance is hopelessly impractical. Recognizing the rights of nonhuman nature would put the brakes on technological growth and plunge us into economic depression.

I remain somewhat agnostic on the desirability of accepting an environmental ethic. But I think that the above two objections can be successfully countered. First a few words need to be said about the concept of *exploitation.*

To exploit someone or something is to treat him/her or it as a mere instrument or raw material for the achievement of one's goals. Two consequences flow from this. First, the exploiter has some control over what the exploitee does. You cannot use something as a mere instrument if you cannot control what it does, and you cannot use it as a raw material if you cannot transform it the way you want. Second, the exploitee always provides a service for the exploiter. Y cannot be exploited by X unless there is a flow of goods from Y to X. These two factors – control and services – are interconnected. What X has control over, in exploiting Y, are the goods and services which Y provides X.

Paul T. Durbin and Friedrich Rapp (eds.), Philosophy and Technology, 187–194.
Copyright © 1983 *by D. Reidel Publishing Company.*

Now it is true that the exploiter often provides services for the exploitee. You must take care of your instruments if they are to work well for you. Nonetheless, when X exploits Y, X usually expects to get more from Y than X gives Y. There are some exceptions to this. Consider the case of the pampered house slave whose master showers him with gifts and amenities. No matter how high the remuneration, this house slave is being exploited so long as he keeps the legal status of a chattel and is treated as such.

From the above, it should be quite clear why exploitation of human beings runs against the basic tenets of Western liberalism. On the political side, exploitation is an authoritarian relation and thus conflicts with our ideals of political equality. On the economic side, it usually involves unequal exchanges between two parties, and thus conflicts with our ideals of distributive justice.

By refusing to extend rights against exploitation to non-human nature, Western industrial culture in effect has been telling us that there are no moral restrictions on the amount of domination we may exercise over nature or on the amount of unpaid surplus we may extract from it.

Let us turn now to the objection against environmental ethics which says that it is hopelessly impractical to grant nature any significant moral or legal rights against exploitation.

From the outset we must concede that it would be hopelessly quixotic to grant rights to nature against all exploitation. Our moral obligations to nature simply cannot be as extensive as our obligations to human beings. We cannot survive if we are not allowed to process raw materials or to use instruments in order to create objects of consumption. Similarly we cannot apply technologies without processing some materials or using some instruments. Technology is inherently knowledge for processing materials or using instruments. Even those technologies which deal primarily with the organization of labor aim ultimately at the more efficient processing of materials or use of instruments. Technology is inherently knowledge for exploiting. We are inherently a technological species. Hence, we are inherently an exploitative species.

We hear talk sometimes of non-Western and non-industrial societies which have not been exploitative of nature or which have not used exploitative technologies in their dealings with nature. This is nonsense. There clearly have been societies which have been less exploitative than ours, or which have accorded nature certain immunities against exploitation, but none which have completely desisted from exploiting nature. Certain parts of nature may have been made off-limits to exploitation (as in the case of sacred animals, mountains, streams, or trees), or might only have been allowed to be exploited with gentleness and care. In some cases, the potential exploiters (e.g., woodcutters) may have been obliged to seek permission from nature, or to apologize to nature, for the performance of their exploitative task (e.g., cutting down sacred trees). Finally, in other cases, the exploiters may have been enjoined to pay back nature what they took away from it. But again paying back nature is compatible with using it as raw material or as instrument, and subjecting it to the sort of domination implicit in these uses.

There is therefore no denying that it is impractical to prohibit morally all exploitation of nature. But is it impractical to impose limited but significant moral constraints on such exploitation? The Reagan administration and other spokespersons for American industry would say quite definitely that it is. Their usual line goes somewhat as follows.

America, we are told, is undergoing a severe productivity crisis. Rates of productivity growth in this country have declined from 3.2% per year during the 1948-1968 period to 1.9% per year during 1968-1973, and to 0.9% per year during 1973-1979. But productivity growth is the most reliable measure of technological growth. Thus, despite impressive advances in microprocessing and other select fields, American technology seems to be stagnating. This in turn is undermining the economy, contributing both to intolerable rates of inflation and unemployment.[2]

What is the cause of this? In the Reaganomic scenario, one of the primary culprits is governmental regulation to protect the environment. Industry is forced to divert funds from investment in new technologies to pollution abatement and resource conservation. Apparently we have gone too far in the latter direction. If this trend is not reversed soon, Reaganites predict, we will find ourselves mired in economic depression. We must put a damper on our enthusiasm for protecting the environment.

Present environmental law is not based on any recognition of rights of nature against unlimited exploitation. It merely protects the interests of certain groups of humans against the encroachments of others. When courts ordered Milwaukee to change its methods of treating the sewage which it dumps in Lake Michigan, they were protecting the rights of the city of Chicago and not of Lake Michigan. There is no doubt that environmental law would be considerably strengthened and stabilized if it were based on the ascription of legal rights to nature. This means that if the Reaganites are right in their assertion that present environmental regulations are having a chilling effect on technological growth, then *a fortiori* the recognition of rights of nature would have a devastating effect on technological growth. So it would seem quixotic to work for the social acceptance of an environmental ethic. Is this really so?

However fraudulent it may otherwise be, there is some point to the Reaganite position. But, quite surprisingly, this point works in favor of rather than against the ideal of an environmental ethic. "Productivity" can mean many things, depending on what benefits one wants to maximize at the expense of what costs. The typical modern state fixates almost exclusively on growth in labor productivity, that is, in the value of goods produced per unit of labor. Labor productivity usually is increased by increasing the capital-labor ratio, that is, the amount of hardware and energy used in the production process for every unit of labor. This strategy will work only if the capital goods (and the energy and mineral resources used to manufacture these) remain relatively cheap.

But here is the rub. Any social order that grants nature significant rights against exploitation will thereby put significant restriction on the extraction of mineral and energy resources and the manufacture of capital goods. Everything else being equal, the prices of capital goods will increase, and the rate of growth in labor productivity will decline. So there is definitely a tension between environmental ethics and growth in labor productivity.

But this is a blessing in disguise. Even in the absence of significant environmental regulation, it will probably become increasingly difficult and frustrating to pursue growth in labor productivity. Within a century, we will have used up virtually all the solar energy which it has taken the earth millions of years to store in fossil fuels, and we will have squandered untold amounts of high-grade mineral resources. The odds are that the new technologies of the future, however exotic, will not yield the same returns in the extraction of resources and energy that the more modest technologies of this century did with considerably less effort and money. The upshot is that by continually pushing to increase the capital-labor ratio, the world industrial system is beginning to run its head against the walls of nature. It is on a collision course with a resource crisis of unprecedented proportions in the none-too-distant future.[3]

Hence, it appears considerably less quixotic to grant nature rights against unlimited exploitation than to stake our priorities on increasing the rates of labor productivity growth. By recognizing the rights of nature now, we would be husbanding our resources. We would be carefully preparing ourselves to ease gracefully into an era of near-zero growth in labor productivity. Surprisingly it appears to be the eminently practical course to turn to some form of environmental ethics.

This does not mean however that technological growth must come to an end. Technology is too malleable for that. There are as many ways for technology to grow as there are ways of being productive. In an environmentally benign society we would pursue growth in environmental productivity at the expense of growth in labor productivity. To increase environmental productivity is to increase the ratio of social benefits (including clean air, uncongested highways, healthy food) to social costs (including pollution, resource depletion). Any technological growth which increased environmental productivity would be desirable.

It is true that technology is inherently exploitative of nature. But there are different ways of exploiting nature, and some are less disruptive and demanding than others. Technology can teach us how to exploit nature in a way that is compatible with nature's survival and integrity. It can teach us how to work with nature rather than how to radically transform it. It can help us repair the damage we have already done to nature. Recognizing rights of nature against exploitation does not turn us against technology *per se.* It turns us toward the "soft" technological path (solar energy, organic gardening) and away from the "hard" technological path (nuclear energy, agribusiness technology). The growth of soft technologies will not necessarily give us more power over nature. What it will do is increase our ability to muddle along within the confines of the walls of nature. Such technological growth is indeed not only compatible with, but also stimulated by, a growing recognition of nature's rights against unlimited exploitation.

Let us turn now to the objection that the concept of rights of nature makes no philosophical sense. Imagine how someone might make this case. "Please don't confuse us with the Reaganites. We are not against the legislation of extensive environmental protection. But to legislate in favor of the environment is not necessarily to give it any moral or legal rights. In fact, we don't know what it could possibly mean to recognize rights of nonconscious beings against unlimited exploitation. If something has some moral or legal right against my exploiting it, then it has claims against me. But it can't have claims against me unless it has interests that need to be protected from me, and it does not have interests unless it has wants, that is, unless it has consciousness. So for us nature cannot have any moral standing. We do not deny that nature has any value. Of course it is of use to humans, and in this respect it has instrumental value. What we deny is that it has intrinsic moral value. The instrumental value of nature is quite sufficient to ground whatever legislation we need to protect the environment. That is, protecting the environment is nothing more than a matter of survival for us."[4]

This argument pretty faithfully summarizes the case against environmental ethics usually made in the philosophical literature. I find it quite unconvincing. First, very few even of us who are in favor of abortion would say that it is conceptually impossible for fetuses to have rights. We would allow that they have interests, although we have no evidence that they have wants.

Furthermore, it falsifies the process of rights-ascriptions to say that we first find out what beings have interests and then determine from this which of these should have rights. Actually, it is sometimes easier to determine something's interests once we have decided what rights we are willing or forced to grant it. If we decided to grant to that biotic system called Lake Michigan the right to be represented in courts as a plaintiff (say, against cities like Milwaukee), we should have no trouble determining what its interests are. The sorts of (morally or legally relevant) interests it is deemed to have will depend on the sorts of moral or legal rights against exploitation which we are persuaded or forced to recognize in it. The rights we bestow on something determine the interests we attribute to it. Suppose we were so silly as to grant to 1967 Ford Mustangs the right to exist in good health. Then their interest would obviously be to be protected from the junk heap and to be kept in good running order. Anything can have an interest to which we are willing to ascribe rights. It is rights that are interest-making, not interests which are right-making.

Finally, those who deny that nature can have intrinsic moral value are laboring under the naturalist illusion that intrinsic value is a property (like color) that one discovers in things. This illusion dissipates if one speaks about intrinsic value in the verbal rather than substantive mode. We value some things intrinsically, others instrumentally, and some both. To say that something has intrinsic moral value is to say that there are good reasons for valuing something as an end-in-itself.

To deny that nature can have intrinsic moral value while allowing that it has instrumental value is to assume that the reasons for valuing something intrinsically must be of a radically different sort than those for valuing it instrumentally. This is false. The same sorts of reasons, though perhaps more compelling or pressing, can lead us to value something intrinsically which lead us to value something instrumentally. Otherwise we could make no sense of utilitarian or contractarian ethics. By appealing to welfare or the greatest happiness for the greatest number, utilitarians are giving us instrumental reasons for valuing people intrinsically, as moral ends in themselves.

Or, consider the scenario leading to the establishment of a Hobbesian contract. The contracting parties agree at the outset that killing must be prohibited on pain of severe punishment if the condition of war of everyone against everyone is to be terminated. But someone rises to say that this is not enough. The issue of killing is too important she says, and the chances of offenders getting away with it too great, for society to depend exclusively on the deterrent power of the state. Society must supplement its legal prohibitions against killing, she concludes, by inculcating in its population a deep veneration for human life. People must come to view others as ends-in-themselves. They must see human life as intrinsically worthy of being protected. Only then can society be assured that all but a few will refrain from trying to kill others.

Could one object here that we are being given an argument for disseminating the illusion that human life has intrinsic value rather than an argument that establishes the intrinsic value of human life? Not if we construe value in verbal terms. For what this member of the Hobbesian community is arguing is not that humans have some peculiar property (intrinsic value) but that we should treat human life in a certain way (value it intrinsically as an end in itself). Thus we have an instrumental justification of intrinsic value.

For the same sorts of reasons, we could argue for valuing biotic systems intrinsically. We could say that the war between us and the environment has become nasty, brutish, and intolerable. To get out of this state of nature between ourselves and the environment, we need not only to prohibit coercively those acts which undermine the integrity of our biotic systems, but also to inculcate in people a reverence for that integrity. People must come to view biotic systems as ends in themselves if they are effectively to desist from excessive environmental despoliation. Why must we appeal not only to people's sense of self-interest (that is, the instrumental value that biotic systems have for them) but also to their sense of reverence (that is, the intrinsic value that biotic systems have for them)? For one thing, the integrity of the environment is of instrumental value for distant generations as well as for this one. But this is something very difficult for the present generation, in its everyday struggles, to make allowances for, if its only considerations are matters of instrumental value. Learning to value the environment intrinsically will inculcate in this generation that extra restraint which will save the environment for future generations. The rights of present and existing biotic systems are much more palpable than the rights of distant and yet-to-be-existent generations. There is another reason for inducing in people that extra restraint that comes from valuing nature intrinsically. Because of our ignorance of the complexities of biotic systems, we often are unaware that we are dangerously undermining our environment. We must sometimes hold back even when the instrumental value of the environment for us does not seem threatened by our activity. We cannot always rely on the city of Chicago to obstruct the city of Milwaukee's pollution of Lake Michigan. Chicago may not be aware of the dangers of such pollution or may not be sufficiently concerned for its long-term intergenerational effects. Thus, on instrumental grounds, we may have a strong case for coming to view Lake Michigan as an end-in-itself with moral and legal rights of survival.

To conclude. Anything can be valued as an end in itself when there are strong instrumental reasons. Some important types of ethical theory are premised on this. There seem to be good instrumental reasons for valuing the environment not only instrumentally but as an end-in-itself. But if this is so, there are good reasons for granting rights to nature against exploitation. Anything that can be intrinsically valued morally has moral standing, and can be the subject of rights. In the way we grant nature rights, we will determine what interests of it we are supposed to protect. So, to the question, "Why should we recognize rights for nonhumans against exploitation?" our answer would be quite simply, "Because it is eminently practical to do so." That is, our response to the charge that environmental ethics is quixotic paves the way for our response to the charge that it is philosophically confused.

University of Wisconsin/Milwaukee

NOTES

[1] Aldo Leopold, *Sand County Almanac* (New York: Ballantine Books, 1966); Christopher Stone, *Should Trees Have Standing*? (Los Altos, CA: W. Kaufman, 1974); Mark Sagoff, "On Preserving the Natural Environment," *Yale Law Journal* 84 (1977): 205–267; Tom Regan, "The Nature and Possibility of an Environmental Ethic," *Environmental Ethics* 3 (Spring 1981): 19–34.

[2] See, for example, the special issue, "The Reindustrialization of America," *Business Week* 2643 (June 30, 1980).

[3] For support of this sort of view, see Hazel Henderson, *The Politics of the Solar Age* (Garden City, NY: Anchor Press, 1981).

[4] For good surveys of this literature, and good analyses of arguments of this sort, see Regan, *op. cit.*, and Scott Lehmann, "Do Wildernesses Have Rights?" in *Environmental Ethics* 3 (Summer 1981): 129–146.

HANS LENK

NOTES ON EXTENDED RESPONSIBILITY AND INCREASED TECHNOLOGICAL POWER*

"It is not the solution of technological problems, but that of the ethical problems which will determine our future," thinks Sachsse (1972, p. 122). In his book on "technology and responsibility" Sachsse is one of the few authors who explicitly deals with ethical problems of technological progress – without, however, being able really to present such solutions. Indeed, it would be presumptuous to hope for neat solutions in advance, while the new ethical dimensions of technology and applied science have just loomed in our range of vision. The reader may be lured into modifying Sachsse's statement, exaggerating it for didactic reasons. Instead, we could say: Not only the solution of technological problems, but also of those ethical problems connected with technological progress and its worldwide application, will (along with other things) decisively stamp the future of mankind. The point is, we cannot afford even today, but particularly in the future, to ignore or neglect the pressing ethical problems of technology and applied science.

Today much more than hitherto ethical and moral problems evolve in connection with the extended technological power of man to dispose of our nonhuman environment, of "nature"; the power to manipulate and temper life, even human life itself. Because of man's tremendous technological effectiveness and the huge power of technological action, a new situation for ethical orientation is evolving which requires new rules of behavior – therefore also a new ethics in the strict sense? Even if the basic principles of goodness remain constant, the "executive" rules applying ethics to conditions must be developed further, must be adapted to new possibilities of behavior, action, and side effects. This must not be a mechanical adaptation of new rules of behavior judged in the light of constant basic ethical values (perhaps interpreted in a new way), but must be an adaptation in the light of predictable, calculable consequences and of critical/pragmatic discussion of details.

More precisely: What new situation is determined by technological development and ever accelerating technological progress? Doubtless the situation is characterized by the fact – among others, but not only – that certain moral and legal concepts do not fit the new technological phenomena and processes. Sachsse shows, for instance, that the process of transmitting information cannot be interpreted simply as an exchange of goods since the

Paul T. Durbin and Friedrich Rapp (eds.), Philosophy and Technology, 195–210.
Copyright © 1983 *by D. Reidel Publishing Company.*

seller still possesses the object of the bargain after the exchange. Information does not, like material objects, obey simple rules of addition and subtraction. Our moral concepts of property, theft, and just exchange – all of them oriented towards the category of substance – are not applicable to information (pp. 134–136).

But the fact of new technological phenomena and processes, by itself, is not – it seems to me – the only characteristic of the new situation generating a new kind of ethical problems related to technological progress. The most decisive new perspective for a new interpretation or a new application of ethics is, beyond any doubt, the immense growth in the technological power of man. Leading to specific risks, it requires new ethical perspectives:

(1) The number of people affected by technological measures or their side effects has increased tremendously. Those affected frequently do not directly interact, in the same context of action, with the intervening agent.

(2) Natural systems are now an object of human action – at least in a negative way. Man can permanently disturb or destroy them by his technological measures. Without doubt, this is an absolutely new ethical situation. Man never before possessed the power, globally or even regionally, to destroy all life in an ecological system, to debase it decisively by technological manipulations. Since some of these encroachments are not controllable and may be irreversible, nature, both as an ecological system and in terms of particular species, should gain a new ethical relevance within the new dimensions of technological power. In the past, ethics was essentially anthropocentrically oriented around actions, interactions, and their consequences among men; however, ethics must now gain an ecological relevance, giving significance to other forms of life (as for example in Schweitzer's "ethics of reverence for life"). Taking into account possible irreversible harmful effects (e.g., changes of climate, injuries by radiation, technological erosion), the fate of man is also at stake – but not at all only his fate.

(3) In view of the increased possibilities of effects of manipulation in the biomedical as well as in the ecological context, the problem of responsibility for unborn individuals and generations is getting new emphasis.

(4) Not only in social engineering and in the manipulation of the unconscious, but in any experiment with human subjects (be it pharmacological-medical or social scientific) man himself has become an object of scientific research. And this is a new special ethical problem (cf., e.g., Jonas, 1969, and Lenk, 1979).

(5) Specifically in genetic engineering man has gained the power to change hereditary stock, to generate new living species by directed mutations,

perhaps even to genetically manipulate or change man himself. This certainly is a totally new dimension for ethics. Can man carry the responsibility; has he the right to produce (and change) artificial species of life; to eugenically alter himself (even if toward the better)?

(6) Man does not seem to be an object of technology only in genetic manipulation; he has become an object of collective and individual manipulation not only by pharmacological means and in terms of mass suggestion and subconscious influences, but also in terms of pharmacological and medical self manipulation (e.g., tranquilizers).

(7) Can we observe a progressive trend called "electronic technocracy"? This would be technocracy in combination with bureaucracy and a progressive development of microelectronics, computer-aided systems organization, and automated administration. The development of applied computer technology and of electronic data and information systems has certainly brought about the possibility of total technocratic control of persons *via* collected, stored, and easily retrievable personal data. Personal privacy seems to be endangered. The secrecy of data leads to the legal problem of protecting against commercial and social exploitation of personal data – a question of considerable moral significance.

(8) However, technocracy also displays another very important component. Teller, the so called "father of the hydrogen bomb," stated in an interview in "Bild der Wissenschaft" (1975) that the scientist or technologist "ought to apply everything he has understood" and should not put limits on that: "Whatever you can understand, you should also apply." These statements constitute an overstated ideology of technocratic feasibility; it turns Kant's old moral dictum "ought implies can" into a reverse "technological imperative" (Lem, 1976): "Can implies ought " (Ozbekhan). Whether or not man is allowed to or ought to make, apply, produce, initiate, or carry through everything he can, certainly comprises a specific and precarious ethical problem. And it may not be as easily answered in the affirmative as Teller thought. According to Ozbekhan, the slogan seems to describe empirically the guiding orientation of technological progress or technological developments in general; technological feasibility seems obviously to have gained such a fascination that it assumes quasi-normative force – the almost automatic requirement that it be applied and carried through. Examples, including the moon landing program, genetic manipulation, and atomic bomb explosions, are familiar.

(Some think counterexamples of earthshaking significance are to be found in the U.S. decision not to proceed to production of the supersonic transport or in the molecular biologists' Asilomar moratorium on dangerous gene research – with the subsequent development of legal restraints. It is also wrong, literally, to say that everything that could be produced will be produced. Only about five percent of patents, on the average, are subjected to technological innovation and implementation. Nonetheless, the slogan is telling enough and does express an attitude.)

Jonas, in his book *Das Prinzip Verantwortung* (The Principle of Responsibility, 1979) with the significant subtitle, "Toward an Ethic of Technological Civilization," explicitly takes up the challenge modern technology offers to the moral orientation of human action. What he develops is a theory of extended responsibility.

According to traditional ethics, in epochs of relative technological powerlessness, nobody was responsible for the "unintended later effects of a benevolent, well conceived and well performed act" (Jonas, 1979). This has decisively changed with the immensely enlarged technological power of man, with the occurrence of many unintended or uncontrollable side effects of applied technology. To be sure, "The old prescriptions of the ethics of brotherhood – those pertaining to justice, charity, honesty, etc. – are certainly still valid in their intimate directness for the day-to-day sphere of human interaction"; however, they are to be supplemented by a "new extended ethics of technological and collective action within which agent, act, and effect are no longer the same as in the sphere of immediate social proximity." Due to the excessive nature of man's technological power, this realm would receive "a new, never dreamed of dimension of responsibility" (p. 26). We have gained a negative power over the planet's biosphere: we could irreversibly pollute it (or at least some subsystems of the planet) – be it by radioactivity, smog, or other effects.

"The critical vulnerability of nature to man's technological interventions" shows "that the nature of human action has *de facto* changed" – insofar as nature as a whole has become an object of human action and human responsibility; "this is a *novum* which ethical theory has to ponder" (pp. 26–27). Irreversibility and cumulative effects go along with this, passing the narrow limits which traditional ethics obeyed, focusing on problems of face-to-face actions between men. Jonas thinks that "no earlier ethics (outside religion) prepared us" to perceive nature and the "biosphere as a whole and in its parts" as having its own moral claims and rights under the trusteeship of man. Here he is wrong; Schweitzer's comprehensive "ethics of reverence toward life" already did. However, Jonas is right in stating that the world view of the natural sciences did not include such a trusteeship with respect to nature.

Today, therefore, not only scientific predictions and technological knowledge gain a new altered ethical significance, but also metaphysical reflections on nature (pp. 28–31). "The collective agent and collective action" require ethical imperatives of "a new sort" involving total responsibility for nature and for coming generations (pp. 32 ff.).

Kant's categorical imperative pertained exclusively to "logical compatibility" of intentions with actions; it was thus merely a formal principle addressed exclusively to individuals, to the "subjective character of self-determination" of the acting person (pp. 35 and 37). Now a new imperative has to pertain to the future existence of mankind and "the future integrity of man" as object and objective. The new categorical imperative has to display content; it cannot be formal any longer: "Act in such a way that the effect of your action is compatible with the permanence of genuine human life on earth"; or, expressed in the negative: "Act so that the effects of your action do not destroy the future possibility of such life"; or simply: "Do not endanger the conditions of an indefinite future existence of mankind on earth"; or, again positively: "Include in your present decision the future integrity of man as an object of your will" (p. 36). Because of asymmetry and irreversibility, the categorical imperative of future ethics has to have content comprising the existence of mankind as such and the perspectives of a possible future of mankind. (Note: Kant also spoke about the existence of man and mankind, as well as of reason being an objective in itself [AA IV, 428ff.] . This is certainly compatible with the conditions of collective actions, future time perspectives, and the plea for a future existence of mankind in the sense of Jonas's "new" categorical imperative. In other words, the principle of a responsibility directed towards the totality of mankind is not as new as Jonas thinks. Kant's statements are at least capable of being interpreted or slightly modified so as to comply with Jonas's ethics of extended responsibility. The "duty to provide for an existence of mankind in the future," "the imperative *that* mankind be," and the responsibility toward "the *idea* of man" [Jonas, 1979, pp. 86, 90ff.] may, as a metaphysical principle of practical reason, easily be deduced from Kant's approach, too. So Jonas's approach is not anything really deontologically new.)

More important is the revision of the concept of responsibility as a function of power and knowledge; Jonas thinks that responsibility within traditional ethics has always been interpreted as causal with respect to completed actions, referring exclusively to legal and moral responsibility (pp. 172ff.). By contradistinction, Jonas thinks it is necessary to develop a "totally new concept of responsibility . . . pertaining to the determination of what is to be done; according to this, I must feel responsible not primarily for my behavior and its effects, but for any object which requires my action"; "the object comes to be mine, because power is mine and has a causal reference toward this object. The object dependent on me in its own right comes to be the commanding instance; what is powerful in its causality comes to be the object of obligation" (pp. 174ff.). Due to power, "my control over something at the same time includes obligations with respect to it . . . ; the execution of power without observation of duty is then irresponsible or a neglect of responsibility" (pp. 176 and 178).

Jonas thinks that the new responsibility for being, for something, for an object, primarily for the existence of mankind's future – and only after that for the ideal of the good life – is the essential feature of the new concept of responsibility (p. 186 and *passim*). According to the dynamics of changing life circumstances in the wake of technological development; according to the immensely increased technological power and the extension of the scope of action and effects (including barely controllable side effects and sometimes irreversible encroachments into natural contexts), "time spans of responsibility as well as of planning based on knowledge . . . have been extended to an unforeseen degree"; this has led to causal effectiveness greater than with previous knowledge which was always doomed to incompleteness in complex systems regarding side effects, especially synergistic and cumulative effects (p. 220). Earlier, one could be relatively sure of a fairly constant order of nature which man could disturb not at all or at most ephemerally. Nowadays, after the increase in the power of technology – according to Jonas – "action has assumed aspects which had not been included in earlier ideas of it. . . . Responsibility for the historical future in terms of action" is consistent with the fact that the power of mankind, his capacity of doing, "engenders the *content* of the ought" (p. 229). *De facto* power is, so to speak, the "root of the ought of responsibility" (pp. 230ff.). Factually and morally it is becoming man's destiny. The ought "derives therefrom as self-control with respect to his intentionally used power"; it pertains to the very being of that power, but particularly also to the being of future mankind and other creatures depending on man's power. Man becomes "the trustee of all other objects which somehow come under the rule of his power" (p. 232). This change in the scope of responsibility and time perspective is, according to Jonas, what is really new in the "ethics of responsibility for the future" that is needed for the technological world (p. 175).

Jonas thinks that in an era of "almighty" (*modo negativo*: potentially all-destructive) technological civilization the first duty of collective human behavior is the future of mankind; "it obviously comprises, too, the future of nature as a *sine qua non*." However, independently, an essential metaphysical responsibility is included once man has become dangerous not only for himself but for the whole biosphere (p. 245). The "community of destiny of man and nature" and "the proper dignity of nature" have been re-discovered. Therefore, man is – in proportion to his power to manipulate, disturb, or even destroy – taking over responsibility for the state of nature, "the state of the biosphere, and the future existence of the species of man" and other creatures (pp. 246 and 248). "The *no* toward *non-being*, first to the non-being of man," seems to be the most important basic principle for "an emergency ethics of the endangered future" which is necessary for placing limitations on technological power which sometimes "goes wild" when faced with apparently apocalyptic situations or catastrophes. Only "the greatest degree of political and social discipline" can be an effective agent subordinating short-term advantage to the long range commandment of the future (pp. 250 and 255).

Jonas sees as the only alternative "an ethics of responsibility which, today, after several centuries of post-Baconian, Promethean euphoria (also prevailing in Marxism), has to put a bridle on galloping ahead"; if not, nature, later on, "will take revenge in its own dreadfully harsher manner" (p. 388). Only "together with the evil . . . does the good to be rescued become visible." "Fear for the basically vulnerable object of responsibility" is "becoming a duty . . . but only if there is also hope." "Fear becomes the first obligation before all others of an ethics of historical responsibility"; it requires the "courage of responsibility" to act in spite of uncertainties: "Responsibility is the obligation to acknowledge care for another being . . . ; what will happen with *him*, if *I* do *not* care for him?" (pp. 391ff.).

The main idea of Jonas's ethics for technological civilization is this: Facing an immensely increased human technological power and a galloping dynamics of life in the industrial world, and also facing dangers to nature and creatures (including man himself) stemming from side effects of the industrial process, an extension of the concept of moral responsibility is necessary – namely, passing beyond a concept of individual causal responsibility to that of human trusteeship, a stewardship responsibility of man. That is, from a regressive attributing of *ex post facto* responsibility to a prospective orientation, a "care-for" responsibility; from a past-oriented responsibility for the results of actions to a future-oriented responsibility for being which demands the capacity and feasibility of control as well as restrictions of power – to a responsibility involving preservation and prevention, so to speak.

Indeed, if we are confronted with cumulative effects and synergistic combinations of consequences, the concept of a responsibility oriented merely at a single agent responsible for completed actions no longer suffices. Individual ascriptions of responsibility cannot be applied to collective processes. We are not, moreover, permitted to leave non-ascribable, non-manipulable processes simply to random development or destiny. This would indeed be irresponsible. At the same time we must be able to ascribe – under the concept of a responsibility of preservation, prevention, trusteeship, and stewardship with respect to ecological systems, nature, and life in general – collective responsibilities aimed at preventing disturbances or destruction. Omissions – particularly intentional omissions – according to the new concept of responsibility should be attributable to some agent, whether individually or collectively. Traditional ethics had considerable difficulty in coping with the moral judgment of omissions. (Analytic philosophy of action tried to interpret intentional omissions as a sort of action of its own.) Every man within his system of interconnected actions and life conditions has to bear his part of this extended responsibility in proportion to his power; this includes the almost everywhere present negative power of disturbing or destroying highly interconnected and therefore highly susceptible systems.

Regarding Jonas's basic approach, one has to correct or add at least one general perspective: there is no real passage or crossing over from traditional responsibility for action results to a responsibility of prevention and preservation; the traditional responsibility for completed acts continues to exist with respect to the causalities of action – even with regard to the technologically extended scope of action if individually attributable. However, with respect to side effects which are difficult to survey and may be unintended, this responsibility is more difficult to sustain and sometimes not easily or not at all to be attributed to an individual agent – as mentioned before. Instead of speaking of a passage or change from the one type of responsibility to another, one should think of *two* concepts of responsibility at the same time: a stricter and narrower causality-oriented one, and a more refined and wider one that includes the mentioned orientation toward prevention and preservation. If there is a transition or a passage, it is at most to be seen in the fact that, according to the changed situation in the technological era, ethical thinking cannot restrict itself to the stricter and narrower traditional concept of responsibility alone; it must also be oriented toward the new wider concept without ignoring or substituting for the traditional responsibility for action results.

All of this certainly has a number of general consequences for ethics. The traditional, exclusively individualistic ethics of moral obligation has to be extended in the direction of a future-oriented ethics of collective agents with wider time perspectives, an ethics for holders of power – even, and most notably perhaps, when these people do *not* act. In a world of progressively evolving systems interconnections; of growing economic, political, social, and ecological dependencies; of increasing susceptibility to technical manipulation and risks, side effects, cumulative and synergistic effects, no mere ethics of "love thy neighbor" (as it has perhaps evolved from phylogenetic and especially historical experiences of face-to-face actions between men) can any longer suffice. Future ethics has to be based, more than ever, on a fundamental responsibility for the whole of mankind – including nature and future generations. In addition, ethics must become more future-oriented with larger time perspectives; more social, more cooperative, and more pragmatic regarding situational dependencies and mixtures of power (including technological power). Ethical responsibility has to be extended to collective agents or holders of power and responsibility under a wider concept of trusteeship and stewardship as well as under the obligation for preservation and prevention. Finally, an ethics which is oriented toward pragmatic conditions of application in an ever changing world cannot remain static; it must confront changing possibilities of effects as well as potentials of side effects in the realm of the technologically feasible – without merely mechanically adapting to technological change. Basic ethical impulses that are constant can and must be pragmatically related to the present situation of *Homo faber technologicus*, taking into account the responsibility of preservation and a more sensitive moral assessment of side effects which may possibly not be foreseen or controlled. Therefore, stricter and more cautious judgments are necessary, without avoiding all kinds of risks. Even if the basic ethical impulse may hardly have changed, the conditions of application have drastically changed under the perspectives of a systems-technological era. As far as ethics refers to the acting man, in particular the man who produces new artifacts and situations and changes the world, ethics has to be continuously developed further regarding dynamic developments in the world. Ethical thinking cannot stay where it is; it has to be dynamized pragmatically, for "new possibilities of actions and an extended power actualize wider and modified responsibilities," as has been stressed by the author before (Lenk, 1979, p. 73). This pragmatic orientation is easily compatible with discussions in analytic ethics (cf., e.g., Frankena, 1972): A realistic and pragmatic modern ethic can only be a mixed theory involving rule-utilitarian and deontological components. This is true whenever "morality is created for man, not man for morality" (Frankena, 1972, pp. 64 and 141).

The insights mentioned can easily be transferred to the problems of technological progress in the narrower sense. Here this can only be sketched out briefly. Technological progress is a multi-dimensional construct, a phenomenon resulting from a permanent interaction with other realms of influences and actions; the phenomenon displays a great complexity with respect to individual contributions, different areas, and basic social factors (e.g., the "status of societal achievement" – Bolte). The probability of improvements and changes is dependent on the actual state of development, which is resulting in the almost law-like basic form of exponential technological progress – particularly the rate of acceleration.

With respect to moral judgments, there is a result similar to the one mentioned earlier regarding synergistic and cumulative effects; namely, that causal responsibility cannot usually be attributed to a single individual, or sometimes even to a single area of activity if development and acceleration are dependent on a multiplicity of mutually escalating interactions. However, in the wider sense mentioned earlier, individuals who are in the game take on a responsibility for preservation and prevention – that is, technicians, engineers, and, generally speaking, members of the so-called technological intelligentsia have to bear this certain co-responsibility without any one of them *alone* being able to bear the total moral responsibility for the application of a discovery or an invention with possibly harmful applications which even they could not have foreseen. (This is the problem of the individual responsibility of technical people and scientists in applied research; it cannot be discussed here in detail.) Weizsäcker's distinction between the "discoverer" and the "inventor" ("the discoverer usually cannot know anything about possibilities of application") seems to be plausible, at first sight. It is, however, plausible at most in an ideal-typical sense, hypostatizing conditions that are too simple. Almost all technical developments (e.g., the development of the internal combustion engine or the production of dynamite) naturally show the ambivalence of positive and destructive applicability. In addition, basic research and technological development are no longer as easily separated and distinguished as the ideal-type distinction between "discoverer" and "inventor" postulates.

With respect to the severing of individual responsibilities from the almost unsurveyable ramifications of decisions and developments, society and its representative decisionmakers take on a collective responsibility for the application of developed technologies – but especially for the development of "big technologies" such as the Manhattan Project (if no thesis of an autonomous, quasi-natural technological process of development is defended). In the last analysis it is acting man who constitutes technology and its development, albeit in a multiramified synthetic combination. Together with the extension of the concept of responsibility mentioned earlier, men take on – individually and as members of an acting collectivity – the responsibility for preservation and for prevention against abuse or misuse. This is particularly true for individuals in strategic positions within large systems.

SUMMARY AND APPLICATIONS

(1) Power and knowledge create obligations – this is as true for technological (superpersonal) power as anywhere else. The creation of new dependencies creates new moral responsibilities of a personal and superpersonal sort. A rather utopian situation of extended power of technological feasibility – with respect to time perspectives and the scope of actions and effects, including sometimes unforeseen and uncontrollable side effects – generates:

(2) extended responsibility: Beyond the traditional responsibility for what he causes, man has to take on a "caring" responsibility for preservation and prevention.

(3) This responsibility applies not only to the well being of our neighbor and to a humane survival of mankind, but also to the preservation of and care for nature (including its systematic conditions of ecologic functioning), as well as for non-human fellow creatures (e.g., animal species). Nature as a whole and in its parts has come to be a moral object – at least in view of man's negative technological power (his capacity for disturbing or destroying).

(4) This extended responsibility is mainly oriented toward the future, toward the future existence of mankind; succeeding generations will have to acknowledge their moral rights with respect to a humane life in an acceptable environment, but also with respect to the future of nature and fellow creatures. A legalizing of the rights of succeeding generations and fellow creatures could and should occur.

(5) The responsibility for preservation and prevention cannot be attributed only to individuals. With respect to the effects and dangers of synergistic and cumulative effects and large technological projects with thousands of co-workers, a collective responsibility on the part of the collective agent and of anyone who disposes of possibilities of encroachments has to be borne. Concepts of team responsibility, of responsibility on the part of the whole generation, responsibility of experts and specialists, have to be developed further.

(6) The responsibility of scientific and technical experts at strategic positions is part and parcel of this responsibility of preservation and prevention. (Think of a strike of the chemists in charge of water supply or something of that sort – instead of merely a strike of air traffic controllers.) The responsibility for prevention can be attributed individually (*modo negativo*) to persons in strategic positions.

(7) The responsibility of the scientist and technologist obtains whenever and wherever detrimental effects can be foreseen and avoided – e.g., in immediately applied technological projects. A personal causal responsibility may be ascribed then. A general, strict responsibility of causality, however, cannot be defended with respect to the ambivalence of collective research and development (in particular basic research). All the more important is preventive responsibility. The distinction between the type of "discoverer" (pure scientist) and "inventor" (technologist) may be useful for a preliminary, overall orientation, but is only an ideal-type model. All kinds of mixtures occur and engender mixed responsibilities within the general responsibility of provision and care.

(8) Scientists and technologists conducting experiments with human subjects, in addition to the responsibility of the expert, must also take on the normal interhuman action responsibility for subjects (particularly in non-therapeutic experiments). Despite declarations of the World Medical Association and various psychological associations, the legal situation of experiments with human subjects is far from being clarified (Eser).

(9) Man certainly must not produce everything which he is technically capable of producing; and he must not apply everything which he can produce. "Can implies ought" is not an ethical imperative; it should not be a categorical technological imperative either. On the other hand, the innovative creativity of technological man should not be restricted more than necessary – all the more so since technological developments, as ambivalent, can be and must be positively used as conducive to the well-being of man and to the preservation of nature. Mankind has become dependent on technological progress and can only dispense with it at the price of catastrophes. Today, mankind canot afford to stop technological progress (as H. Marcuse proposed) or even to devaluate and hamper it. (This does not mean that mankind is dependent on an exaggerated fetishism of industrial growth, on a "technological imperative" to produce or innovate everything feasible.)

(10) What is convenient for man and the cosmos or for ecological subsystems has changed the course of history, with greater dependence on systems conditions. (Anticonception problems, for example, did not occur

in times of population scarcity.) Ethics, therefore, must dynamically and pragmatically take into account the historical situation. Notwithstanding the constancy of basic ethical impulses, ethical analysis has the task of refining capacities, of coping with new technological challenges.

(11) A special challenge is the tendency towards "systems technocracy," a syndrome in which all the trends of bureaucracy, role-segmentation, functionalization, technical perfection, automation, and computerization come together. Legal and ethical problems of protecting data are aggravated.

(12) There is no general ethical recipe beyond the constant basic responsibility for mankind, for fellow men and future generations, as well as nature and its creatures. A necessary condition of coping with future ethical challenges is to foster moral conscience, particularly in professional contexts. The development of professional ethics, both as a field and in terms of educational programs, is flourishing. Yet so far hardly a medical student takes courses in medical ethics; and technicians and scientists are nowhere – or hardly anywhere – introduced to the ethical problems of their disciplines in pragmatic and practice-oriented concreteness, as far as I can see. Ethics should not only be required as a subject in school, but also as a professional "moral guardian discipline" – as postulated by the Mount Carmel Declaration on Technology and Moral Responsibility (1974).

Indeed, we will only be able to exercise moral control over technological progress if we do not stay with the superficial moral policy of avoidance behavior, either blindly relying on or blindly surrendering to the apparently autonomous dynamics of technology.

University of Karlsruhe

NOTE

*This is a revised version of a paper presented to the Xth World Congress of Philosophy of Law and Social Philosophy, Mexico City 1981.

REFERENCES

Albrecht, V. 1969. "Die Werturteilsfrage in der Technik," manuscript prepared for VDI-Ausschuss, "Philosophie und Technik." Karlsruhe.

Barber, B. 1976. "The Ethics of Experimentation with Human Subjects," *Scientific American* **234** 25–31.

Beecher, H. K. 1959. *Experimentation in Man*. Springfield, Ill.

Beecher, H. K. 1966. "Ethics and Clinical Research," *New England Journal of Medicine* **274** 1354–1360.

Belsey, A. 1979. "Scientific Research and Morality," paper prepared for the 6th International Congress for Logic, Methodology, and Philosophy of Science. Hanover.

Birnbacher, D. 1980. "Sind wir für die Natur verantwortlich?" in D. Birnbacher (ed.), *Ökologie und Ethik*. Stuttgart, pp. 103–139.

Bodnar, J. 1979. "Die Ethik der wissenschaftlichen Forschung," paper prepared for the 6th International Congress for Logic, Methodology, and Philosophy of Science. Hanover.

Bolte, Karl M. 1979. *Leistung und Leistungsprinzip*. Opladen.

Born, M. 1979. "Die Zerstörung der Ethik durch die Naturwissenschaften: Überlegung eines Physikers," in H. Kreuzer, (ed.), *Literarische und naturwissenschaftliche Intelligenz*. Stuttgart, pp. 216–224.

Byrne, E. 1979. "The Normative Side of Technology: Philosophy and the Public Interest," in P. Durbin (ed.), *Research in Philosophy & Technology*, vol. 2. Greenwich, Conn., pp. 91–109.

Chain, E. 1970. "Social Responsibility and the Scientist," *New Scientist* **48** 166–170.

Diener, E. and Crandall, R. (eds.), 1978. *Ethics in Social and Behavioral Research*. Chicago and London.

Eser, A. and Schumann, K. F. (eds.), 1976. *Forschung im Konflikt mit Recht und Ethik*. Stuttgart.

Ferkiss, V. 1970. *Der technologische Mensch*. Hamburg.

Frankena, W. K. 1972. *Analytische Ethik*. Munich.

Gehlen, A. 1957. *Die Seele im technischen Zeitalter*. Hamburg.

Gehlen, A. 1961. *Anthropologische Forschung*. Hamburg.

Ginsburg, T. 1980. "Die Verantwortung des Wissenschaftlers heute," in M. Grupp (ed.), *Wissenschaft auf Abwegen*? Fellbach-Oeffingen, pp. 90–103.

Grau, G. 1976. "Die 'besondere' Verantwortung des Technikers für die Gesellschaft," unpublished manuscript.

Grupp, M. 1980, "Gefährliche Wissenschaft?" in M. Grupp (ed.), *Wissenschaft auf Abwegen*? Fellbach-Oeffingen, pp. 68–77.

Heisenberg, W. 1969. *Der Teil und das Ganze*. Munich.

Heisenberg, W. 1979. *Quantentheorie und Philosophie*. Stuttgart.

Hersch, J. 1980. "Die Verantwortung des Wissenschaftlers in der Sicht der Philosophie," *Universitas* **35** 1291–1296.

Hoffmann, R. 1975. "Scientific Research and Moral Rectitude," *Philosophy* **50** 475–477.

Humber, J. M., and Almeder, R. F. (eds.), 1976. *Biomedical Ethics and Law*. New York and London.

Ingarden, R. 1970. *Über die Verantwortung*. Stuttgart.

Jantsch, E. 1967. *Technological Forecasting in Perspective*. Paris: OECD.

Jantsch, E. (ed.) 1969. *Perspectives on Planning*. Paris.

Jonas, H. 1969. "Philosophical Reflections on Experimenting with Human Subjects," *Daedalus* **98** 219–247.

Jonas, H. 1979. *Das Prinzip Verantwortung: Versuch einer Ethik für die technologische Zivilisation*. Frankfurt.

Kadlec, E. 1976. *Realistische Ethik*. Berlin.

Katz, J. 1972. *Experimentation with Human Beings.* New York.

Kropp, G. 1948. *Die philosophische Verantwortung in der Physik.* Berlin and Hanover.

Kurtz, P. 1977. "The Ethics of Free Inquiry," in S. Hook, P. Kurtz, and N. Todorovich (eds.), *The Ethics of Teaching and Scientific Research.* Buffalo, N.Y., pp. 203–207.

Leinfellner, W. 1974. "Wissenschaftstheorie und Begründung der Wissenschaften," in Eberlein (ed.), *Forschungslogik der Sozialwissenschaften.* Düsseldorf.

Lem, S. 1976. *Summa technologiae.* Frankfurt. (Polish original, Cracow, 1964; rev. ed., 1967).

Lenk, H. 1979. "Zu ethischen Fragen des Humanexperiments," in H. Lenk (ed.), *Pragmatische Vernunft.* Stuttgart, pp. 50–76.

Lenk, H. (ed.), 1973. *Technokratie als Ideologie.* Stuttgart.

Lenk, H. (ed.), 1974. *Normenlogik.* Munich.

Lenk, H. and Fulda, E. 1981. "Zur ethischen Problematik von Humanexperimenten in der sozialpsychologischen Grundlagensforschung," in L. Kruse and M. Kumpf (eds.), *Psychologische Grundlagensforschung: Ethik und Recht.* Bern.

Lenk, H. and Ropohl, G. 1976. "Praxisnahe Technikphilosophie," in W. C. Zimmerli (ed.), *Technik oder wissen wir was wir tun?* Basel and Stuttgart.

Lenk, H., and Moser, S. (eds.), 1973. *Techne, Technik, Technologie.* Munich.

Lipscombe, J. and Williams, B. 1979. *Are Science and Technology Neutral?* London.

Lübbe, H. 1980. "Wissenschaftsfeindschaft und Wissenschaftsmoral: Über die Verantwortung des Wissenschaftlers," *Berner Universitätsschriften "Wissenschaft und Verantwortung."* Bern, pp. 7–17.

Mendelsohn, E., Nelkin, D., and Weingart, P. (eds.), 1978. *The Social Assessment of Science.* Bielefeld.

Mount Carmel Declaration on Technology and Moral Responsibility, 1974. Haifa.

Obermeier, O.-P. 1979. "Darf der Mensch alles machen, was er kann?" *Politische Studien* **30** 565–574.

Ottmann, H. 1980. "Praktische Philosophie und technische Welt," *Zeitschrift für philosophische Forschung* **34** 157–178.

Ozbekhan, H. (n.d.) "The Triumph of Technology: 'Can' Implies 'Ought,'" mimeograph, Systems Development Corporation, Santa Monica, Calif.

Pappworth, M. D. 1968. *Menschen als Versuchskaninchen.* Zurich.

Pfeiffer, W. 1971. *Allgemeine Theorie der technischen Entwicklung als Grundlage einer Planung und Prognose des technischen Fortschritts.* Göttingen.

Popper, K. R. 1977. "Die moralische Verantwortlichkeit des Wissenschaftlers," in K. Eichner and W. Habermehl (eds.), *Probleme der Erklärung sozialen Verhaltens.* Meisenheim, pp. 294–304.

Rapp, F. 1978. *Analytische Technikphilosophie.* Freiburg and Munich.

Rapp, F. 1980. "Technikgeschichte und die Grenzen der Machbarkeit," Loccumer Protokolle. Loccum.

Rapp, F., Jokisch, R. and Lindner, H. 1980. *Determinanten der technischen Entwicklung.* Berlin.

Ropohl, G. 1979. *Eine Systemtheorie der Technik.* Munich and Vienna.

Sachsse, H. 1972. "Ethische Probleme des technischen Fortschritts," in H. Sachsse (ed.), *Technik und Verantwortung.* Freiburg, pp. 121–148.

Sachsse, H. 1976. "Der Mensch als Partner der Natur," in G.-K. Kaltenbrunner (ed.), *Überleben und Ethik.* Freiburg and Munich, pp. 27–54.

Sachsse, H. 1978. *Anthropologie der Technik*. Braunschweig.

Sachsse, H. (ed.), 1972. *Technik und Verantwortung: Probleme der Ethik im technischen Zeitalter*; Freiburg.

Schuchardt, W. 1980. "Zur Bedeutung aussertechnischer Werte und Ziele," *VDI-Zeitschrift* **122** 421–429.

Schuler, H. 1980. *Ethische Probleme psychologischer Forschung.* Göttingen.

Spaemann, 1980. "Technische Eingriffe in die Natur als Problem der politischen Ethik," in D. Birnbacher (ed.), *Ökologie und Ethik*. Stuttgart, pp. 180–206.

Stork, H. 1977. *Einführung in die Philosophie der Technik.* Darmstadt.

Teich, A. H. (ed.), 1977. *Technology and Man's Future.* New York.

Zihlmann, R. 1976. "Auf der Suche nach einer kosmosfreundlichen Ethik," in G.-K. Kaltenbrunner (ed.), *Überleben und Ethik*. Freiburg and Munich, pp. 17–26.

KLAUS MICHAEL MEYER-ABICH

WHAT SORT OF TECHNOLOGY PERMITS THE LANGUAGE OF NATURE? CONDITIONS FOR CONTROLLING NATURE-DOMINATION CONSTITUTIONALLY

Economic developments since World War II have brought hitherto unknown wealth to the industrialized part of the world. In the meantime, enthusiasm for scientific and technological achievements has given way to the consciousness that other qualities of life have to be sacrificed if those achievements are to be shared by everybody. If, for instance, everybody drives a car, then either traffic problems or their solutions have to be accepted – and both reduce the quality of urban life. And the diversity of nature has already been so severely diminished in congested industrial areas that only a concurrent degeneration of our perceptive faculties can limit our dismay.

Even a philosopher as averse to criticizing our civilization as John Passmore – he considers "Western science perhaps the greatest of man's achievements" and maintains that "man can live at all only as a predator" (Passmore, 1974, pp. 175, 178) – cannot evade the general conclusion that "what is at fault . . . is the total system, the scientific-technological-industrial complex" (p. 190). His conclusion is that "there is something wrong with any measure which suggests that it constitutes 'economic growth' when parks, shops, hotels, theatres, neighbourhoods, are destroyed in order to substitute banks and insurance offices" (p. 191). This being so, we have to look for the measure the neglect or insufficient observation of which in our actions has brought about the present crisis; that measure, therefore, should be brought to bear on the industrial economies in the future.

To burden the environment with wastes, poison, and noise so that the conditions of life are impeded or destroyed, to some extent violates the principle not to live at the expense of one's fellow human beings. Since, however, it is not only the conditions of man's life that are affected but also those of other animals, of plants, indeed of nature generally, the question therefore arises whether – and which – other principles can be applied here. Approaching the environmental crisis along these lines is said to present a problem, namely:

- whether the traditional criteria of social consideration between human beings "for our own sake" are sufficient to control human domination in nature,
- or additional criteria "for nature's own sake" have to be taken into account in relating ourselves to nature.

Paul T. Durbin and Friedrich Rapp (eds.), Philosophy and Technology, 211–232.
Copyright © 1983 *by D. Reidel Publishing Company.*

The first thesis may be called *anthropocentric*; the idea is that everything must be related to ourselves and to our own presumed interests – a thesis which in modern times has replaced the geocentric conception of the world. The second thesis may then be called *physiocentric.*

This problem, whether man or nature is the measure, arises in Western ethical thought as it has more and more narrowed itself down to human relations and disregarded man's being part of nature. Put this way, however, the problem is as deceptive as the traditional question, whether the (human) individual should consider in his or her action only individual interests or also those of society. I call this question deceptive because it depends on an idea of humanity which neglects that man is a social being, and also because an individual who is not conceived as embedded in society, as to an essential degree constituted by social relations, will obviously not have to consider such relations. What we have to or are allowed to do depends on who (explicitly or implicitly) we think we are; and whoever presupposes an egoistic individualism cannot give a positive answer to the question whether in his actions other people are to be considered for their own sake. For the same reason one should not be surprised that so many authors whose conception of humanity does not take into account man's belonging to nature come to the conclusion that human interest is to be accepted as the primary measure in relating ourselves to the rest of the world.

What an individual may do depends on assumptions about one's own status and about the status of other beings as well as about the relational context of one's own existence with respect to that of other beings. In relating ourselves to nature, the question is, therefore, not which side matters and provides the measure for acting with respect to the other side. Rather, the real problem is how man's relation to nature is to be conceived. Since the relation involves dependency at both ends, both positions, the "anthropocentric" and the "physiocentric," presuppose conceptions of humanity as well as of nature (different in each case) which deserve consideration – especially anthropocentrism.

Francis Bacon found a very suggestive formula to describe man's relation to nature in politics: The most elementary political ambition, he said (in his *Novum Organum*, 1620; part I, section 129), is to get power over one's fellow human beings. A second and more noble stage is to seize power over other countries. The highest and most excellent level of power, however, is, according to Bacon, man's domination over nature. Accepting this hierarchy of political ambitions, we must admit that legitimate forms of exercising power have so far only been fairly established at the lowest of the three levels. An adequate international order, although desperately needed, is not yet in sight, and the same is true with respect to man's domination over nature, whether in national or international regimes (e.g., the utilization of the oceans). Power, however, should only be exercised constitutionally, as we have learned from human history. Also, progress that has been made in the past at the first level is endangered today by technological progress at the third level of power (Winner, 1977).

Man's domination over nature as exercised nowadays politically may be called "absolutist." This absolutism will be the topic of the first section of this paper. In human history, absolutism has given way to the constitutional state, and this is what I propose as an adequate determination of our relation to nature in the second section. The third section, finally, presents an outline of holistic principles on which such a constitution could be based.

1. ABSOLUTISM IN MAN'S RELATION TO NATURE

The traditional legal systems of Western industrialized countries afford protection for nonhuman entities only insofar as they are human property. Originally nature was considered *physis* – the emergent – but Lucretius called his book *De* RERUM *natura*, and Descartes left only *res extensa* as the basic determination of the nonhuman world. To determine nature, in the tradition of Roman Law, as *res* – in particular as *res extensa* – and to conceive man as the ruler to whose regulations nature (*res*) is subject, while the relation is one of owning with respect to being owned, is the core of what I have called the *anthropocentric* system. Even if this term is misleading – insofar as it really means a certain determination of man's relation to nature and, therefore, refers to nature as well as to man – nature here has lost every character of her own and has turned into what mankind determines her to be. Taking into account the different levels of political ambition Bacon pointed out, the relation between man and nature which is implied by anthropocentrism may suitably be called absolutism. "*L'état, c'est moi*" and "*La nature, c'est l'homme*" express the same kind of monarchy.

Absolutism in man's domination over nature is not only the basic character of our relation to nature in the industrial economies; it is also approved as legitimate by philosophers as well as theologians. Passmore, for instance, deliberately as well as cheerfully accepts being called a "human chauvinist" for maintaining that "an 'ethic dealing with man's relation to land and to the plants and animals growing on it' . . . would have to be justified by reference to human interests" (Passmore, 1974, p. 187). And the dominant strain of theological thought – humanistic as it has come to be – is no less Cartesian than economics and the sciences. Moreover, acknowledgment is given to absolutism even by those who feel that it is based on human *hubris.* As Laurence Tribe has pointed out, "Environmentalists often feel disingenuous when they seek to rationalize their position in terms of a want centered calculus" (1976, p. 78). To pretend to be pursuing economic rationality while actually following one's conscience, however, may put a drain on one's non-absolutistic emotions with respect to nature; ultimately these emotions may not come forth any more. The reason is that *homo economicus* does not perceive nature as anything more than a resource; in principle no qualms of conscience can justifiably arise.

Paul A. Samuelson's textbook, *Economics*, may perhaps be taken as a representative of the dominant or leading conception of nature in economics. The book has a very detailed index (30 pages), but there is no reference to nature – irrespective of the fact that the economic process is rooted in nature at both ends (resources in/garbage out). Instead of "nature," between "NATO" and "needs" only the heading "natural resources" is found – and this only with the reference, "See resources." Looking up "resources" then, one is referred to the chapter on underdeveloped countries. Still, in those parts of the world, nature has to be taken into account even if "there's nothing wrong with any poor country that discovery of oil can't cure" (p. 753). One also learns from Samuelson that there are two kinds of resources, natural and human, and that health and nutrition programs are necessary "*both* to make people happier and to make them more productive workers" (p. 752); thus are the implications of Cartesianism fully realized. (Changes in public awareness may be responsible for the fact that the heading "human resources" is omitted in the next edition of Samuelson's book; however, the text is unchanged, and the same seems to be true for man's position within the economic process.)

Production functions in neo-classical economics confirm the impression given by Samuelson's textbook. They depend only on man's effectiveness and are carefully designed not to reflect the natural context in which the economic process is rooted – not even "limits to growth." Nature, again and again, only provides the material – and in this sense is a resource – which is shaped or formed by production processes into utilities, status values, etc. By contrast, in early economics nature was not neglected; it was considered an important factor of production corresponding to its role in agriculture. In the course of time, however, the natural stocks of "nonrenewable resources" turned out to be inconceivably large; also, the prediction made by Malthus and Ricardo – namely that land returns will increase with population while capital returns correspondingly decrease – did not turn out to be justified. Although some economists, from time to time, endeavored to reconsider the natural basis of life and of economic theory, the dominant strain of thought in economics unbrokenly retained its roots in Locke, who thought that economic production is ninety-nine percent a result of human labor and only one percent due to nature. "To emphasize that even the noblest achievements on earth depend only on man's appropriate education, technology, and organization, flatters human pride and the haughtiness of our civilization" (Schmoller, 1920, vol. I, p. 139).

The conception of nature dominant in economic thinking can be questioned along two lines. On the one hand, to consider nature as only a resource not only deemphasizes any limitations; it explicitly assumes the unlimited availability, forever, of resources – so that the idea of "limits to growth" must appear to be a revolutionary discovery when, as the conditions of life are jeopardized, it is finally perceived. These problems can be taken into account by environmental economics – based on the polluter-pays-principle – and by an economic theory of depletable resources; and they can probably be solved as far as these theories go. On the other hand, before economists became aware of environmental problems, of limits to growth, of the need to finally reconsider nature's role in economics, nature was conceived as nothing but material, not good in itself but only in relation to human purposes. However, to take nature as nothing but material is an inadequate conception because "man, human society, and the economy, are part of the biosphere on the surface of the earth" (Schmoller, vol. I, p. 128). One cannot consistently conceive the whole as a resource while making an exception of one part that is of particular concern.

Are there economic schools of thought which consider the part coherently with the whole? Doubts with respect to the economic conception of man have given rise to socialist as well as conservative and liberal ideas; the aim is to reconcile man's serving the economic process with the economic process serving man. However, nature has generally not been thought of even in these approaches. Marxist economics, which began with a redetermination of man's role in the economic process, is not at all more advanced than neo-classical economics with respect to deemphasizing nature and considering technical progress as the leading factor in economic production. It is as if Marx's Parisian Manuscripts – and whatever Bloch, Adorno, Horkheimer, and Herbert Marcuse have thought and written – simply did not exist. In fact, it is not unjust to state that the main stream of Marxist thinking – even taking into account official positions of Eastern countries – still looks forward to a land of milk and honey free of domination, which is to be brought about by political emancipation from human history and by technological emancipation from nature – all at the expense of nature which is to be mercilessly exploited. Marxist economics has remained humanistic only in the sense of addressing social relations, while leaving out nature.

Turning nature over without compassion to "instrumental practice" (Habermas, 1973, p. 176) while mankind contentedly occupies itself on a cloud of unconstrained communication confirms Ortega's caustic remark that the humanities have been invented so that the "*caballeros del Espíritu*" (Ortega, 1947, vol. VI, p. 26) may consider man more noble than nature. The humanities are based on the idea "that man eliminates himself from experience to constitute nature as a lawful order," so that a "second center" of knowledge is established where "the same man turns around from her [nature] . . . to life, to himself" (Dilthey, vol. VII, p. 830); but this can only lead to an inhuman experience of nature and an unnatural experience of man. To separate life from nature conceptually also shares the responsibility for killing nature economically. Dilthey's "two centers," therefore, are one center too many.

With respect to nature's being conceived and treated as a resource, we are inevitably faced with the question how this understanding of nature fits our belonging to nature. Is not man subject to the same laws as all material systems, and do we not gladly recall and recognize this in case of medical treatment? The Cartesian answer to this question shifts our affiliation to nature to a blind spot between the natural sciences and the humanities; we may thus perceive ourselves as "*caballeros del Espíritu*" on the cloud of unconstrained communication, enjoying social freedom on the basis of technologically suppressed nature.

However, we can not succeed in collectively saving our souls as long as we are living, because life is rooted in nature. Absolutist control over nature hinges on and fails in accord with our link to nature. Although economists may disregard this, philosophers cannot. Rationality being one of the leading passions of our time, philosophers like Passmore can maintain with the enthusiasm of the Enlightenment that only human interests matter and that humans require nothing but clever management of resources. But do not the following words of Passmore show much more sensuous than intellectual sensitivity?

It is as if the smoke of the industrial revolution had destroyed men's eyes, ears, noses and sense of touch, or as if only by seizing upon and making their own the familiar

> tenets of what had been a minority puritanism could men justify the ugliness which they were creating around them. A more sensuous society could never have endured the desolate towns, the dreary and dirty houses, the uniquely ugly chapels, the slag heaps, the filthy rivers, the junk yards which constitute the 'scenery' of the post-industrial West and which it has exported to the East. Only if men can first learn to look sensuously at the world will they learn to care for it (Passmore, 1974, p. 189).

Passmore feels a strong need to defend science and technology, democracy, and free enterprise against "mysticism, primitivism, authoritarianism" (p. 173). It is true that those who feel that the primary goals of the fifties have been achieved and that today's main issue is the goals of the eighties frequently present and advocate their ideas in a style that dissociates themselves from earlier achievements. However, there is no general need for this; after all, missing music at a good meal is quite different from complaining about the meal. Passmore's problem seems to be that certain effusions of environmentalists are even more against his grain than the desolateness of industrial economy. Certainly, however, this does not warrant a conclusion that he should defend the paradigm of the dominant industrial economy; those who "care for the world" may travel together at least this far. They may even take my next step, at least as open for discussion.

2. LAWS OF NATURE AND LAWS OF JUSTICE; HUMANITY BEYOND MANKIND

That man belongs to nature or is part of nature means that he has emerged from natural history like the animals, plants, and minerals around us; we are historically related to them. Since Cartesianism in knowledge and in scientific institutions fails to recognize this fact, its converse may be taken as a starting point in moving toward non-absolutist relations between man and nature.

The kinship between, for instance, man and flowers which follows from common roots in natural history does not mean that men are flowers or that flowers are men. So it would be mistaken to treat flowers like men; flowers are flowers and they are not men. But insofar as flowers and men are related through natural history, they are the same – namely, living beings as opposed to minerals. More than that, they are also different from other living beings; men and plants – together with animals – are eukarionts and not prokarionts (bacteria). The kinship of men with horses, dogs, cats, and other mammals may seem to be more obvious than the one to the flowers, because in natural history man came into being with the mammals; but still the same is true as in our relationship to plants: dogs are not men but dogs, and it would be inappropriate as well as entirely mistaken to treat a dog like a man. However, insofar as dogs and men are naturally related, they are the same, being equally mammals; and as far as this equality goes, dogs and men should be treated corresponding to this equality – that is, equally.

That two subjects or cases should be treated equally according to their similarity or sameness and differently according to their dissimilarity or distinctiveness is a basic legal principle. Since equality before the law historically followed equality before an absolutist ruler, absolutism with respect to nature might, at Bacon's third level of power, again give way to legal order. The equality principle which I propose – equal treatment according to equality and different treatment according to difference – may appear to be a suitable extension of the principle of equality before the law from one species to many species. To define nature in legal terms in fact corresponds to the dominant concept of nature in one respect: the term *res* is taken from the legal tradition. The fact that our legal system has not adequately considered the natural environment, that environmental disruption has taken place – indeed the fact that the economic order and political stability would be jeopardized by limits to growth – may also be held up against this common background. The task is not for the first time to introduce legal terms into nature; it is to get rid of a disparity between the legal system as applied in society and our traditional legal understanding of nature.

The idea of causality, historically is rooted in the concept of responsibility or guilt (*aitia*); and the same tradition connects laws of nature and laws of justice. Nowadays we tend to think that this relationship is "merely" historical, that laws of nature really have nothing to do with laws of justice. This lack of coordination between the two kinds of order, however, may indicate that our distinctions, between guilt and cause and between the two categories of laws are misleading. The time may have come to remind ourselves of what is common, of what connects the two sides.

Since the legal *res* does not – at least not any more – provide an appropriate understanding of nature, we should become aware of alternatives. Generally if a Roman tradition turns out to be questionable, the best idea is to refer to its Greek origins; there, as a rule, matters have been discussed on a much higher level. This is especially true in the case of nature. One basic idea of Plato's *Politeia*, for instance, is that truth as it appears in nature does not differ from truth as it is observed in human action. Far from being a naturalistic fallacy, the claim is rather that we have not understood nature before also recognizing the truth of human action in nature; neither have we learned what we must and must not do before we also recognize nature in the ethical measure of human action. Less elaborately, this same idea of unity has already been put forward in the very first philosophical statement which has come down to us:

> Into those things from which existing things have their coming into being, their passing away, too, takes place, according to what must be; for they make reparation to one another for their injustice according to the ordinance of time (Anaximander B1; from Simplicius, *Physics* 24, 18 [J. M. Robinson trans.]).

Here the universe is conceived as a "legal community involving everything" (W. Jaeger, 1954, vol. I, p. 218) – as opposed to Hesiod for whom there was no legal order in nature; otherwise animals would not devour one another.

Anaximander's great idea, in other words, is that truth in court is the same as truth in nature. (Along the same line, the Jews concluded that Yahweh, who led them into political freedom, must also have created the universe.) While in the now-dominant interpretation of man's relation to nature, only man has rights with respect to the rest of the world, the idea suggests itself from the Greek tradition that on Bacon's third level of power rights should also be recognized in the nonhuman world. So far this has been discussed mainly for animals. Since animals – like children or incompetents – cannot themselves look after their rights in court, animal rights must also be entrusted to men in whom nature is becoming aware of herself by way of language (see section 3, below; also Feinberg, 1974).

To limit human power over the animal kingdom by constitutional principles is a thought as old as the French Revolution. Nevertheless, this part of bourgeois liberation is still unfinished; the claim has come down to us, but the challenge still stands: "Our conviction . . . is that *we* require *now* to extend the great principles of liberty, equality, and fraternity, over the lives of animals. Let animal slavery join human slavery in the graveyard of the past!" (Corbett, 1971, p. 238). Of course, freedom has only gradually been extended, from white men to women, Negroes, Jews, and gypsies; it may well be that in our time the circle will just have to be enlarged to include animals.

Animal liberation has aroused no less mockery than women's liberation. Equality claims with respect to demands for equal rights especially have given rise to jokes. For instance, Henry Salt's claim that animals should be conceded the right of personal development does indeed seem to take animals to be like men even where they are not. To impose nineteenth-century ideals of personal fulfillment on a turtle or a chicken, while it may have been meant to compensate for former exploitation, can certainly be judged a most extreme form of human absolutism. I may feel that a turtle, for instance, is an individual and has interests and should receive some convivial attention from time to time; yet to expect personal development, as far as I can see, would be to neglect, even to disregard the nature and character of a turtle's existence.

The basic question, therefore, is which equalities determine equal rights, the equal treatment of equals. The first answer to this sort of question was given by Jeremy Bentham, the utilitarian; he was also the first to proclaim the extension of human constitutional principles to the animal kingdom. Referring to the abolition of human slavery, he wrote in his *Principles of Penal Law* (1780):

> Why should the law refuse its protection to any sensitive being? The time will come, when humanity will extend its mantle over every thing which breathes. We have begun by attending to the condition of slaves; we shall finish by softening that of all the animals which assist our labors or supply our wants (part III, chapter 16, 1962 ed., vol. I, p. 562).

In his *Introduction to the Principles of Morals and Legislation* (1789) Bentham later added the criterion of *suffering* to that of *sensitivity*. While for him this was not meant to be a constriction, in our time the criterion of sensitivity has been narrowed down to "all consciously sentient beings" (Frankena, 1979, p. 5).

We may pause for a moment to consider Kant's well known objection from the "episodic" section of his *Metaphysik der Sitten* (1797; *Metaphysische Anfangsgründe der Tugendlehre* §§ 16–18, A 106–109). Kant proposed a distinction between duties *towards* (*gegen*) and duties *with respect to* (*in Ansehung von*) a subject, and he came to the conclusion that abstention from cruelty to animals is only an indirect duty *with respect to* animals – although it is derived from a direct duty *towards* ourselves. To identify the rights of nonhuman beings as reflective of man's relating himself to himself is a typical Kantian argument; yet in this case Kant seems to have stepped back too quickly. My objection is this: reason's determination of free will constitutes duties towards reason (and nature – which "intentionally attached reason as a government to man's will"; *Works*, vol. IV, p. 20; *Grundlegung zur Metaphysik der Sitten* A4). Thus, from duties *towards* reason (and nature) duties *with respect to* men, animals, plants, and minerals must be inferred. To conclude, however, that the boundary between direct and indirect duties is the same as that between men and animals – since duties towards reason are not necessarily duties towards those who also have duties towards reason – is not justified.

Rights of nonhuman beings, therefore, may not be denied on the basis of

Kant's "episodic" reasoning. Apart from sensitivity, equal *interests* may also provide a basis for equal rights. In our times especially, a common interest is felt to be the conditions of life. Interests are the basic determinations from which, according to Leonard Nelson, duties towards animals (as well as towards plants, in my opinion) can be derived. Nelson's conclusion is that "duties towards animals exist and . . . they are immediate duties, not being derived from duties towards men" (1970, p. 163). Yet these ideas will be convincing only to those who recognize a sameness between man and the other animals; this is, as Schopenhauer correctly observed, because "the alleged rightlessness of animals is based on the assumption of a complete distinctiveness between man and animal which clashes with all the evidence and has been maintained most resolutely as well as stridently by Descartes – this being a necessary implication of his errors" (*Works* IV, 2, p. 238).

If rights of animals, as I propose, can be justified from the basis of a principle of equality, it becomes evident that according to the same line of reasoning duties towards plants have to be recognized. Indeed the relationship between men and plants gives rise to many equalities – so that again equals should be treated equally as far as the equality goes. Rights of plants can also be defended on the basis of Bentham's and Nelson's criteria of sensitivity and interest, respectively, because they can be deprived neither of sensitivity nor of interest. That the "liberation movement" – after that of the bourgeois, Jews, Negroes, women, gypsies, and animals – cannot halt without going on to plants is also becoming apparent from the flimsiness of the arguments which are given to justify the denial of rights to plants. As a representative contemporary example I can show this in Joel Feinberg's analysis (1974); like Nelson, but without referring to him, Feinberg very carefully argues that rights should be granted to all beings which have interests, yet nevertheless he comes to a stop short of plants. Feinberg suggests that:

> trees are not the sorts of beings who have their "own sakes," despite the fact that they have biological propensities. Having no conscious wants or goals of their own, trees cannot know satisfaction or frustration, pleasure or pain. Hence, there is no possibility of kind or cruel treatment of trees. In these morally crucial respects, trees differ from the higher species of animals (p. 52).

Some people feel alien or unrelated to animals while others do so with respect to plants. Those who feel a sensitivity for plants, however, experience plants as having their "own sakes" like we do; they can be healthy or sick, and in some places or situations they prosper much better than in others. A good gardener will always notice when a plant suffers. Feinberg himself may not have been too happy with his verdict, because in order to exclude plants he required, additionally, an "at least rudimentary cognitive equipment." That a philosopher who is fairly unconventional in claiming rights for animals afterwards falls back into darkest Cartesianism with respect to plants may have something to do with the fact that only very few people seem to be as sensitive to plants as to animals. But can this be an excuse for the following statement?

> An automobile needs gas and oil to function, but it is no tragedy for it if it runs out – an empty tank does not hinder or retard its interest. *Similarly* to say that a tree needs sunshine and water is to say that without them it cannot grow and survive. But unless the growth and survival of trees are matters of human concern, affecting human interests, practical or aesthetic, the needs of trees alone will not be the basis of any claim of what is "due" them in their own right (Feinberg, p. 54; italics added).

Overlooking the first sentence's premise – that a situation for a car can be tragic or against its interest – it is monstrous to say that a car without gas and a tree without light and water are in a "similar" situation; the tree's dying is completely ignored here as if it were nothing. A car without gas can be suitably preserved, but a tree without light and water dies. Can we be justified to consider the death of a tree different from that of a human being without being accused of lacking all moral judgment and sensitivity? I am not arguing that the death of a plant is to be considered the same as the death of a man. Plants are plants and men are men, so that a plant dies a plant's death and a man dies a man's death. But how do we dare simply ignore that sameness according to which, in both cases, a living creature is dying and going back to the earth? Moreover, there are such intermediate cases as "human vegetables."

In death what we cannot hold passes away. But to acknowledge plants only insofar as we can hold them precludes experiencing their death. To acknowledge only what we can hold, moreover, generally means to miss the natural context of life or the world's animation. As Fechner said, "Not what is available but what is not available in a subject makes up its soul" (1899, p. 17). The time might come when those philosophers who still consider plants similar to machines will refer to Fechner's book on plants' souls (*Nanna, oder über das Seelenleben der Pflanzen*, 1849) as a proof that philosophy has not always disregarded these most sensitive beings.

Frankena has pointed out that different types of ethics can be classified according to the extent to which "moral patients" – different kinds of beings that we have duties to – are recognized. The basic possibilities that everyone considers are:

– nothing but oneself,
– oneself as well as other human beings,
– oneself as well as other human beings and all consciously sentient beings,

– all living beings, or
– all beings

as morally relevant subjects (Frankena, 1979, p. 5). Leaving out distinctions with respect to ethical justification which Frankena adds to his list, my argument using the equality principle has now led us to the fourth possibility. Generally this corresponds to Albert Schweitzer's "ethics of reverence for life." Schweitzer realized that "an ethic which deals with man's relation to creatures leads to a dispute with the philosophy of nature the outcome of which cannot be foreseen" (Schweitzer, 1947, pp. 227f.). It seems to me that this conflict cannot be responsibly avoided any more.

Moving toward a "practical philosophy of nature" (Meyer-Abich, 1979), Schweitzer's reasoning is exemplary in neither falling into raptures nor committing naturalistic fallacies. It may be observed that this is not true in many other cases where the dispute between ethics and philosophy of nature is experienced. So Mill's essay on nature, trivial as his arguments are, still deserves consideration. What I want to emphasize is that Schweitzer postulates reverence for life fully realizing that this life which he demands to be experienced with reverence sustains itself on the basis of eating and being eaten (p. 242). Nature is not idyllic. Nevertheless, man is that part of nature in which she becomes conscious of her suffering.

That nature is suffering and expecting from man a manifestation of hope and freedom – as Paul wrote to the Romans (chapter 8) – may never, till now, have been realized, or really believed, in Christian ethics. It seems to me that it is our task ultimately to become conscious of nature's expectation for the first time in the history of mankind, although it is not so much a new ethic as the redemption of parts of the old one. If we do all this, guilt will be identified as the basic character of human existence not only with respect to our fellow human beings but to nature as well. In this world of eating and being eaten, we cannot avoid an increase in suffering. Far from taking this inevitability as an exculpation – "good conscience is an invention of the devil" (Schweitzer, p. 249) – the acknowledgment that we invariably live at the expense of other species' suffering changes the situation practically as well as spiritually. Practically, an animistic apology before cutting a tree did not save the tree from being cut, whereas, since one cannot honestly apologize for useless consumption as an end in itself, other beings of nature may be saved from being turned into economic resources. This is how far a legal order might go. Spiritually, and morally, although the five biblical virgins with candles did not look very different from the other five, the very small difference between the two groups was decisive.

Having gotten this far, I do not see any conclusive reason to refrain from taking Frankena's last step, from considering all beings as morally relevant subjects. We have so much in common with the mineral world that the equality principle here again requires careful consideration. A stone's "soul," for instance, can become emergent by the activity of a sculptor, and generally speaking one cannot honestly apologize for grinding stones for concrete, the industrial paradigm of *res extensa.* And, finally, whoever hesitates before this last step should recall Goodpaster's objection that "morality cannot be formed from the rib of egoism" (Goodpaster, 1979, p. 25): "The *last* thing we need is simply another 'liberation movement' . . . ; the mere enlargement of the class of morally considerable beings is an inadequate substitute for a genuine environmental ethic" (p. 29). This being so, application of the proposed equality principle has led us to the whole, encompassing particularistic liberation movements which in turn give way to a holistic approach (J. C. Smuts, A. Meyer-Abich). I will conclude with a short outline of how in my opinion man and nature and the relations between them should be conceived in a holistic way so that the equality principle can become the basis of a constitutional order at Bacon's third level of power.

3. HOLISTIC PRINCIPLES FOR A HUMANE DOMINATION OF NATURE

The universe could be expected to "burst out into jubilation," as Goethe put it, if it were brought into its own by man's living life to the full (*Works*, Hamburg ed., vol. XII, p. 98). As opposed to that, the industrial economies treat nature as a bag of material, as "resources." The crown of creation, however, should crown not only itself but the world; it ought to be conceived as worth crowning as a whole. That nature is not a resource is most conspicuous in medicine; taking a sick man as material to be restored is to miss his humanity. Does this not indicate that we miss the essential life of nonhuman animals, plants – perhaps even of minerals as well – when we conceive and treat them as inanimate? Considering man's affinity to the rest of the world, it might therefore be a good idea to treat other beings humanely instead of treating men like inanimate beings, i.e., as mere resources.

While living in the USA during the Second World War, Ernst Bloch reconceptualized hope, reminding the industrialized societies that technological hope did not originally concern things (*natura naturata*) but life in and with them (*natura naturans*). "The working substrate of nature, that which otherwise has been called effective power or seed, is not referred to. But as soon as a technology becomes concrete, this problem of reference becomes the task of technological hope" (1959, p. 779). Bloch was inspired by Schelling who took the "revival of nature with freedom" as the task of the philosophy of nature:

> Philosophieren über die Natur heisst, sie aus dem toten Mechanismus worin sie befangen erscheint, herausheben, sie mit Freiheit gleichsam beleben und in eigene freie Entwicklung versetzen – heisst, mit anderen Worten, *sich selbst* von der gemeinen Ansicht losreissen, welche in der Natur nur, was geschieht – höchstens das Handeln als *Faktum, nicht das Handeln selbst* im Handeln – erblickt. Die ursprüngliche Produktivität der Natur verschwindet in der gewöhnlichen Ansicht über dem Produkt. Für uns muss das Produkt über der Produktivität verschwinden ("To philosophize about nature means to raise it out of the dead mechanism in which it appears mired, to animate it with freedom, as it were, and to transpose it into its own free development. It means, in other words, to break away from the common view which sees in nature only simple happening – at most sees its action as *fact*, and *does not see the acting itself* within the action. The originating productivity of nature disappears in the common view concerning the product. For us, the product must take a back seat to the productivity") (*Works*, Schröter ed., II, 13f.).

As opposed to this, nature for us has degenerated from the nature of beings and things to the things and beings of nature (Meyer-Abich, 1979). Nevertheless, there is a stream of thought – from Anaximander, Heraclitus, Plato, and St. Paul, to Goethe, Schelling, Heidegger, Bloch, and contemporary holism – in which man and nature, as well as their mutual relations, are conceived in other than absolutist fashion, so that conditions for controlling human domination constitutionally can be defined.

Human practice with respect to the whole of the world depends on our interpretation of humanity. Industrial society so far has been based on the concept of man as the conqueror of nature. Our will be done, industrial society prays. If man is the measure, however, there is no way to limit human domination constitutionally. Human autonomy, as absolute, is a contraction of freedom. We should be aware of nature not only physically as an object or bag of resources but also responsibly as something we have to care for in freedom, and not in autonomy. Which claim does this responsibility respond to?

Kant discovered that we cannot but accept the moral law because we give it to ourselves (*Grundlegung zur Metaphysik der Sitten*, BA 73; *Works*, vol. IV, p. 65). This is the Prussian way to overcome absolutism. At the same time, reason must be considered a gift nature has given us to govern our will (*loc. cit.*, p. 20). This is made explicit in the second formulation of the categorical imperative according to which we should act as if the maxim of our action would become a general law of nature: "Handle so, als ob die Maxime deiner Handlung durch deinen Willen zum allgemeinen Naturgesetze werden sollte" (p. 51). Human absolutism, therefore, can be overcome by observing nature's intentions as revealed to us by reason. Even if Kant was thinking strictly along the lines of classical physics – where the scientific paradigm matches the productive paradigm of the industrial economy – a way to overcome the highhandedness of mankind might be drawn from his basic approach along the line which I have indicated. A general answer to his question of what nature's intentions are is given in his *Idea of General History* (1784); he there suggests that these intentions ought to be acknowledged in history, the history of mankind being the execution of a hidden plan of nature to bring about a perfect political constitution:

Man kann die Geschichte der Menschengattung im grossen als die Vollziehung eines verborgenen Plans der Natur ansehen, um eine innerlich – und, zu diesem Zwecke, auch äusserlich – vollkommene Staatsverfassung zu Stande zu bringen, als den einzigen Zustand, in welchem sie alle ihre Anlagen in der Menschheit völlig entwickeln kann ("One can view the history of the human species at large as the carrying out of a hidden plan of nature, in order to bring about an inner – and, to this end, also an outer – perfect political constitution as the sole condition in which it can fully develop all the natural abilities within humanity") (*Idee zu einer allgemeinen Geschichte in weltbürgerlicher Absicht*, A 403; *Works*, vol. VI, p. 45).

Indeed, human history is only one thread in the evolution of nature. Though it may be a red thread and for mankind certainly is the one we hang on, it is just one thread among others in the course of life. As Johann Wilhelm Ritter put it, representing German *Naturphilosophie* at the beginning of the nineteenth centure: "Nature as a whole is the real genus. In every generation it is nature as a whole that generates" ("Nur die ganze Natur ist die Gattung; bei jeder Begattung zeugt die ganze Natur"; 1810, § 659).

In natural history man is the *Zôon lógon échon*: the animal whose biological constitution implies the faculty of speech. Nature emerges to express herself in the human species by way of language. That elements of speech also may be found in other species does not change the situation. The question remains: Which task or responsibility should the faculty of speech respond to? Many million years after the emergence of man in natural history, nature begins to call herself what she is, namely *phýsis*, and finds a way to express herself in *lógoi*. Aware of the *lógos* – listening to *phýsis*, as Heraclitus put it (fragment 112) – we perceive human history in nature (the natural context of human life) as nature becoming aware of herself by way of language. Emergence and suffering are the earliest characters experienced by the human mind in natural self-awareness. Human thought, according to Plato, is the dialogue of soul with itself; to identify nature as a partner in this dialogue leads to overcoming the absolutist (or anthropocentric) view: Nature – that's me. This is an acknowledgement that we belong to nature, that she is "impelling herself with us," as Goethe said, so that finally the universe may burst out into jubilation:

Natur! Wir sind von ihr umgeben und umschlungen – unvermögend aus ihr herauszutreten, und unvermögend tiefer in sie hineinzukommen. Ungebeten und ungewarnt nimmt sie uns in den Kreislauf ihres Tanzes auf und *treibt sich mit uns fort*, bis wir ermüdet sind und ihrem Arme entfallen ("Nature! We are surrounded and embraced by her – unable to step out of her and unable to enter deeper into her. Unbidden and without warning, she draws us into the circle of her dance and *impells herself with us* until we are exhausted and fall from her arms") (Tobler-Fragment; *Works*, Hamburg ed., vol. XII p. 45; italics added).

As I have argued elsewhere (Meyer-Abich and Schefold, 1981), current problems with the environment and in medicine are the result of basically erroneous understanding of nature. Reversing Marx's eleventh thesis against Feuerbach, we have changed the world too much and should re-examine the interpretation or understanding which guides our actions implicitly or explicitly – and especially our understanding of nature. The same conclusion can be derived from the problems of consumerism.

Industrial products generally express no more than has been impressed and reflected on them by way of production and advertising. Industrial production is a self-exhibiting of mankind, ultimately intending everything to look like itself. The copy does not match the original, however, because in the most advanced stages we are left with Cartesian impacts of pure forms by way of power (energy) in homogeneous time on something which by itself is considered nothing but matter or material resources. It is as if the theory of ideas had really failed in the fifth of the ways Plato described in the *Parmenides* (133allff.), and had given way to a technological solution. But apart from soul intelligence cannot be present in a lively way in anything; thus in Plato's *Timaeus* the Demiurge, when he framed the universe, "fashioned reason within soul and soul within body, to the end that the work he accomplished might be *by nature* as excellent and perfect as possible" (*Timaeus* 30b4-6). This is what happens in art; it is why art may be considered liberation or salvation of nature (cf. Paul's letter to the Romans, chapter 8).

Art, while it is hard work, is different from the industrial process. There is no Cartesian art; the work of art comes into being from the soul of the artist, whereas industrial production tends to replace soul by power machines – which explains the strategic role of the energy system. Works of art, therefore, are not made but generated, and "what has been generated (*genómenon*) has come to be as a whole" (Plato, *Sophist* 245d4). Now the artist does not take matter as material or as a resource which is nothing by itself and has to be shaped according to human arbitrariness. The successful sculptor, for instance, freely exposes, or uncovers, what is the stone's innermost best. He is working on the stone and not making it into something. In contrast to that, the construction industry destroys trees and stones to recast them into building material – concrete or pressed wood – which then is really nothing more than *res extensa.*

Something which has been generated so that its innermost best is nonviolently exposed will, by itself, tell us something. I think, however, that our artificial environment will cease to be dumb (except for the impersonal, mocking reflection of social relations, as in advertising) only when industrial production is fully renounced. This is not to say that there are no works of art in steel or synthetic materials. Many of those who are advocating the "soft" path nowadays are committing what may be called a "Miltonian fallacy"; that is, they are striving for a lost paradise. But this is only to turn from Ahriman to Lucifer. In fact, a "soft" enthusiasm does no more than a "brave new world" to acknowledge what is essentially not soft in human life – something the Kantian dove also overlooked when dreaming how fast it might fly when not retarded by the air's resistance. In fact, the development of traditional technology (recall Francis Bacon) was also inspired by the idea of recovering paradise by overcoming effects of the biblical Fall. (By now, it seems that this development has led into a Golden Cage.)

In my opinion the controversy between those who defend the "hard" path and those who plead for the "soft" path should be considered a question of an adequate balance – so to speak – between a skeleton of industrial production and blood circulating in living tissue. The political problem of defending such a position is that usually those who simply want, for their own advantage, to preserve the status quo also favor a "balanced position" – presupposing that what is convenient for them is balanced. This is not the case in our actual situation, as far as I can see. We have lost our balance, falling to the "hard" side, and we keep on falling further in that direction.

Nature's exultation at being brought into its own by man's living life to the full is, in my judgment, also at the core of the energy controversy. In fact, energy, as part of the industrial triad, form/energy/matter, is thus the *tertium comparationis* between *res cogitans* and *res extensa* in a Cartesian economy. Energy's function here corresponds to that of Descartes' proof of God's existence, which allows immaterial spirit to find its way back into unspirited matter, or to that of the artist's soul in letting nature come into its own. Expanding industrial society's energy budget would have the decisive effect of tipping the balance more and more toward the "hard" side so long as energy is used to transform other beings into resources or to recast those resources so that they become an image of industrial man.

When asked whether the puppy should be raised (*gross gezogen* in German) the child answered: No; just let it grow. Obviously, energy is also required in the process of growing. A being that grows, however, is not passively moved, not violently shaped by energy or power; it makes use of energy in moving or developing *itself*. This distinction between moving oneself and being moved (again referring to Plato's philosophy) is especially obvious and applicable in the critique of contemporary medicine.

Another good example of "natural" technology, in the sense of letting something be what it is by itself, can be seen in domestic energy systems. In terms of energy, a building is a device for keeping the inside space at temperatures different from those outside. In higher latitudes, moreover, to do this generally requires an additional heating system, and this is what we call "the" heating system. The better a house is adapted to its locale, however, the smaller the additional heating system can be. For example, since the oil crisis we have become aware that in Germany no more than about a fifth of the heating energy which was formerly used is really needed to serve this purpose; more than four fifths could economically be substituted for by architectural changes together with advanced technological knowledge – with the result that houses would become comfortable "by themselves" to a much higher degree than before.

In his *Critique of* [aesthetic] *Judgment* (1790) Kant introduced "being natural" as a criterion of art: "Nature was beautiful when at the same time it looked like art; and art can only be called beautiful when we are conscious that it is art and it still looks to us as nature (*und sie uns doch als Natur ansieht*) . . . ; i.e., beautiful art must be perceivable as nature (*muss als Natur anzusehen sein*) (*Kritik der Urteilskraft*, A 177; *Works*, vol. V, p. 405). If, for these reasons, a particular technology today is to be called good – that is, if it is to be considered art – the criterion of being natural should be applied to it. The result would be that some technologies will turn out to be "natural" and others will not. That such a criterion is exactly what we need with respect to environmental disruption is easily seen. For example, nowadays we call "natural" the green world that we see (or miss) when looking out the window, whereas we use the term "social" for the invisible relations between men. Exactly the realm which is most problematic with respect to man's relation to nature, that of technical artifacts, fades out of our perception. If we were able to judge the natural character of a technical system as we are able to judge that of a garden, we would be much better off in our decisions about the acceptability of technologies.

In Plato's philosophy, as well as in everyday language, "natural" is what something is by itself. An opinion, for instance, is called natural when it occurs to us spontaneously – as in Plato's dialogues – and is not forced upon us; unconstrained or casual behavior is also called natural. This same conception of nature is implied when beings that develop by themselves – plants and animals, for instance – are called nature. From this it can be seen that the things and beings of nature are still called nature because the nature of things (*natura naturans*) comes into being here and lets them be what they are (*natura naturata*). Virginal nature is natural in this sense, but domestic animals and plants are too insofar as breeding them from their wild ancestors can be considered agri*culture*, the development of their innermost best – like education. Natural conditions prevail where something can grow by itself or is nonviolently assisted to develop its inherent qualities. Natural, therefore, is not only what is not artificial – indeed the artificial may be more natural than what has wildly grown – but not every technology fulfills the criterion of being natural.

The criterion of being natural, if further developed, would still probably not allow clear-cut or exclusive distinctions. But this is true for most normative judgments and does not warrant renunciation of the criterion. I refrain from discussing examples in this paper, but I would not be surprised if some of Illich's "convivial" technologies turn out to be more natural than the nonconvivial ones. (Conviviality, however, was introduced by Illich as a social criterion, irrespective of man's relation to nature.)

If somebody strives for the good by himself, he is considered free. To behave naturally and to behave freely are really synonymous here. With respect to other beings, however, we do not care enough for their nature and not at all for their freedom. Why not? Perhaps the environmental crisis will teach us to accept freedom as the measure of being natural not only for mankind but also for the whole of nature of which we are a part. "Dissension of the will to life," as Schweitzer called it, is responsible for missing this freedom again and again. Thus in the relation of humans to nature we are indeed experiencing Bacon's third level of power, nature being turned into history; in Bloch's words, "Ultimately manifested nature, like ultimately manifested history, lies in the horizon of the future" (1959, p. 807).

ACKNOWLEDGMENT

Paul Durbin has kindly adapted my German philosopher's English to make it clearer for the reader. Wherever my argument remains less clear than the subject deserves, the responsibility is entirely my own.

University of Essen

REFERENCES

Anaximander Fragments, 1951. See H. Diels and W. Kranz, *Fragmente der Vorsokratiker*. Berlin.

Bentham, J. 1962. *Principles of Penal Law* and *Introduction to the Principles of Morals and Legislation*, in *The Works of Jeremy Bentham*, J. Bowring (ed.), vol. I. New York.

Binswanger, H. 1979. "Natur und Wirtschaft," in K. M. Meyer-Abich (ed.), *Frieden mit der Natur*. Freiburg, pp. 149–173.

Bloch, E. 1959. *Das Prinzip Hoffnung*. Frankfurt.

Corbett, P. 1971. "Postscript," in S. Godlovitch and J. Harris (eds.), *Animals, Men, and Morals*. London, pp. 232–238.

Dilthey, W. 1923–1931. *Gesammelte Schriften*. Leipzig and Stuttgart.

Fechner, G. T. 1899. *Nanna, oder über das Seelenleben der Pflanzen*. 2d ed. Hamburg and Leipzig.

Feinberg, J. 1974. "The Rights of Animals and Unborn Generations," in W. T. Blackstone (ed.), *Philosophy and Environmental Crisis*. Athens, Ga., pp. 43–68.

Frankena, W. K. 1979. "Ethics and the Environment," in K. Goodpaster and K. Sayre (eds.), *Ethics and Problems of the Twenty-first Century*. Notre Dame and London, pp. 3–20.

Goethe, J. W. von. 1981. "Winckelmann" (1805) in *Hamburger Ausgabe*, vol. 12, pp. 96–129. Munich.

Goethe, J. W. von. 1981. "Die Natur; Fragment aus dem Tiefurter Journal" (1783), in *Hamburger Ausgabe*, vol. 13, pp. 45–47. Munich.

Goodpaster, K. E. 1979. "From Egoism to Environmentalism," in Goodpaster and Sayre (eds.), *Ethics and Problems of the Twenty-first Century*. Notre Dame and London, pp. 21–35.

Habermas, J. 1973. *Erkenntnis und Interesse*. 2nd ed. Frankfurt.

Heraclitus Fragments, 1951. See Diels and Kranz. *Fragmente der Vorsokratiker*. Berlin.

Illich, I. 1973. *Tools for Conviviality*. New York.

Jaeger, W. 1954. *Paideia*, 3 vols. Berlin.

Kant, I. 1956–1964. *Werke (in sechs Bänden)*. W. Weischedel (ed.). Frankfurt and Darmstadt. Citations include page references, A, to original edition.

Meyer-Abich, A. 1934. *Ideen und Ideale der biologischen Erkenntnis*. Leipzig.

Meyer-Abich, K. M. 1979. "Toward a Practical Philosophy of Nature," *Environmental Ethics* 1 293–308.

Meyer-Abich, K. M. 1982. "Geschichte der Natur, in praktischer Absicht," in E.

Rudolph and E. Stöve (eds.), *Geschichtsbewusstsein und Rationalität*. Stuttgart, pp. 105–175.

Meyer-Abich, K. M. and Schefold, B. 1981. *Wie möchten wir in Zukunft leben: Der "harte" und der "sanfte" Weg*. Becksche Schwarze Reihe, vol. 242. Munich.

Mill, J. S. 1970. *Three Essays on Religion*. New York. Original, 1874.

Nelson, L. 1970. "System der philosophischen Ethik und Pädagogik," vol. 5. in *Gesammelte Schriften (in neun Bänden)*. Hamburg.

Ortega y Gasset, J. 1947. *Obras Completas*. Madrid.

Passmore, J. 1974. *Man's Responsibility for Nature*. London.

Ritter, J. W. 1969. *Fragmente aus dem Nachlasse eines jungen Physikers*. H. Schipperges (ed.). Heidelberg. Original, 1810.

Rolston, H., III. 1975. "Is There an Ecological Ethic?" *Ethics* 85 93–109.

Salt, H. S. 1907. *Die Rechte der Tiere*. Berlin. English original, 1892.

Samuelson, P. A. 1970. *Economics*. 8th ed. New York.

Schelling, F. W. 1927. *Werke*. M. Schröter (ed.). Munich.

Schmoller, G. 1920. *Grundriss der allgemeinen Volkswirtschaftslehre*. 2 vols. Munich and Leipzig.

Schopenhauer, A. 1950. *Preisschrift über die Grundlage der Moral*, vol. 4 in *Sämtliche Werke*. Frauenstädt and Hübscher (eds.). Wiesbaden.

Schweitzer, A. 1947. *Kultur und Ethik*, in *Kulturphilosophie* (II. Teil). 6th ed. Munich.

Smuts, J. C. 1927. *Holism and Evolution*. London.

Tribe, L. H. 1976. "Ways Not to Think about Plastic Trees," in L. Tribe, C. Schelling, and J. Voss (eds.), *When Values Conflict: Essays on Environmental Analysis, Discourse, and Decision*. Cambridge, Mass., pp. 61–91.

Winner, L. 1977. *Autonomous Technology: Technics-out-of-Control as a Theme in Political Thought*. Cambridge, Mass.

PART IV

METAPHYSICAL AND HISTORICAL ISSUES

DON IHDE

THE HISTORICAL-ONTOLOGICAL PRIORITY OF TECHNOLOGY OVER SCIENCE*

I. INTRODUCTION

The thesis I wish to explore in this essay is that *there is a significant sense in which technology may be seen to be both ontologically and historically prior to science.* There is, of course, an obvious and trivial sense in which this claim may be regarded as true. If technologies in the broadest and most concrete sense involve humans and their uses of tools and artifacts, then at the least one can say that technology in this sense is both universal and was probably involved at the time of the emergence of the human species. There are no instances of societies, cultures, or human groups which do not use tools and artifacts in their relations with the natural environment.

And if science centrally involves a theorizing about things in a systematic and hypothetical sense, then it should be apparent that the practiced and skilled uses of technologies long precede the kind of self-awareness implied in science. In the most general sense then, *praxis* precedes explicit theory.

I wish, however, to suggest that there is a more specific sense in which technology, particularly in its more recent developments, is the *condition of the possibility of science.* I argued in *Technics and Praxis* (1979) that science, in its contemporary sense as an experimental science wedded to specific meanings of measurement, is *necessarily embodied* in its instrumentation. Indeed, one of the chief differences between modern science and Greek contemplative science lies in the development of instrumentation both for measurement and for actual investigative purposes. I showed how instrumentation extends and embodies perception.

Historically, of course, even Greek science in actual practice engaged some measurement technologies. But the lack of a specific technological impetus also doomed Greek science to its primarily speculative attainments (witness the odd ideas about the shapes of atoms as causes of sweet, bitter, or sour tastes in Democritus. Lacking any means of investigation of such micro-phenomena, the speculation had to remain just that.) This lack of appropriate instrumental technology determined the limits of a primarily contemplative science.

Here I wish to push the essential interlocking of science and technology

Paul T. Durbin and Friedrich Rapp (eds.), Philosophy and Technology, 235–252.
Copyright © 1983 *by Don Ihde.*

further by arguing for the historical-ontological priority of technology as a condition of the possibility of science. I shall develop three unequal stages in this demonstration: First, I shall briefly describe what I take to be the standard and dominant theory of the relationship between technology and science. Second, I shall pay my debts to two important intellectual predecessors of the view I shall elaborate. A philosophical debt is owed to Martin Heidegger who may be said to have originated and solidified what has become the philosophy of technology for the twentieth century and who argued most explicitly for the ontological priority of technology over science. A historical debt is owed to the large body of work done by Lynn White, Jr., who made us aware that there was a virtual technological revolution in the Middle Ages which preceded and laid the groundwork for the rise of modern science in the Renaissance and through the Enlightenment.

The third step will then be an examination of certain aspects of the historical technological lifeworld. I shall develop this account along phenomenological lines.

II. THE STANDARD THEORY

There are a variety of conceptual possibilities which could account for the relationship of technology and science, but two extreme cases – I shall call them the "idealist" and "materialist" interpretations – have the advantage of posing the issues most starkly.

What I shall call the idealist view is the interpretation which holds that science precedes and founds technology. It is an interpretation which holds that requisite for creating a (modern) technology, one must have insight into the laws of nature, a conceptual system at the formal and abstract level, and the ability to *apply* this knowledge to the material realm, thus creating a technology.

In this interpretation, technology follows from science, both ontologically as an application of scientific knowledge and historically as the spread of this insight into ever widening realms of material construction.

The standard view is accompanied by an interpretation of the history of modern science and technology which may be characterized as follows: After a long dark period in European history, there emerges a revival of the Greek scientific spirit which emerges within and animates what we call the Renaissance. Europeans regain an interest in nature and speculations about nature and evolve a method of understanding nature which we call Modern Science. Historically this movement becomes dramatic and fulfilled in such figures as Galileo, Kepler, Copernicus, and eventually becomes fully systematized with Newton.

The rise of Modern Science is a development which includes (a) the discoveries of more sophisticated mathematics; (b) a gradual move away from religious and theological notions and a move toward a more mechanistic and materialist metaphysics; (c) a method which diverges from the more speculative ancient roots and moves in a more experimental and verificational direction; and (d) a movement which results in the rise of physics as the primary science or at least the science which is first among equals.

Only after this historical development of science does there arise technology in the modern sense. The Industrial Revolution of the past century and a half and the explosion of the current "high technology" are plausibly dependent upon the precondition of scientific theory. Technology in the contemporary sense seems to spin forth almost directly from science itself.

In this essay I am not interested in a further exposition of the implicit metaphysics of this interpretation, nor am I going to undertake a direct attack upon its presuppositions. As an interpretation of the relationship between science and technology it has both plausible and implausible aspects. I shall point out some of these, but I shall do so indirectly by way of elaborating a certain necessary strategy which this view must entail.

What must technology be; how must it appear if this view is correct? First, what will pass for technology must in the paradigm case be a technology which is obviously dependent for its shape upon scientific-theoretical considerations. Thus the best examples are what we call today high technologies. While I do not intend what follows to be exhaustive by way of definition, I suspect a high technology must be characterized as a technology which must include: (a) a complex and interlocked system; (b) workings which are understood only by way of scientifically derived theories; (c) components which contain esoteric compounds and units, themselves the result of complex and scientifically determined processes; and (d) microscopic machine tolerances, internal organization, mechanical or electronic motions developed from microlevels of manufacture and planned construction. A computer is an obvious case of such a high technology, but there are dozens of other examples which could do as well.

In contrast, "low" or, better, traditional technologies would be those which are simple, arrived at through a process of trial and error, which contain only rough interrelations of parts, and are understandable by any mechanically inclined person. A waterwheel is an example of such a technology.

That there is an apparent and even dramatic difference between the computer and the waterwheel seems clear. But just what and how that difference is to be accounted for is precisely what needs note. However, at the level at which I am developing the case, we need to be aware that the idealist position which holds that science is the condition of technology must accentuate a sharp difference between a presumed pre-scientific and scientific technology. In short, contemporary technology is seen to be disjunctive with respect to traditional technology.

The reason this tactic is conceptually necessary is that otherwise one would have no way of accounting for the previously noted trivial but obvious historical situation in which all peoples and societies use and have technologies whether or not they have a science in our sense. The historical dependence of technology upon science then becomes a special case of dependence; only *scientific* technology is historically dependent upon science.

The Renaissance and Enlightenment, relative to the medieval period, may be seen to be an instance of the focus upon the assumed priority of science in the modern sense. Put most simply, precisely because the values of scientific knowledge as theoretical knowledge were assumed in the Middle Ages to be higher than so-called practical knowledge, the possibly unique attainment of that period is often overlooked.

III. A MATERIALIST THEORY: HEIDEGGER AND WHITE

A contrary position is possible. I shall construct such a view by combining the insights of Martin Heidegger and Lynn White, Jr.

Martin Heidegger is perhaps the philosopher who has most originally and profoundly rendered the question of technology a central concern of philosophy. The position he developed in "The Question Concerning Technology" is one which argues for the ontological, but not the historical priority of technology over science. The argument is complex and I shall look at only a few elements of it.

Heidegger holds that Technology has always underlain what we have called science in the West, but it has been revealed as the origin of science only recently. Embedded in this complex argument, however, is a deep ambiguity about what shall count as technology. On the ontological level, Technology – more precisely the essence of technology – is a certain way of experiencing, relating to, and organizing the way humans relate to the natural world.[1] On the historical level, at least in the chronological sense, Heidegger seems to grant that technology in its modern sense is later than science. In short, Heidegger accepts in some degree the notion that modern or scientific technology is essentially and distinctly different from traditional technology. I hold that he is wrong in allowing himself to accept this notion and that as a result he weakens his own case in such a way that the usual accusation that he is at least partly "romantic" with respect to technology is given weight. In sum, the Heideggerian position is that Technology, while ontologically prior to science, is historically later.

At the core of the view which Heidegger is espousing lies an inversion of the standard view of the relationship between science and technology. This inversion is most dramatically illustrated by his claim that rather than technology being a tool of modern physics, it is exactly the opposite: physics is the necessary tool of Technology. In the first instance, Heidegger discerns that modern physics is necessarily interrelated with its instruments:

> It is said that modern technology is something incomparably different from all earlier technologies because it is based on modern physics as an exact science. Meanwhile we have come to understand more clearly that the reverse holds true as well: modern physics, as experimental, is dependent upon technical apparatus and upon progress in the building of apparatus. The establishing of this mutual relationship between technology and physics is correct. But it remains a merely historiographical establishing of facts and says nothing about that in which this mutual relationship is grounded.[2]

Then, in a much stronger statement, Heidegger argues that physics is the herald of Technology:

> Modern science's way of representing pursues and entraps nature as a calculable coherence of forces. Modern physics is not experimental physics because it applies apparatus to the questioning of nature. The reverse is true. Because physics, indeed already as pure theory, sets nature up to exhibit itself as a coherence of forces calculable in advance, it orders its experiments precisely for the purpose of asking whether and how nature reports itself when set up this way.[3]

This inversion, clearly evidenced in the way Heidegger views the relationship between science and technology, is one which nevertheless retains at least one partial sense in which science precedes technology. (I am quite aware, with most Heidegger scholars, of the distinction between *Historie* and *Geschichte* in Heidegger's use. However, *Geschichte* serves a specifically ontological function.)

This residual sense in which science historically precedes technology is also a sense which allows there to remain a distinction between scientific and traditional technology. The strongest statement concerning this residual sense is this:

> Chronologically speaking, modern physical science begins in the seventeenth century. In contrast, machine-power technology develops only in the second half of the eighteenth century. But modern technology, which for chronological reckoning is the later, is, from the point of view of the essence holding sway within it, historically earlier.[4]

Similarly, the disjunctive sense which the standard view must maintain and which separates modern from traditional technology, is allowed by Heidegger:

> The revealing that rules modern technology is a challenging, which puts to nature the unreasonable demand that it supply energy which can be extracted and stored as such. But does this not hold true for the old windmill as well? No. Its sails do indeed turn in the wind; they are left entirely to the wind's blowing. But the windmill does not unlock energy from the air currents in order to store it.[5]

And again, Heidegger, as he so frequently does, contrasts the peasant's sense of earth with that of the modern technologist:

> In contrast, a tract of land is challenged in the hauling out of coal and ore. The earth now reveals itself as a coal mining district, the soil as a mineral deposit. The field that the peasant formerly cultivated and set in order appears different from how it did when to set in order still meant to take care of and maintain.[6]

Thus while we have the assertion of the ontological priority of Technology over science as an inversion of the standard view, there is retained a secondary sense in which technology chronologically follows the development of science and a sense in which there is a disjunctive difference between traditional technology and modern technology. Science, in Heidegger's view, stands as the event which finally shows to us what Technology is ontologically; science is the herald of Technology in a (chronological) historical sense:

> The modern physical theory of nature prepares the way not simply for technology, but for the essence of modern technology. For such a gathering-together, which challenges man to reveal by way of ordering already holds sway in physics. But in it that gathering does not yet come expressly to the fore. Modern physics is the herald of enframing, a herald whose origin is still unknown.[7]

What holds this argument together lies in the several ways in which Heidegger uses the term technology.

What may be called the surface definition of technology is what Heidegger calls the anthropological-instrumental understanding of technology, technology as a mere tool of science.[8] This definition is not false; it is merely correct. It does not reveal the *essence* of technology.

A second definition derives from the Greek *technē*, and begins to more nearly approximate the Heideggerian sense of Technology in that *technē* is both a name for the activities and skills of a craftsman and for the arts of both mind and hand, but also is linked to creative making, *poiēsis.*[9] For the Greeks *technē* was a production which was a kind of knowledge.

The third, and ultimate Heideggerian definition of Technology, however, makes of Technology a mode of truth or revealing (*alētheia*). Technology, in essence, reveals a world in a certain way. "Every bringing-forth is grounded in revealing."[10] "Technology is a mode of revealing, Technology comes to presence in the realm where revealing and unconcealment take place, where *alētheia,* truth, happens."[11] The essence of technology allows us to see, to order, to relate to world in a particular way. Nature becomes standing-reserve, a source of energy for human use, and this mode of relating to the world becomes, in a technological era, the dominant and primary way in which we understand world.

I shall not further explore the Heideggerian view, other than to note that only after Technology is discovered to be this way of relating to the world may one begin to understand how science under this mode is seen to be the necessary "tool" of Technology. Science becomes a means of knowledge which gives power; science becomes Baconian. And with this move the inversion is completed: Technology, as the revelation of the world as standing-reserve, is the ontological presupposition and ground of modern science.

Philosophically things would have been neater and clearer were it the case that Technology could be shown to be not only ontologically, but historically prior to science. And this would especially be so if the historical priority were of such a nature as to be understood as an experiential condition of the possibility of modern science.

Such a view would also have the advantage that it would be continuous with the ordinary observation that some form of technology is universal and occurs wherever there are human societies.

I think this is the implicit import of the work of Lynn White Jr., who has clearly caused a revision of the way in which we understand the medieval period with respect to technology.

White's publications concerning medieval technology span two decades of work. The landmark book, *Medieval Technology and Social Change* (1962), shows how technological development was deeply implicated in systems of warfare (the stirrup led to mounted shock warfare, thence to changes in social structure in Feudalism), agriculture (the plough combined with horse power and the development of three-field rotation led to a shift of food production to Northern Europe) and in that increasing hunger for mechanical power which laid the basis for other forms of increased productivity.

By looking at the burgeoning technology of the Medieval Period, White paints a historical picture of a Europe rapidly changing, avidly searching for inventions, and particularly hungry for power. This is particularly the case with the newly invented mechanical devices for extracting power from water and wind. By the year 983, water power is being used for fulling mills; but within a century the *Domesday* census reveals that there are already 5624 watermills in operation in England (a harbinger of the Industrial Revolution centuries later).[12] The windmill is referred to as early as 1180 and is common in much of Europe by 1240. The search for power in the Middle Ages utilizes every source. Inventions from foreign lands are rapidly experimented with in new ways, often hardly practical, but rarely overlooked. This medieval search for power laid the groundwork for later industrial technology but it was also intricately tied to a search for knowledge. Giovanni da Fontana, for example, in 1420, designed the forerunners of our robot measurers in the form of swimming fish, flying birds, and running rabbits, all linked to a plan to measure surfaces and distances in water, the air, and out-of-the-way places.[13]

One dramatic technological development during this period, a development which transformed the human perception of time, was the clock. In White's words, "Suddenly, towards the middle of the fourteenth century, . . . [it] seized the imagination of our ancestors. . . . No European community felt able to hold up its head unless in its midst the planets wheeled in cycles and epicycles, while angels trumpeted, cocks crew, and apostles, kings, and prophets marched and countermarched at the booming of the hours."[14] Time and the movement of the spheres was tied to a mechanical device. And thus by 1382 the universe itself began to be conceived of according to a mechanical metaphor.

> It is in the works of the great ecclesiastic and mathematician Nicholas Oresmes, who died in 1382 as Bishop of Lisieux, that we first find the metaphor of the universe as a vast mechanical clock created and set running by God so that "all the wheels move as harmoniously as possible." It was a notion with a future: eventually the metaphor became a metaphysics.[15]

His more recent works have taken account of the unique intellectual climate which encouraged technological development in Europe. By the time of his publication, "Cultural Climates and Technological Advance in the Middle Ages," White can claim, "The technological creativity of medieval Europe is one of the resonant facts of history."[16] What he finds is that medieval Europe was highly receptive to the use and development of technology and that several factors encouraged this. The organization and climate for order, stemming from the earlier monastic reforms, readily adapted technology. The clock, used first to establish the order of time, agricultural techniques, machines to lighten labor, were all affirmatively valued. Indeed, his survey of the literature of the time finds that detractors from the praise of technology are rare. Contrarily, praise of invention, machines, and their use is the rule.

Prior to our Bishop Oresmes who declares the heavens to be clockwork, one finds praise and prediction concerning a glorious technological future common. "Roger Bacon, 1260, pondering transportation, confidently prophesied an age of automobiles, submarines, and airplanes."[17]

This attitude of fascination and obsession with the technological stands in stark contrast to other areas of even Christian civilization. Whereas the Latin West, from the monasteries on, accepted technology into the precincts of the holy – every cathedral must have a clock – the Eastern regions forbade such inventions in sacred space. Clocks must remain outside the realm of eternity, thus outside the church in the Orthodox lands.[18]

The positive evaluation of inventiveness, linked to a desire for machine power, was also accompanied by the willingness to adapt ideas and artifacts from any culture. What became the bow for our string instruments came from Southeast Asia. A Tibetan prayer wheel may have inspired the windmill, and so the list goes. In short, the Medieval Period was suffused with interest in, desire for, and the development of technologies.

By the late Middle Ages, at the dawn of the time of the rise of modern science, White points out:

> About 1450 European intellectuals began to become aware of technological progress not as a project [. . . this came in the late thirteenth century] but as an historic and happy fact, when Giovanni Tortelli, a humanist at the papal court, composed an essay listing, and rejoicing over, new inventions unknown to the ancients. . . . It was axiomatic that man was serving God by serving himself in the technological mastery of nature. Because medieval men believed this, they devoted themselves in great numbers and with enthusiasm to the process of invention.[19]

In short, what White has established is that by 1500, a period whose image is consolidated by the technological genius of da Vinci, there is a self-awareness of technology, the process of invention, and the desire to master Nature through human artifacts.

By the year 1500, Europe had already developed even some of the instrumentation so fundamental to the very investigative possibility of science in the modern experimental sense. Lenses were invented by 1050, compound lenses by 1270, spectacles by 1285 and, by 1600, Galileo's period, the microscope and telescope. Clocks, essential to measurement, began to be developed in the ninth and tenth centuries and by the

1500s were widespread from cathedral to town hall to individual watches.

On the industrial side one can note that Europe is by this time covered with wind and water mills; the lowlands were being drained by wind-power; there were railways in mines; and the massive, sophisticated architecture of cathedrals, suspension bridges, and other large projects were part of daily life. Yet, in spite of the now reflective obviousness of this pervasive technological achievement of the Middle Ages, White is still probably right in claiming that "the scholarly discovery of the significance of technological advance in medieval life is so recent that it has not yet been assimilated to our normal image of the period."[20]

IV. THE HISTORICAL-ONTOLOGICAL PRIORITY OF TECHNOLOGY

If one combines the claims of Heidegger concerning the ontological priority of technology with those of White concerning the immediately preceding historical technological revolution, one arrives at this essay's thesis. However, to consolidate this thesis I shall develop speculatively something of a phenomenology of daily life, first as it appeared in the European lifeworld, then, as a variation, as it appeared in a different culture, that of the Polynesians. In so doing I shall focus upon spatial and temporal orientations.

IV. A
A "Reconstruction" of an Aspect of the Medieval Lifeworld

My strategy in this reconstruction of a medieval lifeworld will be to focus upon selected experiential components as they are embodied in praxis. It should be obvious by now that in the late Medieval Period, mechanical contrivances were very common and indeed pervasive in many ordinary activities. The world had already implicitly begun to be thought of in terms of mechanical metaphors. But in my focus upon space and time I am concerned with the way these dimensions are *perceived.*

I begin with the well-worked-over example of clocks, which were familiar and taken-for-granted aspects of daily life in the late medieval world. Lewis Mumford in his 1936 book, *Technics and Civilization,* has already noted how the clock was crucial to the development and reorganization of medieval life. According to Mumford, clocks were first commonly used in conjunction with monastic life and the development of disciplined and common order. The keeping of hours for religious exercises and the ordering of work set the pace for public or intersubjective life. Heidegger, too, in *Being and Time* (1927) pointed out the way in which clocks are not mere artifacts, but "take account" of human surroundings and nature. One can say that once clocks are developed, we begin to perceive time through technology.

One should take careful note of the specific perceptual representation of time via the clock. First, until recently, all clocks represent time through a use of moving pointers. This is the case whether one regards the moving shadow of the sundial, the linear scale of the early water-clocks, or the eventual round cyclical face of the cathedral clock. I would point out here that this representation of time is one which has both a focus – the instant of time which is the precise "now" as that point where the pointer "stands" – and a field or duration or span of time within which the instant finds its place. The field or span of time is the spread of the clock face, whether linear or circular. Thus "now" takes its place within a duration of time.

If, now, one begins to reflect upon the evolution of the clock, one can note the following distinct developments. At first the movement of the pointer is crude and relates primarily to fairly large units of time. The earliest circular faces of clocks were marked only into hours and had only one hand. But as clockwork became more mechanically refined, time was divided into smaller and smaller units, a second pointer was added to mark the minutes, and then a third to mark the seconds. Time was more and more quantified. This quantification was gradually more finely divided and the perception of time became ever more open to finer discriminations, to what may be called the micro-features of time. Moreover, these micro-features could be considered atomistically as units which were discrete relative to each other. In short, the clock allows us to perceive time latently as a series of atomized, discrete instants, a representation of what was to become a "scientific" mode of analyzing time. Time is perceived via or through the clock and this perception is a technologically mediated perception.

Historically, what eventually became more and more important was the focal point of technologically mediated time. The instant of its micro-features stands out. It becomes the means for further investigating things and is now essential for contemporary scientific measurement. Simultaneously, but almost unnoticeably, the field of time, which is the background but grounding feature of clock time, recedes and becomes less important. This development reaches a qualitatively different result in the contemporary invention of the digital clock. The digital clock represents only the focal instant of time, the field of time is no longer perceptually represented and in the process the perception of time also changes. The person who awaits the train, who once could glance at his watch and *see* that it was yet ten minutes until arrival time by *seeing* the relation between the pointers and the span, now sees only the number and must infer or calculate the span. This is to say that the mental operation for telling the time changes, even if unnoticeably, with the digital clock. What this portends for us, I shall not predict now other than to observe that if part of the essence of technology is "calculative thought" in Heidegger's sense, then the digital clock is an enhancement of this process.

Clocks were, prior to the rise of science proper, part of the daily experience of medieval humanity. They were an ordinary part of the life-world, the technological mediators of the sense and perception of time. And in a sense, they made possible the very calculations which lay at the (much later) basis of measurement undertaken by the Galileos and Keplers of the early scientific era.

Turn now to a spatially mediated experience and note that the same invariants occur again. One of the most important technologies which allowed the science of the modern era to become truly experimental was optics. Lenses were developed in the tenth century, were already compounded by the thirteenth century, and simultaneously with the first explicit scientific observations, the microscope and telescope were invented.

Vision is embodied and mediated through lenses. What changes is what might be called a shift of focus from ordinary perception to the technologically mediated micro-dimension. Distance is reduced, what is far is brought near, but this is equivalent to saying that what was for ordinary vision a micro-feature is now made present. The microscope brings into view for the first time the small and unexpected creatures found in drinking water; the telescope reveals that the shaded areas of the moon are seas and mountains and craters. The span of space is changed, reduced, and the object is "brought closer." What was previously so distant as to be unperceived, is now perceived in a near-distance of optically mediated space. Again, both what is focal and what was the field of space changes under the tranformations of technologically mediated perception.

This is to say that through the use of technologies experience had already come to be prepared for the scientific experience of the world. A world whose features could be considered as discrete units, a world whose micro-features would fascinate, a world conceived of under the sign of mechanical relations, was a world which was prepared for by the taken-for-granted technologically mediated experience of the Medieval Period.

Late medieval experience of both time and space could be considered to be thoroughly embedded in and often mediated through technologies. One could expand upon these examples in many areas of life. One could also contrast these examples of technologically mediated perception of space and time with cultures which did not have clocks or lenses and note that time and space are differently perceived. But I shall now turn to a more dramatic example of the way experience and praxis are organized and examine a crucial case of long distance spatial orientation, the variant development of a perceptual, and *technologically mediated* perceptual, navigational system.

IV. B.
Variant Long Distance Spatial Orientation: Atlantic and Pacific Navigation

One of the features which stimulated the European development of technology was the availability of ideas and devices from many areas of the world, an availability made possible through the early exploratory trips of Europeans. We are familiar with some of the historical events which were associated with this cross-cultural interchange: the Crusades, the travels of Marco Polo, the centuries of coastal voyages, and only much later, the full spice trade and voyages of conquest for gold and riches which fed the end of the Medieval Period. I shall here focus upon the development of cross-oceanic navigation as it contrasts with the Pacific variant.

Coastal navigation, essentially navigation within sight of land or never far from it, is distinctly different from transoceanic navigation. The principles or practice of coastal navigation and the body of knowledge which goes with it were known from ancient times. Such navigation was largely perceptual and traditional in that observations of currents, animal life, noise and sight of breakers over shoal waters, wind patterns, etc., were necessary for safe coastal piloting. Fears of out-of-sight navigation were not merely those clothed with superstitions about the unknown (monsters, the edge of the world, etc.) but were related to lack of knowledge about how even to return to a known area. In short, what was needed was a means of dependable spatial orientation across the expanse of uncharted ocean.

Early Western transoceanic navigation was successfully undertaken by the Vikings who travelled from Scandanavia, not only throughout Europe and the Near East in coastal raids, but also to Iceland, Greenland, and Nova Scotia in the New World. How these voyages were undertaken lies somewhat obscured by a sparse historical record, except that we know that two features of navigation unique to Northern Europe were already known: a fixed star, the North Star (Polaris) was known and navigational calculations could be based upon this fixed point. And the primitive use of the lodestone which also points to a fixed area, was already common with the Vikings. Thus, although very simple, one can say that the very origin of trans-Atlantic navigation was technological in a most primitive sense. Orientation was secured through a device.

If, however, one takes the voyages of Columbus as more typical, then the technological determination of orientation is abundantly clear. By 1492, at the transition period for our purposes, not only is there a magnetic compass, but measured and careful cartography was known, and a vaster array of instrumentation was also available. Compass, astrolabe for calculating angles to the sun and other heavenly bodies, clocks (although not yet fully useful for ocean voyages), and various measuring devices were used for navigation. Columbus's daring voyage was a voyage undertaken through a technologically mediated orientation to possible space. (Columbus knew very well that the earth was round; that it was of approximately a certain size – although vastly underestimated by his era – and that it could be plotted through calculations via instruments.) His navigation already conceived of the world as a grid-work upon whose surface one moved. And his perceptions were instrumentally mediated. Thus our earliest voyages through the period of world-exploration were voyages which were undertaken through technologies.

When one turns to the Pacific, one finds that the Polynesians and related peoples had, already a thousand years before the Vikings, explored and populated virtually every inhabitable island chain of a much larger ocean. Western explorers were amazed by the two-hundred-foot long catamaran war canoes which speedily navigated the Pacific, yet they did not pick up the secrets of Polynesian navigation at the time. One must conclude, on the basis of praxis, that both Atlantic and Pacific navigation were successful; but, on examination, both were distinctly different systems.

Polynesian navigation is instrumentless; it operates without fixed points (such as Polaris, which is not visible in the Southern hemisphere, nor did the Polynesians have the technological fixed point of the compass). It is, rather, a complex system of perceptual observations carried on through a secret tradition by a school of navigators.[21] I shall not outline all of the features of this perceptual system, but shall point to enough features to illustrate its subtlety:

(i) One key feature of the perceptual system was a highly developed sense of wave patterns. These waves march with regularity across the Pacific and the Polynesian navigators learned to use them for precise directional purposes. By judging the angle of swells relative to the direction the canoe takes, one can maintain direction. Navigators became so keenly aware of this wave harmonic that even when local storms confused the seas, they could detect the swell pattern under and with the storm (often they would sit in the bottom of the canoe to feel this – their claim was that only men could so navigate because they felt the pattern in their testicles). They also were aware of what we would call refraction waves: swell patterns bend when they approach a land mass such as an island and the change in direction was detected and understood as an indication of a distant island.

(ii) Cloud and light patterns were also learned. Far over the horizon a column of cloud, slightly green-tinted skies, and other more dense moisture indications would be read as the presence of an island. Again the indications were perceptual readings of the phenomena.

(iii) Although bird behavior and patterns were not unknown to European coastal navigators, the precision of observation which knew exactly how far each species strayed from land, the knowledge that a direction towards land could be obtained at dusk by returning birds, and even knowledge of which fish inhabited near-island waters enabled the Polynesian navigators to regard the ocean stretches as a familiar, readable world.

(iv) Star paths were learned and conveyed from generation to generation of navigators. Lacking an immovable pole star, the Polynesians developed a highly temporal, dynamic mode of reading star tracks over the horizon with changes of direction timed to moving locations. Indeed, all constants were in effect dynamic and temporally changing in this system.

Here, then, was a navigational system which historically was at least equally successful in conquering transoceanic distances, a system which had if anything more difficult tasks to perform in that small island systems are harder to locate than continental masses, and a system which was thoroughly perceptual and historical. It was a system whose "map" of the earth was based upon perceptually acute readings of the ocean, without either a mathematics – except for a time sense (but no clocks) – or an instrumentation. It was a variant orientational praxis.

One might very well expect that a variant praxis would be sedimented in a variant understanding of the world. And that certainly is the case. The Polynesian view was – if interpreted by Western standards – "animistic." The ocean was not perceived as either alien or strange, although its dangers and threats were clearly appreciated. It was a deity whose many natures could nevertheless be understood. It was the source of nurture and support and thus a voyage upon its face, while it may pose dangers, was not a voyage into the wild nor something over which humans could expect mastery.

The point I am making here should not be misunderstood. I am not claiming that this lifeworld is better than that of the technologically oriented modern. But it is different. Its perceptually focused praxis, while it may achieve similar goals, implicates a different understanding of world. It is a world which does not become "standing-reserve" because the earth's bounties are conceived of differently.

One might also point out that the Polynesian world is one which is disappearing. Its navigational arts, though still extant among a small number of persons, have been replaced by the now highly micro-determined instrumented navigation of the West. Long voyages by islanders are now undertaken on trading schooners or ships. (Although their ability to sense land before the Westerner remains. Trading schooner captains indicate that they have lapsed into only rough navigation because they know that their passengers will begin to sing when approaching their island, long before the Western captain knows it is near.) My point is that two differently patterned praxes implicate two different ways of understanding the world, and ours is and has been historically Technological for centuries, indeed virtually for at least a millennium.

If Heidegger is right, that the essence of Technology shows itself only recently, it is because we have failed to look at what was under our very noses for a long time. But Technology is like a set of spectacles: those who see through them and who have become accustomed to them, do not notice them. Thus that which is closest and most familiar to us, we have failed to notice. Yet what we have failed to notice turns out to be basic, perhaps the most basic thing about the very way in which we see the world.

V. CONCLUSION

I have suggested that there is a significant sense in which Technology is both historically and ontologically prior to science. This priority, I believe, is one which is not contrary to the more trivial sense in which the human use of technologies is both universal and archaic, common to all cultures whether or not they have developed science.

I have also suggested that the way in which this priority operates is at the level of a basic praxis within a lifeworld, a praxis which inclines or predisposes us towards what becomes a scientific worldview. I have developed only some of its features, features which include a technologically mediated basic perceptual experience. This is an experience which harbors invariant characteristics such as transformed foci regarding ordinary and micro-dimensions of experience, a tendency towards discreteness and the atomization of things, and the enhancement of calculative activities. In this sense Technology at the level of familiar praxis precedes and sets the conditions for science.

Science, in turn, becomes the coming-to-self-consciousness of these activities, a self-consciousness which projects the form of life implicit in the praxis upon the universe, and a self-consciousness which becomes increasingly purified of diverse elements. Such a purification, however, is also a purification of the essence of Technology.

Even the Renaissance, enamored of inventions, with its desire to measure and use the world, created its artifacts in the shape of animal and human life. Da Fontana's measuring robots were conceived of in the form of fish, rabbits, and birds. The predecessor of the steam boiler was the *sufflator,* literally "blower," whose shape was always that of a human head whose mouth blew forth the steam which powered various devices. Only gradually did the *abstraction* needed for Technology in its contemporary sense emerge and thus free technologies to be "scientific" in the sense of being embodiments of a purely technological metaphysics.

The gradual movement to de-animate our technologies, to move towards purer *functionalism,* is both latent within technology and a preparation for a scientific worldview. It is a long step from the symbolism of the clock whose movements represented the heavenly bodies to the bare, instantaneous numbers of the digital, but the movement is one towards a more totally technological and scientific representation.

There is one issue still left hanging in this paper, the issue which separates idealist from materialist interpretations of science and technology. But it may begin to be understood in a different way, too. That issue is whether or not and in what sense *scientific* technology may be distinctly different from traditional technology. My answer is that in one sense it is different, in another not.

The sense in which it is not different, is the sense in which technologies have and continue to have the same existential dimensions with respect to the humans who use them. Technologies may embody and mediate experience so that our lifeworld undergoes changes; technologies may be "other" than we as that to which we relate; and technologies may increasingly be surrounding features of our lifeworld. In each case these appearances of technology may be seen to be continuous with even the most archaic technology.[22]

The sense in which scientific technology differs from traditional technologies depends upon the synergistic interaction of a technology made abstract or purified through the self-conscious connection with science. Thus the break from "natural" materials to the manipulation and creation of materials, the gestalts which separate scientific fields, the extrapolations made possible by revolutions in science, could only happen when the essence of Technology had become manifest. But precisely because it has become so, we can now notice more distinctly and clearly that we are wearing eyeglasses and we can begin to reflect upon the implications of that wearing.

State University of New York at Stony Brook

NOTES

* This article appeared in *Existential Technics*. Suny Press, 1983.

1 Technology, capitalized (italicized in the German) indicates a use similar to what Heidegger calls the *essence of technology*.

2 Martin Heidegger, "The Question Concerning Technology," *Basic Writings*, translated by David Krell (Harper and Row, 1977), p. 296.

3 *Ibid*., pp. 302–303.

4 *Ibid*., p. 304.

5 *Ibid*., p. 296.

6 *Ibid*., p. 296.

7 *Ibid*., p. 303.

8 *Ibid*., p. 288: "The current conception of technology, according to which it is a means and a human activity, can therefore be called the instrumental and anthropological definition of technology."

9 *Ibid*., p. 294.

10 *Ibid*., p. 294.

11 *Ibid*., p. 294.

12 Lynn White, Jr. *Medieval Technology and Social Change* (Oxford University Press, 1962), p. 84.

13 *Ibid*., p. 98.

14 *Ibid*., p. 124.

15 *Ibid*., p. 125.

16 Lynn White, Jr., "Cultural Climates and Technological Advance in the Middle Ages"; reprinted in White's *Medieval Religion and Technology* (Berkeley: University of California Press, 1978), p. 218.

17 *Ibid*., p. 219.

18 *Ibid*., "In a separate building outside Hagia Sophia, Justinian placed a clepsydra and sundials, but clocks were never permitted within or on Eastern churches; to place them there would have contaminated eternity with time. As soon however, as the mechanical clock was invented in the West, it quickly spread not only to the towers of Latin churches but also to their interiors"; p. 249.

19 *Ibid*., p. 250.

20 *Ibid*., p. 228.

21 A popular discussion of these techniques may be found in *National Geographic*, Vol. 146, No. 6, December 1974.

22 I call these "existential relations"; see chapter one of my *Technics and Praxis* (Dordrecht: D. Reidel, 1979).

REINHART MAURER

THE ORIGINS OF MODERN TECHNOLOGY IN MILLENARIANISM

I. WHY THE PHILOSOPHY AND THEOLOGY OF HISTORY ARE NECESSARY FOR THE PHILOSOPHY OF TECHNOLOGY

In the context of a criticism of the traditional philosophy of technology, Simon Moser discusses Donald Brinkmann's proposition that "the consciousness of a need for salvation within Christianity combines with a certain religious longing in man in the technical age that salvation can be achieved by the active shaping of reality and that it does not depend on an act of grace from God."[1] According to Brinkmann, a process of secularization has changed the direction of the energy of belief in Christians. Originally it was concentrated on the transcendental; now it is directed toward the technical shaping of this world. This is his explanation not only for the origin of the technical age, but also for the "enormous stimulus" which still gives present technology its energy.

Moser remarks that Brinkmann's proposition is related to Max Weber's reflections on the connection between the Protestant ethic and capitalism.[2] Still, he has fundamental misgivings about Brinkmann's supposition that an originally religious longing for salvation should have turned into a desire for self-redemption through technology. He feels this is such an essential perversion that it would mean Christianity had lost its dogmatic and ethical character completely. Moser finds it almost impossible to believe that this has happened, although, for instance, Kierkegaard and Nietzsche had diagnosed a similar perversion of Christianity.

Moser is only willing to agree with Brinkmann up to a point; he says, "And even if the process of secularization of the belief in revelation were an historical condition for the origin of modern technology, that by no means proves that the longing for salvation is an intrinsic feature of modern technology." This is certainly correct, if one interprets "intrinsic feature" as the methodological, operational structure of the application of science to the problem of technical control over nature. But the human, subjective, and intersubjective (social) motivating forces which develop and improve this control over nature could very well be influenced by the "longing for salvation." In comparison with the factors which directly play a role in

Paul T. Durbin and Friedrich Rapp (eds.), Philosophy and Technology, 253–265.
Copyright © 1983 *by D. Reidel Publishing Company.*

scientific, experimental, technical, and industrial procedures, these motivating forces admittedly appear "irrational."

This impression, however, could be dependent on the fact that human reason and its objectives disappear behind the powerful systems of technical and socio-technical rationality. (And one does not really know if rationality is only instrumental or whether it pursues substantial, immanently reasonable goals.) In any case, it seems clear which general goals these powerful means serve or should serve: the satisfaction of human needs and in consequence the preservation and development of man. This is apparently a goal which is reasonable, since human self-assertion is beneficial, and at the same time a goal which has nothing to do with salvation in the religious sense.

Really? The apparently completely "rational" goal, the satisfaction of human needs through technological control over nature, could be intermingled with "irrational" hopes of salvation – a mixture of apparent opposites similar to that which Löwith assumes when he writes: "Historical materialism is the history of salvation (*Heilsgeschichte*) in the language of political economics."[3] If this originally religious motivation is important for the development and further growth of technological control over nature – a possibility which even Moser cannot completely deny – then it would be a declaration of bankruptcy on the part of the philosophy of technology if it could not think of a better description of that motivation than the defensive, helpless label "irrational."

In that case, the philosophy of technology would have no answer to the question of why modern science and technology developed in Christian Western Europe, why technical civilization, which has today embraced the whole earth, comes from this area, why "the development of the powers of production happened in Christian, not Muslim or Buddhist countries,"[4] as Ley says. Ley, in this context, is criticizing Moser's criticism of Brinkmann, and he alludes to personal experiences in India and Algeria. He found that the people in those countries are not satisfied with the negation of Christianity in favor of modern, atheistic rationality; rather, they want to know why, for instance, in the countries under the influence of Islam it has not been possible to achieve a comparable success in technology.

Ley the Marxist believes he has the answer to these questions. It is not the religious differences between Europe and the rest of the world that are the important ones but the socio-economic and social class structures. But at the beginning of the modern world, when scientific and technical innovations began, conditions in Europe were by no means so different from those in other higher cultures – neither the socio-economic conditions nor the class structures nor the development of the means of production and other technical aspects of the control over nature. Therefore these conditions cannot be considered a sufficient explanation for the appearance of those innovations. Other feudal societies with largely agricultural production had similar socio-economic and class structures. And, technologically, other cultures were to some extent further developed than the Western European.

Apparently, the secularized longing for salvation which Brinkmann speaks of did play an important role in the development of scientific-technical civilization. At least it represents a factor which was not manifest in other cultures, or at least not as strongly as in Western Europe. But Brinkmann treats the subject far too vaguely to allow a real understanding of what happened. The way he goes at the problem, one could assume that he really means "irrational motivations." If, however, the philosophy of technology would call on the help of the philosophy and theology of history in the question of the sources of modern technology – sources which still influence the development – then it might be possible to understand the specific structure of the goals of Western motivation.

Hübner too demands, in the context of the discussion of the problems of the philosophy of technology, that we explore the future of technical civilization by exploring the deeper sources of the goals that are implicit in the technology of today, sources from the past.[5] Only in the historical context can anything be truly understood – and that means at the same time that we can call a halt to the ingeniousness with which we still consider technology.

This paper is intended as a contribution to the investigation of the sources of technology – a contribution combining philosophy of technology with philosophy of history in which I attempt to clarify terms like "religious need for salvation," "secularization," and "self-redemption," which haunt contemporary discussions of technology.

II. A CURRENT EXAMPLE OF THE USE OF TECHNOLOGY TO ACHIEVE MILLENARIAN GOALS

By way of introduction, the following example from Jürgen Habermas is an interesting instance of the connection between the Judaeo-Christian idea of history and the drive toward technological control over nature. To be sure, this connection is not to be found in orthodox Judaism or Christianity but only in heretical undercurrents. The secularization of a transcendent anticipation of redemption – i.e., the "enlightened" renunciation of the transcendent aspect in favor of an appeal to man to bring about by his own means his own salvation at the end of history – had a long preparation. In this connection Habermas notes: "It is only one step from mystical heresy to the Enlightenment."[6]

Habermas connects these originally Jewish speculations, through Schelling's ideas, with the Christian doctrine of the Incarnation in man of God through Jesus Christ – although Schelling's further ideas, as Habermas remarks, go in a completely different direction. According to the Jewish heresy – which has an obvious similarity with Gnostic speculations[7] – man must bring about his own salvation and no longer trust in God to direct history. Here man himself is considered to be a God who makes his own history, but this God is different in one essential regard from the original religious God of creation: for him "the connection with nature has been disrupted and control over nature has been lost." With this, man has become unfree. His *inner* incentives, which issue from his needs, meet with *external* impediments and he is forced to create external limits on his freedom because the lack of goods that results from his deficient control over nature makes institutions – above all the state – necessary to manage the resulting scarcities.

The goal of man's creation of history[8] is to overcome these scarcities through technological control of nature. Only by this means can mankind – as the other God – arrive at the point where the religious God has always been. In this context Habermas asks: How can mankind "overcome the power of the external things over the internal life other than by confronting the external things externally," by directing its efforts to overcome nature through the creation of an external foundation for life by technical-industrial labor?

If mankind is successful in this effort, then "the theogonic veil would drop from the historical process" (Habermas) or, in other words, the incarnation of God in man (which "is still in its beginnings" – Schelling) would prove to be man's becoming man, or man's becoming God, something he had attained on his own through technology. Here the goal of historical development is a final condition of perfect satisfaction of needs for all, without the exercise of power by man over man. The technological control over nature will have enabled man to eliminate the imbalance "which has existed historically up to now between powerlessness in control over what can be controlled, on the one hand [i.e., over external nature] and power in control over what cannot be controlled, on the other [i.e., over a mankind which is destined to freedom]." At the end of time we would have "a mankind become God . . . united in love with God become man" – that is: mankind would unite with itself in love in the eschatological world community of a millenarian empire. This would be the triumph of the Christian commandment of love brought about by technological control over nature. The two main causes of human misery would have been eradicated: (1) an incompletely controlled external world, that is, the scarcity that prevents satisfaction of material needs; (2) oppression and lack of freedom as a result of the exercise of power by man over his fellows.

III. CHRISTIANITY AND MILLENARIANISM

From the perspective of established Christianity, as it became dominant in the Christian West (above all through the ideas of Augustine) this conception would have to be condemned as millenarian[9] – all the more so since it represents a secularized form of millenarianism. We find the idea of a thousand-year empire preceding the end of history in chapter 20 of John's *Apocalypse.* At that time the true Christians are supposed to live together with Christ and rule over the earth. According to Augustine, it would be too worldly an idea to imagine this thousand-year empire as an ideal condition at the end of history, as a Kingdom of God on earth. He understands the thousand years more as an expression for an indefinitely long historical period which began with the coming of Christ. Christ guarantees salvation and the true church lives according to Christ, but final salvation does not occur as long as history lasts. The Kingdom of God is not of this world[10] and as long as history lasts, *civitas dei* (the community of God) and *civitas terrena* or *civitas diaboli* (the earthly community or community of the devil) are mingled and in conflict with each other, even within the church as a worldly institution. The idea that man himself, not Christ as the Messiah, or an angel sent by him, could found the Kingdom of God, would have been a horror for Augustine. Had he encountered it, he would have condemned such an idea as a product of the human original sin of pride or arrogance (*superbia*).

How is it possible that the religious longing for salvation could have provided a motivation for the technological control of the world in a Christian culture if the dominant theology of history in Christianity related man's history to a transcendent goal? The following are several possibilities:

(1) First of all the universalization and unification of our many histories in a single world history created the ideal framework for a uniform self-understanding of mankind which in turn made collective action possible.

(2) This integration was made possible by the theological orientation of all human history towards a single goal: the Kingdom of God. For the

Christian theology of history – as conceived by Augustine and accepted for centuries within Christianity – the final goal of history was transcendent: the Kingdom of God could only come about *after* the end of world history. Essential, however, to the Christian theology of history was that it was *teleological* – i.e., consciously directed to an end. With this, human history had acquired a goal, at least for those who believed in it – first of all for Europe, which was dominated ideologically for centuries by the Christian church or churches.

(3) The original transcendent goal became historically immanent through secularization. In 1798 Schlegel wrote, in this connection, "The revolutionary wish to actualize the Kingdom of God is the dynamic origin of progressive ideas and the beginning of modern history."[11] On the one hand this revolutionary wish could be taken to mean political revolution; on the other it could be interpreted as a scientific-technical transformation of man's relation to the world. In Hegel's philosophy of history – which summarized the world history of Europe at the time it was finally about to pass over into a world history of the whole earth – the political aspect is in the foreground. But the industrial-economic aspect, which was subsequently dealt with at length by Marx, is also present.[12] As is well known, Hegel was familiar with the political economy of his time (Ricardo, Smith), which reflected the actual historical conditions. But Hegel understood this development at a deeper level, describing the conflict-laden emancipation of a worldly self-awareness from the Catholic Church, which until then had claimed and, for the most part, actually possessed an ideological monopoly. With regard to the early modern era, the time of the great commercial cities, of the discoveries and inventions, he writes: " . . . For this reason the Church recedes behind the World-Spirit [i.e., the secular consciousness], because the latter has come to understand the sensual as sensual, the external as external, to act in the finite world in a finite manner, and to be independent in this action as a valid, justifiable subjectivity." In this connection one also finds in Hegel: "The technical makes its appearance when there is a need for it."[13] But where, precisely, in Christian Europe does the need for secular action in a finite world – which emerged generally at that time, ready for cooperation – come from?

(4) That suddenly at that time a process of secularization set in which turned human efforts from otherworldly goals would be a poor explanation. It would have to be based on irrational forces, if it could not at the same time explain why it was precisely Christian civilization that adopted a technical worldly view. *One would have to indicate what tendencies in that direction*

could be found in Christianity from the beginning. And this can be done; see the remaining points, 5 – 8.

(5) First, one can refer to the Judaeo-Christian concept of an otherworldly God to whom man bears a special relationship despite his being a part of the world. This anthropocentric relationship estranges man from the rest of the world and makes it the domain he is to rule over. (*Genesis*, I, 28: "Fill the earth and subdue it.") Nature is no longer a realm filled with demons and gods, in part friendly, in part hostile, that man can ignore only at his peril. The connection between Judaeo-Christian monotheism and modern technology is described as follows by Kamlah: "It is in Christianity and not in the ancient world that God is seen as thoroughly 'beyond' the world and as its 'creator.' Against paganism Christianity proclaims the radical loss of power of all 'powers,' of all gods and demons. . . . This despiritualization of the world was successful in the history of the West and made possible modern science and technology. . . ."[14]

(6) Second, this special relationship to an otherworldly God, which removes man from everything else around him, can result in indifference towards the world – as indeed it did up to the Middle Ages. But this was not the only essential characteristic of Christianity: the theoretical indifference with regard to the world was joined with the commandment of love for one's fellow men, that is, with the *practical* devotion to one's fellows, who are also capable of having that special relationship to an otherworldly God. For this reason the church went so far as to burn people's bodies in order to ensure the salvation of their souls. But the separation of body from soul in Christianity was not as clear as this would imply. Augustine, for example, emphasizes the significance of the body against the Platonists.[15] The commandment to love one's fellow men, in the New Testament, plainly includes attention to their physical needs. It would seem, then, that a society such as bourgeois society (which Hegel called a "system of needs"[16]) in which, as much as possible, *all* members would provide for the needs of all by labor, would satisfy the practical demands of Christianity. This could even include democratic decision-making by all members of society. If man's labors are more successful by the application of science and technology, then this greater success is to be understood in terms of Christian morality. This could even be the case when society, as a system exercising technological control over nature, leads to a reduction of personal responsibility and warmth – which seem to be made superfluous by the state's welfare measures.

(7) Third, according to Christian dogma the otherworldly God, in Christ, has become man and with this a part of history. This intervention is supposed

to have concrete consequences for history, even if mankind's final salvation from the misery that dominates history will transcend history. It was only a short step to Schelling's idea (quoted above from Habermas) of the continuing incarnation of God in history, which could at the same time be brought about by man's following the Christian commandment to love one's fellows. A similar thought can be found in Lessing's idea of a gradual education of mankind by the continuing influence of the "revelation" and the once actual, historical presence of God.[17] Even the early Augustine and, more clearly, several other early Christian thinkers tend in this direction.[18] The value gained by history as a result of the incarnation of God in Jesus Christ – compared with the ancients' understanding of history – could at least be seen as a gradual approach to the Kingdom of God, even if this goal of history remains essentially transcendent. In Christianity the relation between immanence and transcendence is essentially dialectical, a tension, where a stronger emphasis of the one or the other side, and the search for mediation between them, is possible.

(8) Fourth, it is true that the medieval church, despite its politicization, adopted the Augustinian, transcendental-eschatological (i.e., *non*-political, *non*-social) interpretation of the Bible's thousand-year dominion and the Kingdom of God. But the *immanent* understanding found support among various groups which, during the post-Augustinian era, went against the official church. A millenarianism can be discovered here which proclaims a worldly goal of history: an ideal condition at the end of history which was to encompass the elect, or those who deserved it, or even everyone, once the enemies of salvation had been annihilated or re-educated. This millenarianism had a political, or at least an ideological, quality: it was possible to act to bring about the ideal final condition of freedom, equality, justice, and fraternity by overcoming material need; it was possible to follow inspired leaders, or to battle against the enemies of a worldly salvation. Occasionally there were spectacular outbursts of mass millenarian movements. Authors such as Cohn associate them with modern totalitarian movements.[19] Voegelin does the same, although he finds that it was not Judaeo-Christian millenarianism but Gnostic undercurrents of Western culture that gave rise to such movements.

IV. MILLENARIANISM AND SCIENTIFIC-TECHNICAL-INDUSTRIAL-ECONOMIC CONTROL OF NATURE

It can, therefore, be said that the millenarian diversion of man's relationship to an otherworldly God – which is a problematic possibility in Christianity – gave an initial integrative impulse to efforts aimed at collective human self-salvation. Millenarianism is understood here in an extended sense to mean the orientation towards a historically immanent final condition. This condition is often called "utopia," where "utopia" designates a state which, though it cannot be found anywhere, is attainable in the future with sufficient effort.

At first the millenarian goal inspired *political* action, where one then calls the path to this goal "revolution" or "emancipation." This does not designate where it leads to, but where it is meant to lead away from is clear: from conditions of unfreedom, of oppression, of injustice. Marx argued most clearly that this political emancipation can, indeed must, remain formal and partial. There must be a further emancipation embracing man as a totality – that is, for Marx, an emancipation which is social; in other words, one that is fundamentally politico-economic.

What is meant by "self-salvation," what mankind wants to save itself from, is quite clear: material need and political oppression. To what extent human suffering would be overcome here is another question. Miscalculations are possible here and have actually occurred. In any case, not only Marxism but Western pragmatism sees in the practical overcoming of material need through the scientific-technical-industrial control of nature and the politically just distribution of the goods produced the means of fulfilling (and thereby making superfluous, wholly or partially) religious hopes for salvation. The actual point of departure for an essential improvement of mankind's condition seems to be progress in the control over nature. Without this progress political changes are superficial. It is only through technology and social engineering that a concrete path to the millenarian goal can be found; moral and political ideals have shown themselves historically to be unsuccessful.

But why should it be necessary to speak of "millenarianism" at this point? Why is the concept of a Kingdom of God on earth necessary, if it is only a question of satisfying vital human needs that belong to man by nature and which push naturally for satisfaction? The answer is that, although the needs have always demanded satisfaction, it is only in Europe and in Western civilization (at least originally) that they have found the means of mass satisfaction. Precisely here lies the problem: *Why did this happen in Europe*?

That a particular culture was the origin only becomes clear when one realizes that human needs are not simply given by nature, but are always culturally influenced. They always demand satisfaction but seek it in a particular, collectively effective fashion only at a time when their diffuse pressure is drawn together by the image of a goal. *In Europe it is precisely millenarianism which provided this general orientation.* Thus it created out of the comparatively amorphous nature of human need *that* need for whose satisfaction modern technology "made its appearance" (to use Hegel's phrase; see above).

This was possible because its immanent, future-oriented teleology of history activated and removed the limits from another specific impulse of Western culture: the techno-rationality which has come down to us in the philosophy and other sciences of ancient Greece. I am not claiming that the ancient Greek *technē* – as it was interpreted philosophically above all by Plato and Aristotle – is basically the same as modern technology. This impulse of Western culture which was inherited from ancient Greece was transformed radically under the conditions of the Christian desacralization of nature (see above). Only by its separation from the Greek teleology of nature and its connection with the millenarian teleology of history has it become an instrument of a control over nature which at first seemed unlimited but which has now begun to meet up with *ecological* limitations.

With this we have the explanation for – or, more precisely, we become able to understand in terms of a historico-social goal dynamics – why it was in Europe that the human, subjective, and intersubjective forces were able to develop which led to a technical civilization that now spans the world. The motive force lay in the Christian theology of history. The possibility of its being activated and made immanent by millenarianism was in embryo in Christianity, and its emergence was only a question of time. Why it was precisely at the beginning of the "modern" era (which in part received its name for this reason) that this occurred still remains to be investigated.

At least the following seems to be clear: a further religious impetus played a role. One would have to explain in terms of instinctual economy why it was that fundamental human energies were no longer released in the longing for another world, that they turned away from otherworldly distant goals and towards worldly distant goals. One would have to explain the willingness to renounce an imperfect, but present satisfaction of needs for the sake of a future perfect satisfaction (a willingness which at present seems to be in decline, at least in certain subcultures). The development of the modern scientific-technical possibilities for the control of nature, and for their techno-economic realization, required a uniquely extensive collective intellectualization as well as a work discipline which required the renunciation of the immediate satisfaction of needs. The pathway to this was also prepared by Christianity: the Christian turning away from an incomplete, permanently imperiled earthly happiness for the sake of a perfect otherworldly happiness had demanded a collective asceticism which only required a bridging principle for it to be turned to use for worldly purposes.

Nietzsche had already recognized that there was a connection between Christian asceticism and the modern scientific mentality, which seems to be directed completely towards this world. In 1887 he said that the manner in which science had developed up to that time "was not the antithesis of that ascetic ideal but, much more, its most recent and eminent form."[20] By "science" Nietzsche meant here that orientation towards the technical control of nature (including man) which Max Weber shortly thereafter called *Zweckrationalität* (instrumental rationality). It was Weber who then discovered the bridging principle in certain Protestant (Calvinist and Puritan, not Lutheran) attitudes concerning commercial exploitation of the technological control of nature. With the interpretation of success in business as a sign of Christian divine election, Puritan asceticism (the denial of instinctual satisfaction to that end) acquired a high value. Christian asceticism, which was originally directed towards a transcendent goal, became "worldly." Once this secularizing transformation had been completed, the system of industrial-economic-instrumental rationality acquired its own momentum, compelling that worldly asceticism and work discipline from all[21] and developing further – even if that asceticism is in part no longer practiced and the (partially parasitic) enjoyment of the fruits of the technological control of nature has become more important.

V. CRITICISM OF MILLENARIANISM

This "mechanism that determines us all with an overwhelming force" will run on, perhaps, "until the last ton of fossil fuel has been burned" (to quote Weber[22]). By making use of the especially dangerous potential of atomic energy for the control of nature it could run on even further. If on the other hand it is to be contained or diverted, we have first of all to become aware of the basic powers that drive it – as we have tried to do by answering the question, "Why did this happen in Europe?" If the thesis that a continuing millenarianism lies at the core of this movement is true, then it is here, at the root, that criticism and the attempts to change direction have to begin – that is, if it is not only symptoms that are to be treated as is the case in current popular "ecological" movements.

One has to ask what is questionable in a millenarian belief in having finally found, in the scientific-technical-industrial control over nature, the means of realizing the age-old dream of mankind of attaining an ideal state at the end of history. To a certain degree this seems quite possible; technology can achieve many things. But not everything. Precisely therein lies the danger of millenarianism: it rests on extravagant hopes which are only meaningful in the context of transcendent belief in a religious God, of hopes for a total salvation which technology cannot fulfill. This hope *could* be meaningful within society and history, but only in the limited sense of following the principle that one must demand the impossible in order to achieve the possible. As the extravagant hopes of a technical civilization, they run the risk of destroying the good life that is possible at present by striving for the impossible – or (and this amounts to the same thing) by striving for ever new future possibilities. Everything which exists at present – not only external nature – is deemed disposable, is considered an instrument towards the excessive goal of a perfect satisfaction of man's needs, a secularized Kingdom of God on earth.

In this, technical civilization might end up subject to the fate that Allah visited upon King Shaddad in *The Thousand and One Nights*: Shaddad, we may recall, "loved to read old books, and, when he once found a description of Heaven in which Paradise was described with all its castles and balconies, trees and fruits and other wonderful things, he wanted to build himself something similar in this world." He ordered the Paradise city of Iram to be built, expecting to move in with his followers when it was finished. But this was not to be: "Happy that he had achieved his goal, he travelled toward the City of Pillars, Iram, until he was only a day's journey away. But then Allah smote him and all the unbelieving heretics who were with him with a God-given punishment from his almighty heaven and destroyed them with great uproar."

We have not yet reached this point, although we have reached a point at which the Persian priest/politician Khomeini can declare that Americans and Europeans (he forgot the Russians!) have been inspired by the Devil with the delusion that "progress and growth contribute to the happiness of mankind. The Devil seduces the Westerners with the thought that Paradise can be achieved in this world by progress and economic growth. The Devil intends to destroy mankind and the West has already become his servant."[23] And since Khomeini is one of the lords of that natural product oil which technological civilization – whether Western liberal or Marxist – still desperately needs, it behooves that civilization to take advantage of that verdict passed on it by another culture to reconsider the principles on which it was founded.

Free University, West Berlin

NOTES

[1] Simon Moser discusses Donald Brinkmann in Lenk and Moser (eds.), *Techne, Technik, Technologie* (Munich: Pullach, 1973), pp. 17ff.

[2] M. Weber, *Die protestantische Ethik und der Geist des Kapitalismus* (Hamburg: Winckelmann, 1975), vol. I, pp. 27ff.

[3] K. Löwith, *Weltgeschichte und Heilsgeschehen* (Stuttgart, 1953), p. 48.

[4] H. Ley, "Zu einigen Erscheinungen bürgerlicher Technikphilosophie," in *Wissenschaftliche Konferenz mit internationaler Beteiligung PHILHIST–78* (Dresden: Technical University, 1979), vol. I, pp. 242ff. and 248–249.

[5] K. Hübner, "Einführung in die Diskussion philosophischer Aspekte der Technik," in Zimmerli (ed.), *Technik oder: wissen wir was wir tun*? (*Philosophie aktuell* 5; Basel and Stuttgart, 1976), pp. 11 ff. and 21–22.

[6] J. Habermas, "Dialektischer Idealismus im Übergang zum Materialismus," in his *Theorie und Praxis* (3d ed., 1974), pp. 199–200. Cf. F. Rapp, "Technik als Mythos," in Poser (ed.), *Philosophie und Mythos* (Berlin, 1979), pp. 110ff.; W.E. Mühlmann, "Heilsverlangen und Unheilsmächte in der Welt von heute," in Hommes (ed.), *Was ist Glück*? (Munich, 1976), pp. 205ff.; and F. Rapp, *Analytische Technikphilosophie* (Freiburg and Munich, 1978), pp. 108ff.

[7] Cf. E. Voegelin, *Die neue Wissenschaft der Politik* (3d ed.; Munich, 1977); also, several of his essays.

[8] For the origin of the idea of "making history," see R. Kosellek, "Über die Verfügbarkeit der Geschichte," in his *Vergangene Zukunft* (Frankfurt, 1979), pp. 260ff.

[9] Augustine, *De civitate Dei* XX, 7 and 9.

[10] *John* 18, 36.

[11] F. Schlegel, *Athenäumsfragment* no. 222.

[12] G. W. F. Hegel, *Vorlesungen über die Philosophie der Weltgeschichte*, Lasson (ed.). (2d ed.; Hamburg, 1976), pp. 854–855.

[13] *Ibid*., p. 871.

[14] W. Kamlah, *Der Mensch in der Profanität* (Stuttgart, 1949), and *Die Wurzeln der neuzeitlichen Wissenschaft und Profanität* (Wuppertal, 1948).

[15] Augustine, *De civitate Dei* XXII, 26–27.

[16] G. W. F. Hegel, *Grundlinien der Philosophie des Rechts*, sections 189ff.

[17] G. E. Lessing, *Die Erziehung des Menschengeschlechts* (1780); see *Gesammelte Werke*, Rilla (ed.). (East Berlin, 1956), vol. 8, pp. 590ff.

[18] Cf. W. Kamlah, *Christentum und Geschichtlichkeit* (2d ed.; Stuttgart and Cologne, 1951), pp. 112–113 and 311ff.

[19] N. Cohn, *Das Ringen um das tausendjährige Reich* (Bern, 1961).

[20] F. Nietzsche, *Zur Genealogie der Moral*, 3 23.

[21] "The Puritan wanted to have a vocation; we must have one" – M. Weber, *op. cit*., p. 188.

[22] *Ibid.*

[23] Quoted in the Berlin newspaper, *Tagespiegel,* 25 January, 1981.

CARL MITCHAM

THE RELIGIOUS AND POLITICAL ORIGINS OF MODERN TECHNOLOGY

INTRODUCTION

The present paper is part of a larger work in progress, "The Philosophical Origins of Modern Technology." Although the economic, scientific, and social origins of technology are commonly discussed at length, the properly philosophical origins are not nearly so well considered.

Enlightenment philosophers, following the lead of Machiavelli, Bacon, and Descartes, sharply distinguished modern science-technology from premodern theory and practice. Beginning with Nietzsche, however, critics of modernity began to argue that modern theory and practice might well have decidedly premodern roots. For Nietzsche himself these roots were twofold: metaphysical (from Socrates) and religious (from Christianity).

Extending but not necessarily agreeing with Nietzsche's analysis, one can identify three basic theories about the philosophical origins of modern technology. These focus on (1) metaphysics, (2) religion, or (3) politics. Martin Heidegger, for instance, who does agree with Nietzsche, argues that modern technology is an outgrowth of Western metaphysics; while Hans Jonas maintains a modified thesis that technology is founded on modern scientific metaphysics. Max Weber, and after him a host of historico-philosophical commentators, including the American historian Lynn White, Jr., have argued that some form of Christian theology has made a decisive contribution to the rise of modern technology. A much less well known theory can be found suggested in the work of Leo Strauss: that it is modern political philosophy, especially as this emerged with Machiavelli, that has exercised the definitive influence on the development of technology. George Grant, a Canadian philosopher, has restated this theory, linking it with Weber's thesis. Another related theory is that of Eric Voegelin, that the modern political realm has become dominated by the drive to conquer the world – using means ranging from science-technology to totalitarian violence – because it has been distorted by various ideologies or *Ersatzreligionen*.

The present paper limits itself to a consideration of the last two theories – as found in Weber, White, and Strauss – and argues for the primacy of political over religious origins for modern technology. At the same time, it

Paul T. Durbin and Friedrich Rapp (eds.), Philosophy and Technology, 267–273.
Copyright © 1983 *by D. Reidel Publishing Company.*

seeks to acknowledge the way in which Christianity has contributed to this political foundation by subtly altering the character of the political realm as such.

RELIGIOUS ORIGINS I

The *locus classicus* for the theory of the religious origins of modern technology is Weber's *Die protestantische Ethik und der Geist des Kapitalismus* (1904–1905). Weber's argument may be restated as follows. Traditional productive or economic activities (as Aristotle pointed out) are not properly conceived as ends in themselves. They terminate in some external or extrinsic goal. They are thus normally pursued for some definite purpose and tend to cease when that purpose is achieved. The traditional laborer who works in order to make a certain amount of money each day will not increase his productivity if offered a higher hourly wage; he might even work less, because less work now produces the same income. The traditional businessman will act in a roughly parallel manner, as he is encouraged to do by traditional Catholic moral teaching regarding the vanity of riches and the superiority of spiritual goods.

However, if productive activity were to take on a purpose internal to itself – as it did for those Christian monks who saw *labora* not merely as a necessary response to human need, but also as a spiritual *ascēsis* for the mortification of human pride – then making could become, as it were, a kind of doing. Doing as distinct from making does not "pass into external matter" but "abides in the agent" (Thomas Aquinas); it is thus not ordered toward any end other than the skilled performance of the activity itself. Considered as a spiritual *ascēsis*, making no longer terminates in some objective end; it no longer exhibits built-in limits. On the basis of such a conception it becomes possible, under certain circumstances, for making to become a religious "calling" – which is precisely what happened, according to Weber, with a number of Christians influenced by the theologies of Luther and Calvin.

POLITICAL ORIGINS I

Weber's argument has been subjected to numerous criticisms of both a historical and philosophical sort. The most philosophical critique has been given by Leo Strauss. According to Strauss, Weber, because of an implicit Kantianism in his ethical understanding, is prohibited from approaching the political realm on its own terms and, therefore, from recognizing its most decisive influences. Like Kant, Weber views the moral worth of an action as inherent in the nature of that action and not dependent on its consequences. With Kant, Weber conceives the rightness of an action as prior to and determinative of the goodness of its consequences, instead of vice versa. Weber is a deontologist rather than a teleologist in ethics. Thus in searching for the moral foundation of the capitalist spirit Weber inevitably insists that this foundation must be of the deontological sort; that is, that modern technological activity be justified in itself – which appears to be done in certain Protestant transformations of the concept of a divine "calling."

Weber's Kantianism obscures the much more likely possibility that modern economic and technological activity could be justified on a consequentialist basis as conducive to the common good. Surely this is today the most widely held argument for the perpetuation of technological practice. According to Strauss, the problem of the origin of modern technology is thus the problem of the emergence of the minor premise in the following practical syllogism: Men should work to contribute to the common good.

> The unlimited production of wealth (= modern technology) is conducive to the common good.
>
> ---
>
> Therefore, we should have modern technological activity.

The major premise is the ordinary presupposition of both ancient and modern political thought. The difference between ancients and moderns is the rejection by the ancients of the minor premise. In popular studies of the history of technology this is readily ascribed to the aristocratic prejudice of a society based on slavery. But in fact the premodern distrust of wealth can be found defended in well-articulated religious and political arguments. The Hebrew prophets, not to mention Jesus of Nazareth – both of whom reflect the views of a non-slave-holding society (or, perhaps, of the lower classes of such a society) – are strongly critical of riches and their temptations. The works of Xenophon, Plato, and Aristotle likewise contain able vindications of the virtues of frugality.

For Strauss the crucial break between ancient and modern judgments on the goodness of unlimited wealth can be traced to Machiavelli's "lowering of the ultimate goal" of political life while simultaneously increasing the possibility of its worldly realization. This paradoxical lowering and raising of the standards of human action can be restated in terms of the technological pursuit of material affluence – or the "conquest of nature for the relief of man's estate" (Bacon) – although Machiavelli does not himself use these terms. Classical political philosophy in its love of virtue and its distrust of material goods is impotent or unable to realize its ideals in any systematic manner because it sets its standards too high. Machiavelli seeks to take his bearings not from how men ought to act but from how in fact they do act; he limits his horizons in order to get results.

However, none of Machiavelli's explicit arguments for the lowering of the standards were unknown to the ancients. For Strauss, Machiavelli simply represents in exceptionally vivid form one of the two major fundamental alternatives in political philosophy which are coeval with man. The question thus arises as to what enabled this alternative, at this particular time in history, to supplant its rival and become the prevailing view. Strauss's answer is that Machiavelli's success rests more on his rhetoric than his arguments – on a subtle way of writing which disarms his critics while encouraging his followers. But Strauss also suggests that even this rhetoric would not have enabled Machiavelli to have such a thoroughgoing impact on the political realm had it not been for the historical contingencies associated with the rise of Puritanism. Puritanism, having rejected Aristotelian philosophy for religious reasons, became the chance "carrier" of ideas of a wholly non-religious sort.

RELIGIOUS ORIGINS II

Both Weber and Strauss assume that modern technology originates in that transition from medieval to modern times which has been called the Renaissance (emphasizing one aspect) or the Reformation (emphasizing another aspect). The medieval historian Lynn White, Jr., has fundamentally shaken this assumption, and thereby the particular arguments based on it, by maintaining that modern technological practice arose in the Middle Ages under the influence of Latin Christianity.

White's revision of the history of Western technology begins with a description of technical changes in warfare, agriculture, and mechanical power, together with their corresponding social transformations, during the medieval period. It continues with an argument concerning a fundamental shift in moral attitude toward technology which White finds clearly present by the ninth century but traces back to as early as the fourth. White concludes with speculations about the religious inspirations behind this shift. The last two stages of his argument are of primary philosophical interest.

With regard to the shift in moral appreciation of technology, White relies on an interpretation of pious customs, homiletic images, and manuscript illuminations. He considers, for example, the iconographic representation of *temperantia.* It was, asserts White, originally one of the less central virtues, and in the early Middle Ages is often found symbolized by a woman diluting wine with water. Yet by the fifteenth century, *temperantia* can be found occupying a predominant position in iconographic representations of the virtues, and its symbol has become a mechanical clock. Given the associations of the clock with a rationally ordered, technologically oriented life, plus the importance of temperance or moderation (gratification postponement) in facilitating capital accumulation in technological development, White argues that this iconographic transformation reveals "that in Europe below the level of verbal expression, machinery, mechanical power and salutary devices were taking on an aura of 'virtuousness' such as they have never enjoyed in any culture save the Western."

As for speculations about the religious foundations of this transformation, White stresses the Semitic approbation of work, which he finds picked up and emphasized in the Benedictine monastic tradition of *ora et labora.* Coordinate with this, White (drawing on the Marburg theologian, Ernest Benz) cites three basic elements of the Christian revelation that seem to encourage technology. One (from *Genesis*) is the idea of the world as a created artifact, of man as created in the image of God and called to participate with God in exercising dominion over this world. The second element (from *Exodus*) is that history is not cyclical but leads to a definite end, toward which man is called to contribute by means of worldly activities. The third (from the Gospels) is the doctrine of the Incarnation and Resurrection; matter is not evil but created for a spiritual purpose and destined for regeneration. To these factors White himself adds the de-animization of nature and Christian moral concern for the suffering of the world.

POLITICAL ORIGINS II

White's argument is open to the following factual criticisms. First, he is simply mistaken to think that temperance was not the central virtue of pre-Christian moral thought. For Xenophon and Aristotle, *egkrateia* (self-control) and *sōphrosynē* (moderation) – the two words most closely related to *temperantia* – are clearly the foundation of all virtue. In Plato's *Republic, sōphrosynē* is one of two virtues distributed among all classes, and thus the foundation of the state. Second, classical authors did exhibit a high regard for work, especially as it could contribute to the cultivation of virtue. See, e.g., Xenophon, *Memorabilia* II, vii-viii, where Socrates argues in effect that "troubles spring from idleness" (compare Benjamin Franklin). Third, the Benedictine tradition is not properly characterized by the phrase *ora et labora*, which was not coined until the nineteenth century in conjunction with the romantic revival of monasticism. In the *Regula* of St. Benedict, the closest conjunction is between *ora et lectio divina.* Fourth, the idea of human dominion over animals, which is all that *Genesis* 1:28 grants, is not unique to the Bible. Compare, e.g., Xenophon, *Memorabilia* IV, iii, 10. Furthermore, the biblical injunction is clearly qualified by being given before the Fall.

The more serious difficulty with White's argument, however, derives from its failure to appreciate the difference between knowledge and opinion, the thought of the few versus that of the many. Like Weber, White can cite no reputable theologian who uses Christian doctrine to rationalize the virtuousness of technology. Hence he resorts to the argument that this takes place "below the level of verbal expression." Is Christianity itself properly held responsible for such a rationalization, especially when the same did not take place in the Eastern Orthodox tradition? Surely the ambiguous stance of other segments of the Judaeo-Christian-Islamic tradition would also call into question any such judgment.

Nevertheless, White does offer genuine historical insights, and in light of them one may venture the following hypothesis. Christianity has contributed more to the origins of modern technology than Strauss at least explicitly acknowledges. But the Christian religion has exercised this influence most decisively by means of a complex and sometimes indirect transformation of the political realm, rather than by a transformation of the economic (Weber) or the social (White).

One may begin to gauge the character of this transformation by recalling Socrates's understanding of the paradoxical dependency and sufficiency of the political. According to Xenophon, Socrates makes two complementary distinctions between divine and human affairs; that is, between the religious and the political. The first distinction is between *whether* or not to undertake some practical action, and *how* to proceed once it is undertaken. The former is a question which depends on the gods, the latter is one in which man is sufficient unto himself. The second distinction is between questions concerning the origin and nature of the cosmos and ethical questions concerning what is good and bad, right and wrong. Here again, the former are properly left to the gods, the latter to be determined by human reasoning – although it is important to note that ethical questions can never be completely resolved. Indeed, their indeterminate character imparts to the political realm a strong measure of caution or moderation.

Against such a background one can observe that the Christian revelation will tend to alter one of the two relationships suggested by these distinctions. If revelation is interpreted (as it has been in the Latin West) to be a definitive disclosure about *whether* or not to undertake certain actions, then it will affect the first relationship – and the provisional nature of all ethical reasonings (*how* to proceed). Whereas if revelation is interpreted (as it has been in the Greek East) to be a definitive disclosure about the nature of the cosmos, then it will affect the second relationship (what is right or wrong as subject to human reasoning). Independent of any particular moral content of the first interpretation, this will necessarily open the political realm to technology by shifting the focus of attention in a decisive way from dependency on the gods to the human search for means or technique. That such a shift should be able to be exploited by men such as Machiavelli, who place worldly success above other-worldly aspirations, should not be surprising.

Polytechnic Institute of New York

REFERENCES*

Leo Strauss. 1953. *Natural Right and History*. Chicago: University of Chicago Press.

Max Weber. 1930. *The Protestant Ethic and the Spirit of Capitalism*. New York: Scribner's. German original, 1904–1905.

Lynn White, Jr. 1978. *Medieval Religion and Technology: Collected Essays*. Berkeley: University of California Press.

NOTE

* Only the main works discussed are cited here; for more complete references, see my "Philosophy of Technology," in P. Durbin (ed.), *A Guide to the Culture of Science, Technology, and Medicine* (New York: Free Press, 1980), pp. 282–363.

WOLFGANG SCHIRMACHER

FROM THE PHENOMENON TO THE EVENT OF TECHNOLOGY

(A Dialectical Approach to Heidegger's Phenomenology)

The philosophy of technology must deal with a phenomenon which will likely decide the survival of the human race. An erroneous judgment on the nature of technology could have fatal consequences. What we know about technology appears to be insufficient. For if we really understood what technology is, we would then be in a position to effectively arrest the rapid destruction of our world through modern technology. A number of technologies, without which we could not live, shape our daily lives. Yet despite this seeming familiarity the fundamental concept common to them all persistently escapes philosophical definition. The traditional assumption is that technologies are means to certain ends, as could be seen in the craftsman's technology. If so, this would mean the present destruction through technology is due to a false use of instruments, to an inadequate determination of technology's objectives. But this plausible explanation is refuted by the fact that neither *new* goals which are set nor *other* applications of instruments have made a change in the course of advancing destruction. The scientific-technical definition of technology as applied natural science is well-founded, and its rational employment for a purpose in the life-world appears to be unproblematic.[1] But such explanations become questionable when they only theoretically prove the life-sustaining function of a phenomenon, whereas, according to all experience, it brings about our premature death. How can a technology be "true" whose successes poison our food, pollute our water, despoil and disfigure our land through poor construction planning, devastate our forests, mercilessly exterminate our fellow creatures and derange our own psyches? By now it has become clear that it is not the individual, isolated error of technology which will kill us; rather, the human race is condemned by our entire way of living, a mode ever more determined by technology.

But is this technology the real technology? Have we not perhaps failed to recognize the true meaning of the phenomenon of technology? In that case we must earnestly begin our inquiries into the essence of technology anew and test all answers hitherto given. In his radical critique of Western philosophy, of metaphysics, Martin Heidegger also dealt with technology as one mode of our customary thinking and acting, and brought out its crucial significance for the present. Heidegger's hermeneutical phenomenology allows

Paul T. Durbin and Friedrich Rapp (eds.), Philosophy and Technology, 275–289.
Copyright © 1983 *by D. Reidel Publishing Company.*

the phenomenon of technology to be comprehended for the first time as the "event of man."[2] This means that man becomes what he is, becomes man in and through technology and only then is he capable of satisfying the universal order in his actions. All prior knowledge of technology is thus radically overthrown. Heidegger's phenomenology demands an entirely altered attitude towards life, it calls for a leap into non-metaphysical thinking. For the time being only very few individuals will be capable of such. But Heidegger's leap will not come without having been prepared; its jump-off base is contained in the theory and praxis which have prevailed up to now. Metaphysics, of which instrumental technology is to be considered a part, can only be overcome when it is understood more truly and intrinsically. Our knowledge of and our dealings with technology both contain the *adequate* concept of technology, as improbable as this may seem in the face of the world's destruction. The present death-bringing technology is, though in a totally distorted way, the selfsame life-sustaining technology of man conceived of originally. Heidegger's "event of technology" expresses therefore not a new phenomenon, but rather the hidden meaning of present-day technology.

In order to be capable of questioning anew we must learn to look at technology differently and not to merely accept it as a fact already recognized. If we can recapture the "foreign" way of seeing – i.e., looking at an assumedly familiar phenomenon as if for the first time, without preconceptions, as we did when it was foreign to us – a number of things become obvious. We see how manifold and divergent the use of technology is in science, industry, and society; how greatly our understanding of it must be differentiated from the actual use; and we see that our theories about technology can again be something quite different. The enumeration and classificatory analysis of this pluralistic occupation with technology alone fills an entire book[3] and yet makes no change in the doubtfulness of such pluralism. This diversity is contradictory, it blocks a common understanding and hinders a meaningful interaction with technology. It creates *de facto* a situation in which each person unthinkingly does what he wants with technology. To counteract this, an attempt should be made to comprehend the phenomenon of technology in a unified sense, as a unity beyond all its contradictions, to be achieved with the help of a dialectical method learned from Hegel. Hegel's dialectics[4] can be defined as a technology of reorientation, of contrary thinking through permanent repudiation of sense-certainty and common sense. Precisely this makes a dialectical procedure possible, and instead of rejecting contradictions, allows for their inclusion and ultimate synthesis as being fruitful for the investigation. The justification of a phenomenon, if only a right to a power of evocation, is still guaranteed, even when it is refuted. The usual black and white picture painted of the world is thereby excluded, just as is the tendency of a lazy mind to prefer an easily enjoyed plausibility to those rarer fruits attainable only through strenuous mental exertion. The labors of the theorists of technology as well as their opposing views are taken so seriously that it is essentially impossible to take sides. Such theories are to be constantly compared with experience, with the practical use of technology. We must not be allowed to avoid looking back at the history of technology and of its concept. It must be noted that those born of a later technological generation are not principally right, nor are present technical definitions superior to those of the past merely because they claim to be more comprehensive and enlightened. The division between theory and praxis with respect to the phenomenon of technology is not to be simply established so that we may then proceed as before; rather, we must draw systematic conclusions from it. Even if the truth about technology is to be found only in bits and pieces and is often barely recognizable in a greatly diverging theory and praxis, this should not give cause for resignation. Each aspect of the technological problem has its own truth which must be "preserved" in all its contradiction in a dialectical critique of the totality of the phenomenon. Dialectics has always been guided by such a wholeness; it corresponds to the wholeness of our spontaneous comprehension of the world. Like this comprehension, neither is dialectics the sum of its individual parts; it is rather a mode of existence, a being-in-the-world as precondition of theory and praxis. Heidegger's presentation of the event of technology disclosed in advance the phenomenon of technology. But perhaps technology can, for our commonplace understanding, only be grasped by following the dialectical path through technological knowledge taken here.

I. INHUMAN TECHNOLOGY: MACHINES

A technological knowledge which, in its analysis of actual technology, never questions the instrumental character of technology and which considers it a means of remodeling the external world possesses the advantage of being largely free from ideological consideration towards man. This explains the fact that the philosophy of technology, which adheres closely to the technical understanding of the engineering sciences, exhibits in this instance exactly the opposite of that which it claims in its basic tenets. In the individual case mechanical technology appears as the phenomenon of experience in its precise immediacy. Practical technology (*Realtechnik*) has fundamentally changed our daily life and has become ever more dissociated from man as we know him in his basic nature. A simple imitation of nature no longer occurs; instead, functional operations of technology replace primal nature. An immediate relation to the products manufactured, as was once possible in the craftsman's technology, no longer exists. According to Rapp's comprehensive analysis, mechanical technology is based on anonymous processes, inorganic homogeneity, and functional reproducibility.[5] Its procedure is indirect[6] and its progress lies precisely in its transferral of work once performed by man to automated machines. Technological feasibility, not human desirability, becomes the ontological standard of measure. Man is considered a system disturbance which must be eliminated as far as possible.

To be sure, it is emphasized that all this takes place solely for the benefit of man himself; but such a wholesale assurance makes no change in the demonstrable "inhumanity" of technology as tool. Nothing new is established with such an observation; but the trivial fact that technology is not identical with man is taken seriously and systematically. The significance that such inhumanity of technology can have for us becomes immediately apparent when we are forced to admit that the human world of today has itself become a machine. But this mechanical system does not rule over other things; instead it has constructed a world in accordance with the machine. The order of such a world has little similarity with the order we know; we vainly seek hierarchical structures. Machine technology is also to be found operative in those areas in which, as administration, it fulfills political tasks and functions in cycles. Nodal points, those central connection points which could be cited as proof of technical hierarchy, are in fact mutually dependent upon their many sources of information. All components of the system are equally essential; integration and not subordination is what is observed. The machine is not an instrument, as we would have it. It was indeed originally directed towards a goal, and applies the laws of nature with a definite intent. But machine technology is functional, not instrumental. From our standpoint this appears to be the same thing, but there is in fact a fundamental difference. To fulfill a function means to be oriented from within, to be defined by itself and its possibilities. To be an instrument means, on the other hand, to be employed from without, to be employed for a purpose which has only a coincidental relation to the characteristic quality of the phenomenon itself. Thus being used as an instrument for something else is a state of being that no one, and nothing, is willing to tolerate by nature. Kant called the use of man as an instrument immoral. It must now be considered "un-technical" to want to employ machines contrary to their function. The use of an automobile to commit suicide occurs more often than one might believe, its own true function being thus disregarded and remaining unfulfilled. Nor does a motor vehicle stuck in a traffic jam fulfill its function, even if rush hours appear unavoidable and the drive to work requires the instrumental use of an automobile. The cities and rural areas destroyed by motor traffic are just as far from being functional in their true sense. Not only is the anger and disappointment over the destruction a factor which delicately impairs the social climate conducive to successful driving, but beyond this, the loss is the infamous "price" we must pay for "progress," a price which is not demanded by the automobile in order to function as a mechanical system. It is a system which of its nature wants simply to function successfully, regardless of purpose. Slashed tires of cars parked on bicycle paths or sidewalks are an unmistakable indication of function failure, or more precisely, improper parking leads to a situation in which the driving function of a car cannot be realized. If the purpose, which has been assigned to machines by the individual or a group does not disturb or disrupt the function of the machine, it will be tolerated. This must not be anthropomorphically misunderstood as the intention of the machine itself; it is merely an objectively establishable fact. For when a functionally inappropriate purpose is obstinately insisted upon, even the best technology has a destructive effect. In a strict sense this is also valid for the misuse of information through technological means in totalitarian states. The function of technical information systems is to produce the most adequate image possible of the world in the form of data. But the use of the most modern computers to tabulate an election result of 99.9% tells those in power nothing about the mood of the voters, and is a senseless waste of time and energy.

Technology in its earlier craftsman's form fulfilled the Aristotelian criterion for those phenomena which are not self-determining. Machine technology however exhibits the characteristics of an indisputable autonomy; its functioning obeys an intrinsic law, is clearly automatic. This in no way makes technology the subject; it means technology is apparently no longer adequately describable within the customary subject-object relationship. Its liberation from human standards of measure is unavoidable, something expressed to no small extent in the explosive increase in the so-called unforeseeable side effects which often eclipse the main purpose of a technological undertaking (e.g., the Aswan High Dam). But technology is "free" only in the sense that even for stubborn humanists the Alpha Centauri is free from "service to man." That this inhuman technology occurs in our midst and all around us, and is for this very reason indissolubly linked with our destiny, leads to the very result that a technology, whose automatic functioning has been disturbed by a purpose, can produce nothing but man's destruction. This means that the increasingly unsuccessful living of our life in the technological world is a relevant and objective indication of a functional disorder of machine technology, for which we have to answer. A technology left to its own resources – and this always means only a mechanical system developed with the help of man and derived from an understanding of its function and purposes which intrinsically correspond to it – would *by its own nature* have developed neither nuclear power plants nor hydrogen bombs, neither lethal poison gases for chemical warfare, nor forty-story apartment houses. Senselessly high risks, potential annihilation of all living things, and housing developments which are unlivable for their intended occupants do not fulfill the basic function of every technology, namely, to be a technology for life and of life.

Technology represents the decisive means of life, of existence for the human race. This remains latently present in the form of experience, despite the sense of foreignness towards and defense against it. We rely on machines constantly, without ever accounting to ourselves for the extraordinary confidence this reliance expresses. Should we ask how it is that this confidence remains unshaken even by technological catastrophes, we would come upon an unexpected characteristic of the phenomenon of technology: in their own way, through reliability, adequacy, and necessity, machines fulfill the ideal of perfection as a matter of course. In the language of cybernetics it makes sense to say about machines that their center is everywhere, that their system is perfectly structured in function as well as in relation. The theological analogy is evident. Thinking *sub specie machinae* has replaced the infallible God as well as fallible man.[7] If man should free himself from the conception of his being unique, and should come to comprehend personality and morality as the improvised solutions of an "imperfect being" (Gehlen), he would be astounded at how closely machines approximate the self-image of man, how they are more human than humans. Automatons outdo their human models in the degree of their dominance over nature; they possess a greater capability of transforming the world and are unsurpassed in their uniform and calculable aspect. Machines are always objective, they have an exceptionally high load capacity, yet are non-directive, as are good teachers. Through their *function,* both machines and teachers make a learning process possible. They allow for advancement and a common experiencing without forcing it. No machine will *by its own nature* ever force man to submit to its function, nor infringe upon our freedom to *prepare our death.*[8] This means that the rockets carrying hydrogen bombs, automatically steering towards their goal, are neither automatic in the true sense, nor do they fulfill a technological function. They are exclusively instruments of suicide, conceived by human beings who have not understood their technology. But even the heart and lung machine in the intensive care ward – doubtlessly a contribution to a technology for life – becomes as hostile to life as a bomb when handled unfreely. The machine's function does not however hinder our potential for death, only the aim of the hospital, an externally imposed purpose, lends itself to the prevention of our dying, when that time has come. Any number of cases of people "technically alive," attesting to such misfunction, can be cited. Modern technology "responds" functionally to its own possibilities and does not impose external objectives. Violence towards nature does not constitute technology, but is instead caused by our existing handling of nature, which in its turn misuses technology. Technology gives an unmistakable indication of its abuse by threatening to become a technology deadly to the human race instead of being one of life.

Human identity in opposition to machines proves to be abstract. This does not mean that man is a machine and should therefore be perceived one-sidedly, from a biologically behavioristic viewpoint. Through machines certain human attributes are fulfilled – those attributes of which man is so painfully conscious, due to their absence, but which are so essential to his wholeness. In this sense the machine is human. It does not replace man; it compensates for his errors. It expands him in his potential, is an organ of man threatened by that quick success which brings failure in the long run. In this respect the machine does not differ from the hand or spleen. The antithesis organic-inorganic, of living and dead matter, has long become untenable. Machines belong to man's being as much as eating, breathing, and thinking – and not by pure chance. The symbiosis between man and machine is the necessary expression of our life-systems, it is a humane mode of being-in-the-world.[9] The important thing is to let machines be machines *through us*, to learn a more expanded way of living from their function as newly disclosed, human relationship with nature. But fearful, we suppress such a mechanical realization of our own potential. Instead we leave it to the irresponsible power of government, to the "objectivity" generated by society, and to the irrational fear of hostile factions to employ scientific-technological developments solely for the establishment of a purpose, with absolutely no conscionable consideration towards technology itself. In our need we, the powerless and potential victims, develop an "ethic of the technological age," and would rather amputate our technological organ in the belief that we cannot prevent its misuse. But we should not be afraid of opposing – *together with technology* – its indiscriminate expansion so indifferent to the environment, and in so doing, to restore to technical realizability its vital meaning. The machine not subjected to short-sighted goals produces the phenomenon, as all nature produces, and at the same time respects their "concealedness."[10] For in any way other than with such regard, technology would preprogram its own misfunctioning and inevitably its own destruction. In that case it would not be functioning according to the one technically just and appropriate understanding. A human technology which is not universally valid remains incomplete, a situation today which we must all too painfully endure. Individual, isolated successes gratify certain interests, but do not count technically. Correctly understood, machine technology produces not individual artifacts, but above and beyond this, *the* valid order for the phenomena of the human world. Technology has no interest in achieving a certain end, it wants to succeed, like every order. The meaning of humanness is in no way altered through technology. On the contrary, such seemingly unfulfillable ideals as gentleness, selflessness, and love suddenly appear to be realizable. A technology which is no longer misunderstood as instrument, but rather comprehended as functional, will not however become a myth which brings forth no works, nor will it revert back to magic practices of the savage mind. It remains a real and working technology, but as an alternative to being an instrument, is now art. Works of art have always been useless, free, creative, and yet of a significant effectiveness. Without intending to, they shape the world and work through images. It is as impossible for man to withdraw from his non-instrumental technology as it is from the Mona Lisa. A fulfilled function is convincing.

II. HUMANE TECHNOLOGY: CYBERNETIC ART

In the synthesis now possible, which conceives of technology as art and which follows almost freely and naturally from the description of the phenomenon of technology, an ancient European knowledge is renewed. Simultaneously, we begin to adequately interpret the confusing phenomenal findings that technology is humane and is not at our arbitrary disposal. Such interpretation is normative, but only in as far as it respects those rules existent in the nature of man, and which cannot be selected by us. "Techne" was the Greek philosophy of life. In that time, art was not yet conceived of separately from technology, from the work of the artist. Not until the one-sided preference for usefulness in the modern age, and the systematic design for domination of nature[11] following Bacon's teachings, did technology become a weapon against nature. And with it the meaning of art was subjugated. Man had naturally invented tools for survival long before the appearance of modern technology. But life was not understood as being synonymous with breaking out of the unity, the sense of oneness with the universe, God, and nature; nor was it taken as a challenge to this unity, or as a struggle against this original order of things.[12] In recent times machine technology has been restoring art to its place in daily life. The art movement represented by the "Bauhaus," which to this date remains exemplary in architecture and functional art, strove for a bond between, a synthesis of art and technology in the "best form." In other words, the contours of a thing and its function should find reciprocal expression in, and be mutually dependent upon one another. Necessity and use, the idols of modern technology, and the pretext for a separation of art and technology, reveal their forgotten reality in the Bauhaus concept, a reality in touch with and correspondent to the environment, one which does not interfere with, but instead suffices for universal functioning according to its abilities. We must not be allowed to willfully destroy if we want to stay alive; and we do not actually need to. What benefit does the human race derive after all from a technology which produces nuclear power plants, closing an alleged energy gap by decades, but at the same time opens an abyss of a millenary threat caused by "final storage" of atomic wastes? Successfully functioning technology does not apply to the individual case; it is oriented in relation to the universe. For only a *truly* successful function in the long run is in the interest of that individual species *calling* itself man and *existing* as technology.

Modern technology bypasses purpose. It does not serve the end of intoxicating us with freedom and power, of producing masters to rule over the earth. Technology answers such misapprehension of its reality with "side effects" the world over. The structural identity of natural and manufactured products becomes apparent in the decay with which both react to inappropriate use. The fundamental condition of technology is also its function of disclosure, of bringing about the world in its realities, letting it occur. Technology also finds its perfection in this function. The condition of technological working is realized only in its complete state of being, in its wholeness, not in any degree of reduction prescribed by individual interests. Architectural-constructional technology is indifferent to the house, but dwelling is essential to it. Transportation technology is concerned only with transporting, not with the means of transportation. Dwelling and travelling are ways in which man gives an account of whether he has comprehended his role in the universe. This is to be understood in a phenomenal manner: "home," and all that this concept embodies, is realized for us in dwelling. Heidegger defined home as the sojourn of mortals among things, under the heavens and upon earth.[13] If we do not inhabit the earth, we remain strangers. Human ethics is expressed in dwelling as well as manifesting moderation and respect for the sensual and the transcendent. Rules and the justification of ethics are without foundation unless there is this experience of dwelling. In driving, on the other hand, we experience the world of others, and understand them as either close to us or distant, in adventure or danger. Of course, we accompany ourselves on every "excursion" (Bloch), but we return home as changed people in the end. The individual technologies of dwelling and travelling are unimportant – as long as their respective functions fulfill the meaning of dwelling and travelling. No technology by its nature "wants" to change these basic modes of living, yet we employ technologies of construction and transportation which flagrantly violate the intrinsic functions of dwelling and driving, and which cause us to stumble towards the brink of existence. Or can we truly state that we feel at home in our despoiled environment and in the "silos" of our metropolises which we call apartment "houses"? Do we learn to live "ethically" – i.e., as humans? Opinion polls do indeed report positive attitudes towards the life-styles in large cities and their green suburbs, but the rampant increase in mental illnesses in both living areas speak another language. Man is an unexcelled master in the art of intellectual self-deception, but it will not protect him from the physical and mental injuries which contradict his self-estimation. Similarly, modern traffic conditions exact a bloody toll of lives "worthy" of a Roman arena, and allow us a paltry second-hand experience. We bring our preprogrammed expectations with us on our travels, and they are promptly fulfilled – from Singapore to Tierra del Fuego: Viennese sausages, Löwenbräu and streusel cake for Germans, wherever they go.

Technological working has nothing in common with such degenerate particularity. Technological functioning is concerned with truth, and therefore the totality, the wholeness of a process is addressed. This also must not be misunderstood as metaphysics, but merely as positive evidence of that which we have always known about technology. As soon as it begins to function successfully, technology is an adequate condition of the world process. How else could it succeed? Technology becomes expressive and tangible in its work, but is not to be determined by its artifacts. It can be reproduced however through philosophical contemplation, and at the same time practiced as a technology of truth. This has always been granted the "useless" arts, and now is just as valid for the most modern cybernetic technology. Microprocessors complement nature in an altogether natural way; or is there a single unnatural component in this technology, human ingenuity included? Technical perfection embodies beauty. In eternal recurrence technology occurs as art.[14]

Cybernetic machine systems do not manipulate the process, but rather, belong to the process, are a part of it. In this way, and in no other, can they be justly described. Man alone is the manipulator. The technical art-work process corresponds to the purest form of process. Information is neither presorted for a determined purpose, nor instrumentally misappropriated. Instead, the function, in the form of an expressed statement, results from significant information having undergone a process – as is the case with a work of art. The sense of function remains flexible and unrestrained with every new discovery and with every new application. In this manner, the cybernetic technologies make possible the rendering of true occurrence, and not willful dominance by man. The more objective technology is, and this also means the less it is determined by external, irrelevant purpose, the better it can attain its humane purpose. The countries of the Eastern Bloc have had to learn the hard way in their attempt to seriously bind science and technology to dialectical materialism, to thereby reject certain technologies as being capitalistic, or to devise an ideological biology of their own. But even our "objective" standpoint is often only an economic one! Instead, cybernetic technologies can be free arts, free to be true, and liberating for man. Modern technology then becomes the art of humanity.[15] It humanizes man by healing him of his anthropocentric violence which he – aided by instrumental technology – is presently carrying to all extremes. The consequence is world destruction. But at the same time it becomes clear that, approached anthropocentrically, technology can *only* be misused. Man does not need to be the center of the universe; he experiences himself as artist in modern technology, as participant in an all-encompassing game which develops his abilities. This technical development occurs incessantly, and is the art of life.

An interpretation of technology as cybernetic art has not only to struggle merely against the fact that such a mode of being has been very seldom realized up to now (and therefore might well be phenomenally inexpressive, might appear to be speculative). More importantly, a destructive contradiction has already been built into its description. For we, as the "artificial," as technical beings by nature, are inevitably destroying the preexistent natural order.[16] With technology, man is apparently withdrawing himself inexorably from the nature within himself, the nature out of which he evolved. The fluctuation between tensing and relaxing, characteristic of all biological processes, is being forced to conform more strictly to technical processes which always remain invariable.[17] Even relaxation rigidifies in homogeneous free time activities treated as technology. We find ourselves part of a "techno-evolution" (Ellul), whose perfection is making us not into technicians, but into mere figures in a technical synthesis of the arts. Through the cybernetic machine, in the end, man and machine will be brought together in a synthesis of a work of art honoring the artificialness of both – but missing the link to life and to our nature. The human world would then become a gigantic "happening," loud and senseless. Because of this the suspicion becomes conclusive that it is precisely the artificiality of our present existence which must be made responsible for the ecological crisis. Inhuman machine technology set free its contradiction – the all-too-human art-technology. In it, man consciously places himself in opposition to nature. He sets himself "free" from nature. The new "second nature" becomes purely human. Only within nature's systems are there recognized successes. Mutual effect is their model reality. The framework (*Gestell*) rules unconditionally.[18] The development towards total artificiality in recent conceptions of a "soft technology" or a "biotechnology," among others, is recognized in being opposed.

III. HUMAN BEING: THE EVENT OF TECHNOLOGY

We must not allow ourselves to be misled by the fact that all life has two sides, life and death; or that, often, from a distorted viewpoint, death is taken for life. The arrogant "No" to nature as well as the violent dream of dominance over it, is indeed a possible mode of being for art-technology, but it is not characteristic. On the other hand, the path leading "back to nature," in its denial of our own artificiality, also leads us astray. Instead, we must learn to live with the realization that *artificiality is the nature of man,* and that technology, in all its so uncommonly diverse forms, is the realized, cosmic mode of being peculiar to our nature and which must be further perfected. Technology is the human way of corresponding to the universe. Considered abstractly, technology could indeed also be "untrue," but if such were the case, how could we still be in existence today, having made such a fundamentally wrong judgment about ourselves? The turning[19] away from artificial death towards technical life occurred at every moment in which we survived. It is essential that we now comprehend this relation of things and see in the merely "trivial" phenomenon of technology the "event of technology" which is of vital concern to us, first and foremost. The substance of things, not techniques, seemed to decide about our life up to now. But in truth, only techniques are real and the "substances" are our irresponsible projections. Men have long been in agreement on the decisive substances – freedom, equality, fraternity, peace – who would dispute them? The way in which we deal with them is what condemns our technologies, not the subject matter itself. But unsuccessful functioning of technology also brings their meaning to light; the only thing that can kill us is that which keeps us alive. Deadly-ness is the reverse side of lively-ness, nothing in the universe is excepted to this rule. Justice is valid for all phenomena. The death-bringing technology is the negative of the life-sustaining technology.[20] Both attest to the "event of technology" in which man becomes himself – or fails himself; in which case the human race will, justifiably, perish. We can meaningfully correspond to all phenomena through technology, and in our actions and thought we can respond to the world in a manner it deserves. The more attentive, the more refined our technologies become – from the practical technology of an engineer to the symbolic-affective technology of the poet – the easier it will be to live. Through our technologies we participate in the universal successful functioning which avoids violence to nature. The "event of technology," for all its differentiation, is simple in its basic sense. The answer from the universe is life or death. Whoever misinterprets his own event of becoming, be it man or stone, shall not become himself. He thereby forces himself into nothingness. The event of becoming is the guarantee of "own-ness" (*Eigenes*) within the whole. It can be monitored anytime through the condition of the particular species. And we know whether the human race will live or die. The interpreters cannot misinterpret forever.

The present destruction of the world reveals itself as the refusal to accept our technology of life. All technologies whose perfection is in accord with the universe correspond to our life pulses. Until now this was impartially admitted only by natural technologies – from breathing to the metabolic process. A mode of living, oriented in relation to the universe, and whose form has been fulfilled by the philosophy and religion teachers of the human race for thousands of years, has been misunderstood for too long as naive idealism. At the same time the universe has been *answering* the so extremely distorted existence of modern man *clearly and appropriately with deprivation and annihilation.* But if man and things are, according to the universal measure, left to be as they are by their nature, then we solve our problems and fulfill our tasks effortlessly, almost unintentionally; phenomena become technically approachable paths to identity. Even in instruments, when appropriately used, the essence of a thing can still be respected. For *man the technician* this means radically rejecting *presence as an accomplished fact*, Being as possession, time as the pressure of time.[21]

Mortality is the essential wager for man. No institution, no tradition survives its defenders; they will all be mere data in an electronic archive in a few technical decades. Every minute people are born who have no "sense of reality" of the age in which we live. For their survival they demand that to which they are cosmically entitled: love, natural unravaged living space, and power for no one. These human beings know intuitively and experience physically that they need nothing more than this in order to live a fulfilled life – in spite of those sufferings which cannot be avoided[22] and which are themselves part of living. A perception, open to everything around us and a characteristic of technical reason, hears in the particular the tone of the universe and is horrified by rigid particularism. This horror, which no one has to evoke since it is in accord with our way of living, is what urges on man the technician. Experienced in fear, trained in differentiated technologies, and with an open eye toward the world, he has no enemy to dread. As long as our instrumental technology does not run amok and lead us into catastrophic destruction, there is every reason to believe we can look forward to the imminent birth of man in the "event of technology."

University of Hamburg

NOTES

[1] W. Zimmerli (ed.), *Technik: oder wissen wir, was wir tun*? (Basel: Schwabe, 1976), p. 110.
[2] W. Schirmacher, *Ereignis Technik.* Dissertation, philosophy, University of Hamburg, 1980.
[3] F. Rapp, *Analytische Technikphilosophie* (Freiburg and Munich: Alber, 1978).
[4] W. Schirmacher, "Dialektik der Technik," *Hegel-Jahrbuch*, 1982.
[5] Rapp, *op. cit.*, pp. 45–78.
[6] H. Sachsse, *Anthropologie der Technik* (Braunschweig: Vieweg, 1978), p. 9.
[7] A. Baruzzi, *Mensch und Maschine* (Munich: Fink, 1973).
[8] W. Marx, "Die Sterblichen," in *Nachdenken über Heidegger*, U. Guzzoni (ed.) (Hildesheim: Gerstenberg, 1980), pp. 160–175.
[9] H. L. Dreyfus, *What Computers Can't Do* (New York: Harper & Row, 1972).
[10] M. Heidegger, *Poetry, Language, Thought.* Trans. by A. Hofstadter (New York: Harper & Row, 1975), pp. 59ff.
[11] W. Schirmacher, "Naturbeherrschung," in *Historisches Wörterbuch der Philosophie*, J. Ritter and K. Gründer (eds.). Vol. VI (Basel: Schwabe, 1982).

[12] M. Heidegger, *The Question Concerning Technology*. Trans. by W. Lovitt (New York: Harper & Row, 1977), pp. 3–49.
[13] W. Schirmacher, "Bauen, Wohnen, Denken," *Philosophisches Jahrbuch* 89, 1982.
[14] T. W. Adorno, *Ästhetische Theorie* (Frankfurt am Main: Suhrkamp, 1971), pp. 153f.
[15] D. Ihde, "Heidegger's Philosophy of Technology," in *Proceedings of the Heidegger Conference*, Villanova University, 1978.
[16] Baruzzi, *op. cit.*, p. 205.
[17] Rapp, *op. cit.*, p. 193.
[18] H. Mörchen, *Macht und Herrschaft im Denken von Heidegger und Adorno* (Stuttgart: Klett-Cotta, 1980), p. 75.
[19] H. Maurer, *Revolution und "Kehre"* (Frankfurt am Main: Suhrkamp, 1975).
[20] M. Heidegger, *Vier Seminare* (Frankfurt am Main: Klostermann, 1977), p. 104.
[21] W. Schirmacher, *Technik und Gelassenheit* (Freiburg and Munich: Alber, 1983).
[22] W. Schirmacher. 1982. "Gelassenheit bei Schopenhauer und bei Heidegger," *Schopenhauer-Jahrbuch* **63** 54–66.

(Translated by Virginia Cutrufelli)

JOSEPH MARGOLIS

PRAGMATISM, TRANSCENDENTAL ARGUMENTS, AND THE TECHNOLOGICAL

When Heidegger remarked, close to the beginning of his famous essay, "The Question concerning Technology,"[1] that "the essence of technology is by no means anything technological [or, perhaps better, 'technical'] ,"[2] he imposed an obligation – doubtless a fair one – on all who might attempt to answer, to characterize technology in terms of man's ultimate relationship to reality. That burden is a heavy one but not altogether unwelcome: initially at least, because it seems to insure the least partisan and most comprehensive approach to the issue one can imagine, and because it permits us to focus the question through Heidegger's important and original perception without betraying any support for Heidegger's own rather extraordinary account; and ultimately, because it permits us to come to grips in an efficient way with a more plausible theory of technology, in fact a theory opposed to Heidegger's, more congruent with alternative conceptions (favored elsewhere within philosophical reflection) of man's relation to reality, less tendentious, and more manageable with regard to the developing record of Western speculation.

One cannot help thinking, in trying to characterize technology philosophically, that one ought not to be careless about whom one cites first. Also, of course, it must surely appear a little strange to be able to turn so directly to say what the true relationship is that human beings should occupy with respect to Reality or Being writ large. And yet, the convenience of selecting Heidegger lies precisely with his equivocal role – in exposing what he takes to be the nihilistic feature of the technological mode of modern philosophy *and* in sheltering the relevance of theories of technology largely opposed to his own.[3] The point of Heidegger's remark is to call for a demonstration, first, that "the instrumental definition of technology" (that "technology is a means to an end" – *merely*, that "technology is a human activity" – *merely*) is correct both in a trivial and (particularly in our own age) a sinister sense and, second, that, though "correct" (and even inevitable), that definition cannot possibly be "true," utterly fails to reveal the "essence" of technology.[4] To assimilate Heidegger's contribution and to reject it at a stroke, we may say, by way of epithets that are somewhat cryptic but perhaps not disagreeably so, that Heidegger pretends to have made a *transcendent* discovery about technology (indeed, about the whole of Western philosophy), whereas the

Paul T. Durbin and Friedrich Rapp (eds.), Philosophy and Technology, 291–309.
Copyright © 1983 by D. Reidel Publishing Company.

best (and entirely adequate) effort that men can hope to make in answering the Overwhelming Question is to offer a *transcendental* proposal about the nature of technology and reality. In fact, Heidegger's transcendent claim masquerades as a transcendental one – which accounts both for its somewhat strident tone and seeming good sense.

The contrast intended is radical. But, properly understood, it captures the dominant contributions of recent vintage of both Anglo-American and Continental philosophy and shows us how to tolerate ideologically alternative conceptions of technology as at least eligible, as inviting and claiming partisan adherence, and as exhibiting in a richer way than Heidegger himself is prepared to admit the connection between technology and ontology. One may be tempted to construe Heidegger's own thesis as a particularly powerful version of transcendental thinking. But to do so completely fails to square with Heidegger's confident discovery of the end of philosophy, that is, of the technological mode of philosophy running from Plato to Nietzsche,[5] and the optimistic, transcendent prospect of once again "letting" Being reveal itself – as Heidegger would say – in that way that permits human beings to be revealed to themselves.[6] Still, in Heidegger's ironic and good-humored way, we may well ask ourselves, "But where have we strayed to?"

I

Stripped of its unusual language, Heidegger's thesis maintains, in part, that technological thinking leads to the result that man himself comes to be viewed reflexively as no more than a technological resource – which he takes to be absurd and dangerous. The corrective offered requires mankind's recovering the deep truth that the technological characterization of reality as instrumental is itself a peculiarly partial and distorted (but not altogether ineligible) way of understanding reality: we need to hold ourselves, it seems, in readiness for the undistorting way of comprehending reality, of which the technological is essentially a black sheep but genuine offspring, so that we can recover ourselves and our proper destiny. In this sense, Heidegger appeals to the transcendent – and, inevitably, violates his own emphasis on the historicity of human existence. This may perhaps clarify the meaning of such characteristically dense remarks as: "The essence [*Wesen*] of modern technology starts man upon the way of that revealing [through Enframing (*Ge-stell*) – roughly, ordering or stacking real things solely to service contingent human interests or ends] through which the real everywhere, more or less distinctly, becomes standing-reserve [*Bestand* – roughly, resources available to order for instrumental purposes]."[7] This is how Heidegger understands the history of the subject-object relationship through the whole of Western philosophy.

That he does not mean actually to eliminate the technological from his would-be reform – but only to avoid its pernicious ontological pretension (its "ontic" restriction) – is reasonably clear from his seemingly naive and idyllic vision of the peasant's work in the era of the windmill and the subsequent transformation allegedly due to modern technology:

> . . . the windmill does not unlock energy from the air currents in order to store it . . . The field that the peasant formerly cultivated and set in order [*bestellte*] appears differently than it did when to set in order still meant to take care of and to maintain. The work of the peasant does not challenge the soil of the field. In the sowing of the grain it places the seed in the keeping of the forces of growth and watches over its increase. But meanwhile [that is, nowadays] even the cultivation of the field has come under the grip of another kind of setting-into-order, which *sets* upon [*stellt*] nature. It sets upon it in the sense of challenging it. Agriculture is now the mechanized food industry.[8]

Two objections are in order here – both instructive about the recovery of the technological: one, that it is difficult to share Heidegger's confidence about the ontologically uncontaminated orientation of the peasant (or of the beer merchants of ancient Mesopotamia, for that matter) or about the unqualifiedly technological sensibility of our own ecologically troubled age; the other, that we must see that adjustments with regard to such considerations cannot plausibly be made to bear the weight of the unusually extreme contrast in ontological reward that Heidegger assigns to a technological ontology and to an ontology committed to his own transcendent vision. Add to this that the transcendent claim is, demonstrably, philosophically indefensible, unnecessary, and arbitrarily assigned to the experience of favored theorists and societies, and one cannot but be strongly inclined to seek an alternative theory.

The trouble with Heidegger's thesis is straightforward enough, though it is due to a variety of faults. It claims to expose the distortion of the prevalent technological ontology of our own age, though it is itself somehow antecedently exempt from that failing. It cannot explain how we confirm the validity of its own claim, though, in exposing the danger of Enframing, it promises a kind of ontological redemption: "The rule of Enframing [Heidegger says] threatens man with the possibility that it could be denied to him to enter into a more original revealing and hence to experience the call of a more primal truth."[9] And it conflates a dialectical and normative criticism of the contingently limited vision and insensitivity of particular expressions of technological thinking with a putatively timeless discovery of a "higher ontological orientation" – that, to be sure, deploys itself in an historically distinctive way. This is the sense in which Heidegger offers a transcendent rather than a transcendental philosophy. We seem to correct in a dialectical way the substantive distortions and inadequacies of prior speculation, but the truth is that we must now suspect all determinate corrections as well. We seem to be productively confined by the contingencies of diachronic reflection, but the truth is that we need no longer be captured by any particular theory thus produced.

We begin, Heidegger says, by "questioning concerning technology, and in so doing we should like to prepare a free relationship to it. The relationship will be free [he claims] if it opens our human existence to the essence of technology [that is, to *Wesen*, to the 'enduring as presence,' to the 'presencing' (*Anwesen*) of Being]."[10] This presencing is occasioned by all the forms of human *technē*. Every such occasion constitutes *poiēsis*, which entails the revealing (*das Entbergen*, or *alētheia*) of what regarding Being is concealed.[11] Here, as is well known, Heidegger means to substitute truth as the revelation of Being (*alētheia*) for truth as correspondence. His fundamental claim is that, "It is as revealing, and not as manufacturing, that *technē* is a bringing-forth (*Her-vor-bringen*)."[12] But though it indeed has something inspirational about it, the argument appears to be both purely formal and arbitrary and wears its pretensions on its sleeve. Sometimes, Heidegger holds that technology is a deficient *poiēsis*, and, sometimes, he holds it is no *poiēsis* at all:

> And yet the revealing that holds sway throughout modern technology does not unfold into a bringing-forth in the sense of *poiēsis*. The revealing that rules in modern technology is a challenging [*Herausfordern*], which puts to nature the unreasonable demand that it supply energy that can be extracted and stored as such.[13]

Alternatively put, we are misled, Heidegger thinks, if we attempt to *correct* (always within the confines of historical existence) our theories of the dense and complex world of encountered things: we must *turn* rather to the "call" of Being itself, that is, to the very source of our prodigal but dangerous ontologies. This is the key to Heidegger's extravagance.

The result is a peculiarly preposterous theory that entails that, however impressive it may be, very nearly the entire movement of Western philosophy from Plato to Nietzsche has gone ontologically astray *and* that Heidegger himself controls its permanent exposure. It is also what may be termed a "sly" theory, because its emphasis on the historicity of human existence and inquiry somehow manages to ignore the concrete history of actual existence and actual inquiry. *If*, however, the transcendent is genuinely inaccessible (not an option at all) and *if* the transcendental permits a certain measure of diachronic and reflexive correction, then it cannot but be the case that what Heidegger dubs the technological mode of ontology provides (even on his own view) the sole avenue for clarifying man's cognitive relationship to Being. It is often remarked, rightly enough, that, already in *Being and Time*, Heidegger's philosophy is essentially "praxical" in nature – in the express sense that, for Heidegger, what we suppose we are doing in theorizing about, or contemplating, the nature of things is ontologically dependent on our first being able to engage things as tools or equipment in the most fundamental activities of life. *Vorhandenheit* presupposes and depends upon *Zuhandenheit.*[14] The theoretical is thus an abstraction from the practical – in effect, as the essay on technology makes clear, an ultimately pernicious and philosophically ideolatrous abstraction. Heidegger seems concerned very nearly exclusively with the formal relation between these two conceptions of things. He does not seem to wish to attend to the provisional validity of particular theories and how they change and must change under the force of a changing *praxis*; he turns rather to the putative structure of Being writ large, which somehow flowers in things manifested in our "dealings" with the world as "ready-to-hand." "Our real theme," he says (and he means it) "is Being . . . not a way of knowing those characteristics of entities which themselves are [say, contemplatively accessible] ; it is rather a determination of the structure of the Being which entities possess."[15] *Praxis*, therefore, is itself peculiarly abstract in Heidegger's philosophy. We are to understand that, being what we are (*Dasein*), we encounter things as "equipment" (*Zeug*); this is how Being parsimoniously reveals itself to us.[16] In the essay on technology, therefore, we are warned not only about the delusional nature of philosophical contemplation freed from praxical concerns but of the seductive appearings of praxical things as well. Hence, we are enjoined not so much to adjust (however we may, within the boundaries of historical reflection) our account of the plural things of the world thus encountered as to dwell on the timeless truth that Being (writ large) ever discloses itself in such appearings. "Enframing [Heidegger declares] comes to presence as the danger. . . . But *the* danger [is that] Being itself endanger[s] itself in the truth of its coming to presence, [hence] remains veiled and disguised."[17] The curious result is that Heidegger quite gymnastically manages to reject human history and conceptual corrections made solely within the movement of human history precisely in the context in which he stresses the historicity – the historically changing openness – of *Dasein* to the disclosure of Being. In this sense, he snatches a transcendent theme from his own transcendental reflection on the praxical contingencies of formulated ontologies. This has often been seen as marking the distinction between the early and later Heidegger, though there is already more than an inkling of this shift in *Being and Time*.

It is impossible to ignore, here, Marx's insistence on construing philosophy and science – all theoretical knowledge – as forms of *praxis*; they are, Marx affirms, conceptually and really dependent on the historical conditions of actual production. In this sense, whether or not we agree with Marx's diagnosis of captialism (or, indeed, of the whole of human history), we cannot fail to see the important sense in which Marx anticipates and (in effect) resists Heidegger's philosophical injunction. Already, in the *Theses on Feuerbach*, Marx declares:

> The chief defect of all hitherto existing materialism (that of Feuerbach included) is that the thing, reality, sensuousness, is conceived only in the form of the object or of contemplation, but not as sensuous human activity, practice [*Praxis*], not subjectively. (Thesis I)
>
> The question whether objective truth can be attributed to human thinking is not a question of theory but is a practical question. Man must prove the truth, i.e., the reality and power, the this-sidedness of his thinking in practice. The dispute over the reality or non-reality of thinking that is isolated from practice is a purely scholastic question. (Thesis II) [18]

There, Marx offers his own account of the abstract subject-object relationship favored in "theoretical" or "contemplative" views of truth. But, unlike Heidegger, Marx resists the temptation of the transcendent and links the discoveries of science and philosophy precisely to what Heidegger regards as the condition for the technological mode of ontology. Connect the Theses mentioned with what may well be one of the central, systematic themes of the *Grundrisse*, and the contrast between Heidegger and Marx becomes entirely transparent:

> Each individual possesses social power in the form of a thing. Rob the thing of this social power and you must give it to persons to exercise over persons. Relations of personal dependence (entirely spontaneous at the outset) are the first social forms, in which human productive capacity develops only to a slight extent and at isolated points. Personal independence founded on *objective* [*sachlicher*] dependence is the second great form, in which a system of general social metabolism, of universal relations, of all-round needs and universal capacities is formed for the first time. Free individuality, based on the universal development of individuals and on their subordination of their communal, social productivity as their social wealth, is the third stage. The second stage creates the conditions for the third. Patriarchal as well as ancient conditions (feudal, also) thus disintegrate with the development of commerce, of luxury, of *money*, of exchange value, while modern society arises and grows in the same measure. [19]

Marx's emphasis on *praxis* preserves the historicity of human inquiry in a sense that is quite in accord with Heidegger's own praxical insistence. But Heidegger's claim is both purely formal and somehow disconnected from the very conditions of practical life that he himself has so often stressed elsewhere. The irony is that Heidegger's reasonable complaint against the abstractness of the correspondence theory of truth (and, in the same sense, against the subject-object relationship within the technological orientation) applies with equal force and with a peculiar vengeance to his own view – at least to the extent that he adheres (as he does, in speaking of technology) to his supposed transcendent vantage. Marx, on the other hand, provides an entirely fresh view of the nature of transcendental thinking itself, within the bounds of historical *praxis* – hence, in a way that enriches rather than abandons the technological sense of the subject-object relationship. Heidegger somehow supposes that we can escape that relationship; Marx merely clarifies the sense in which its distorting pretensions may be avoided and (within the boundaries of practical reflection) corrected – that is, corrected in that sense that is fully compatible with the thesis of historicity and the concrete conditions of historical existence. Within the Marxist tradition, the proper analogue of Heidegger's formal notion of preparing for the revelation of Being (without prejudice to the relative humanity or appeal of potentially opposing views) is Stalinism: for Stalinism is the result of violating the cognitive constraints of historical inquiry *within* the supposed historical limits of science and transcendental reflection; whereas, in Heidegger, the correction prepared regarding the revelation of Being is first informed by the condition of historicity but, then, that correction itself somehow pretends, within historical time, to have escaped its inherent constraints.[20]

The full meaning of these remarks is still somewhat obscure, since they presuppose clarifying the nature and import of the contrast between transcendent and transcendental philosophy. Briefly, what is needed is an account that explains how transcendental reflection functions within the constraints of historicity; how (in so doing) it escapes conceptual anarchy, chaos, radical relativism, solipsism, and the like; and how (further) it precludes the need for transcendent foundations of any sort. Armed with such an account, we cannot fail to see that Heidegger's attention to the historical nature of human existence – and technology – claims an (ontological) privilege in exposing the distorting vision of Western philosophy, that, consistently with its own transcendental discovery, it cannot insure. In short, the essentially historical nature of human inquiry may well be regarded as the single most important *transcendental* discovery of post-Kantian philosophy. It has, however, already taken two centuries to appreciate its full force – for example, as in the indefensibility of any foundationalist account of cognition or of a privileged access to Being or Reality; or again, as in the inherent violation of every historicized inquiry that totalizes the diachronic currents of transcendental thinking – that is, that attempts to reach a transcendent truth about all transcendental disclosures. But in order to formulate this thesis effectively, we must go somewhat further afield; in so doing, however, we shall find it much easier to specify the nature and significance of technology itself.

Nevertheless, we must be clear that we have broached the question of technology in a peculiarly large spirit, in a way in fact that threatens to exclude (from immediate relevance at least) detailed appraisals of quite particular technological changes – for instance, regarding specific industrial dangers or tradeoffs in pollution and conservation. Doubtless, this is true of Heidegger's and Marx's orientation as well. To theorize, just like that, about the conceptual connection between human cognition and technology is to collect our concerns in the most global way possible. No criteria regarding technological assessment, for instance, can be reasonably expected to follow directly from such large reflections – *a fortiori*, no specific findings that would depend on the application of such criteria. This is hardly a loss, however. For, for one thing, our smaller intuitions about the comparative benefit of this or that technological change need to be systematically linked with an appreciation of the pervasiveness of the technological orientation itself; for a second, the very notion of a human ecology requires a grasp of the relationship between the biological and cultural aspects of human existence and the extent to which the technological colors the cultural and affects the biological (or the purely physical features of the human context); and for a third, we must be able to generate the widest variety of views about the human condition as affected by technology in order to entertain flexibly all possible perspectives on the changing direction of technological assessment. It is simply a mistake to think that the humane appraisal of the relevant changes in the quality of human existence can be pursued as a matter of course merely by attending to the piecemeal, practical contingencies we happen to encounter. We need a coherent conception of how human history has altered and must alter the setting within which our very convictions about directing technological change are themselves subject to change under the influence of an unfolding technology. Only the largest reflections about what it means to think of *anything*, under the conditions of human existence, will orient us promisingly. Whatever we take to be a reasonably adequate and informed principle (among others that may be equally plausible, even if incompatible) will require a linkage to such speculation. There is, therefore, a fundamental practical cast to the most seemingly abstract and unengaged reflections. What follows is intended to help sort the complexities involved in this persuasion.

II

Seen through American eyes, contemporary Western philosophy is decidedly pragmatist in cast. This is not to say that the specific doctrines favored by Peirce or Dewey or James are correct or vindicated, or that current theories are returning to their specific views. It is not even to lay much importance on the use of the term "pragmatist." Doubtless, in European quarters, precisely the same tendencies collected as pragmatist by American philosophers will be linked with equal justice to phenomenological, structuralist, deconstructivist, hermeneutic, semiotic, Marxist, existentialist, and similar currents.[21] The fact remains that there are certain salient, determinate, convergent tendencies to be noted in nearly all contemporary speculations about reality, cognition, methods of inquiry, that compellingly preclude Heidegger's partisan doctrine about technology while at the same time remaining congruent with his larger emphasis on the historicity of human existence. Here is the point of noting the essentially equivocal nature of Heidegger's influence on the development of our question. His own thesis has actually quickened the life of contemporary pragmatism, but he himself has somehow managed to prophesy, under that very banner, a return to an ontological capacity incompatible with that thesis.

In order to avoid misunderstanding, let us construe "pragmatism" as a term of art. We shall take any philosophy to be pragmatist to the extent that it distinctly favors three doctrines, regardless of its conceptual sources or variant details. It must, first, oppose foundationalism with respect to whatever cognitive powers it presupposes. Hence, the epistemological claims of Platonism, Cartesian and Kantian certainty, sense-datum theories, Husserlian reduction, totalized systems of the structuralist possibilities of any cultural or cognitive domain, the discovery of the "true" or "single" language of nature, even the discovery of the fixed laws of nature, are all precluded.[22] Secondly, it must compensate for this first constraint by presupposing that, in a sense not epistemically privileged or even fundamentally epistemic in nature, a certain tacit condition regarding the survival of the human species obtains: namely, that our theories of reality, of our cognition of reality, and of our rational methods for constructing and validating such theories may, however divergent or incompatible or changeable over time they may be among themselves, be judged sufficiently grounded in reality that our sustained adherence to them does not as such entail the extinction, near-extinction, or related jeopardy of the human species or of strategically engaged parts of the human species. Among American philosophical pragmatists, this theme is particularly emphasized in quite a variety of ways, as may be seen in the theories of W. V. Quine, Nelson Goodman, and Hilary Putnam.[23] Thirdly, relative to these two constraints, it must provide for the cognitive success of the valid forms of inquiry in terms continuous with, and dependent upon, the conditions of social *praxis*.

Preeminently, both Heidegger and Marx exhibit, in this sense, a pragmatist orientation: Heidegger, more in *Being and Time* than in "The Question concerning Technology" or "The Turning"; Marx, notably in the *Theses* and the *Grundrisse*. But the pragmatist strains in Heidegger are perhaps more clearly developed among those he has influenced – for instance, within the hermeneutic tradition, Gadamer (*Truth and Method*) as opposed, say, to Dilthey; and among the existentialists, Sartre, particularly in his effort to reconcile existentialism and Marxism.[24] Among relatively recent Continental thinkers of importance, one may note that the Wittgenstein of the *Investigations* but not of the *Tractatus* may reasonably be said to be a pragmatist in his treatment of philosophical issues.[25] Husserl's search for the indubitable foundations of knowledge and the characterization of the transcendental ego mark him as a clear target of pragmatism. But descriptive phenomenology, notably in the work of Merleau-Ponty, is obviously capable of functioning in the pragmatist manner.[26] Among the more eclectic thinkers influenced in varying degrees and ways by Marxism, phenomenology, Heidegger, hermeneutics, existentialism, and structuralism, the pragmatist bent is particularly noticeable in Habermas, Marcuse, Bourdieu, and Apel.[27] So-called analytic structuralism, as instanced, for example, in Lévi-Strauss, is a ready target of pragmatism, for much the same reasons that the totalizing tendencies (altogether abstracted from *praxis* and historicity) that appear in the *Tractatus* and the work of such early linguistic structuralists as Saussure are criticized. Genetic structuralism, on the other hand, particularly associated with Piaget and those he has influenced, attempts to restore the primacy at least of the tacit conditions of the developing powers of human cognition, conceding, in particular, the primacy of human action within real-life contexts. Unfortunately, Piaget himself has rather little to say about the social import of *praxis* and treats cognitive (and linguistic) development more in terms of structurally fixed transformations than in terms of actual *praxis*.[28] This helps to explain, for instance, the sense in which Lucien Goldmann, combining Marx (*via* Lukács) and Piaget, is more clearly oriented in the pragmatist manner than either Piaget or Lukács.[29] The possibility, within this eclectic network, of leaning toward radical relativism and conceptual anarchy – though still distinctly pragmatist in orientation – is perhaps best illustrated in the work of Jacques Derrida.[30]

In fact, the principal thrust of current pragmatist developments distinctly favors the following doctrines: first, that in rejecting fundationalism and conceding the historicity of human inquiry, we can no longer meaningfully oppose realism and idealism as independent alternatives regarding the cognitive status of science; and second, that in linking the ultimate realist import of competing theories to the tacit conditions of species survival, we can no longer insure an exclusively adequate or correct account of "what there is." Realism now signifies that the explicit dialectical opposition and replacement of ontologies manage to be suitably continuous with the ecological, essentially precognitive, conditions under which the race persists; and idealism now signifies that candidate theories can be tested only relative to our remembered record of the fortunes of earlier theories and our current conception of how to continue to test theories informed by that record. Constrained by these adjustments, relativism now signifies the tolerance of alternative ontologies that, consistently with what, on the best transcendental evidence, we take to be the requirements of system, meaning, simplicity, coherence, validity, may be fitted to the corpus of the developing data of our contingent and essentially incomplete inquiries. The idea that formulated ontologies incompatible with one another could not possibly be jointly validated (in whatever way ontologies can be validated) presupposes both the correspondence theory of truth and a correlated commitment to the strict demarcation of realist and idealist alternatives. Yet even our conception of the requirements of system, simplicity, and the rest cannot possibly escape a measure of relativization. To avoid those presuppositions, therefore, is to apply the force of skepticism against claims of exclusive truth and to reinterpret the new realist import of well-defended theories as falling within the competence of what we may now more profoundly characterize as the instrumental nature of human inquiry. Instrumentalism need no longer be opposed to realism and need no longer be construed as a form of subjectivism or subjective idealism.[31] Realism is itself the transcendentally imputed import of the results of diachronic inquiry interpreted as improved approximations to the way the world is.[32] Hence, it cannot be freed from its conceptual dependence on idealist and instrumentalist constraints; and it cannot be freed from its effective dependence on our contingently dominant technology and social *praxis*.

The point of this abbreviated survey is to confirm the variety of pragmatist positions and the strong contemporary convergence, despite their obvious scatter, toward the three themes sketched – on both sides of the Atlantic. For, in fixing the import of that convergence with respect to the themes of *praxis* and the transcendental, we find ourselves in an improved position to formulate a balanced conception of technology – with regard, say, to both Heidegger and Marx. We need not, therefore, subscribe to any particular version of the pragmatist philosophy (of the sorts suggested) in coming to terms with the pragmatist orientation toward technology itself.

III

There are, one may say, two decisive conceptual foci for contemporary philosophy – that, as it happens, are decidedly congenial to the pragmatist orientation:

(i) human engagements involving the real world are inescapably and ineliminably technological;

(ii) transcendental discoveries are a species of empirical discovery.

These two theses prove to be nicely complementary, in that they entail or presuppose the historical and social nature of human existence and cognition, the equivocally liberating and confining orientation of changing and partial horizons, the reflexive discovery of our cognitive capacities with respect to alternative horizons, the continuity of precognitively ordered and cognitively informed survival and work, and the viability of nonconverging and even incompatible theories of the nature of reality and of man's cognitive relation to reality. Without any doubt, these are among the most salient and insistent philosophical claims of our day and appear, with equal regularity, in both the Anglo-American and Continental literature.

(ii) draws attention to the unique role of man as the sole agent of ontological and epistemological discovery – hence, to the sense in which (opposing Heidegger's worry) man cannot be reduced to a merely technologically accessible object, *even if* there is a defensible sense in which he remains an object of inquiry: very simply put, (i) itself has transcendental import.[33] The relevant lesson is a dual one. In the first place, it is precisely man's transcendental grasp of the technological grounding of his own sustained survival and cognitive achievement that most plausibly insures his sense of the structure of the real world and of his capacity to discern it. And in the second, it remains an empirical question as to what is actually disclosed about technologically accessible reality – both nonhuman and human. There is no possibility – and no need – to escape a technological orientation; there is no need for a flight to Being writ large; there is only a need to realize that theoretical overviews of reality and human cognition, disclosed reflexively within a particular historical horizon of successful *praxis* and its attendant ideologies,[34] are bound to be continually replaced. But such replacements are rationally constrained: (a) by the need to make coherent sense of the relationship of antecedent overviews relative to their own horizons and the present vantage; (b) by the interpreted record of the continuity of human inquiry itself; (c) by the technological clues transcendentally favored as affording the most convincing overview within any present horizon; and (d) by the inevitable concession that, both diachronically and synchronically, no exclusively valid overview could possibly be constructed. In a reasonable sense, this is to combine Heidegger's emphasis on historicity and *alētheia* and Marx's emphasis on the priority of the social conditions of production over individual consciousness.

This is why such recently fashionable views as Richard Rorty's repudiation of philosophy (particularly, of ontology and epistemology), Paul Feyerabend's radical anarchism regarding science, and Thomas Kuhn's stock insistence on the mere discontinuity of paradigms are decidedly premature and both unnecessary and arbitrary.[35] Rorty ignores or rejects the possibility of transcendental arguments; Feyerabend ignores the methodological import of whatever, even on his own view, proves to be scientifically pertinent[36]; and Kuhn ignores the need to locate relevant discontinuities within a larger context of continuous human inquiry. There is, frankly, something dreadfully funny (and wrong-headed) about historicized reflections on the powers of human cognition that seriously conclude that philosophy is altogether misguided (Rorty) or that science is altogether incapable of rational direction (Feyerabend) or that different paradigms of inquiry commit their adherents to living in altogether different worlds (Kuhn). What is wrong, precisely, is the same tendency (to snatch transcendent from transcendental truths) that we have already noted in Heidegger. Like Heidegger himself, all three exhibit that distinctly contemporary emphasis of pragmatism that relies primarily on the thesis of historicity. In so doing, however, each is nonplussed by the challenge of relativism and the puzzle of recovering a measure of objectivity and methodological discipline in the absence of foundationalist guarantees. Failing to resolve the matter, each announces the impossibility of a solution from a pretended vantage that would itself have required a transcendent privilege obviating his own skepticism. It is perhaps because the Kantian conception of transcendental arguments has led us to suppose that such arguments must enjoy foundationalist grounds of certainty, are impossible under the condition of historicity,[37] and are fundamentally opposed to empirical reasoning, that Rorty, Feyerabend, and Kuhn may have been drawn too early to their own discouraged conclusions. But those conclusions are extravagances in any case – and altogether unnecessary.

The pivotal issue remains how to recover transcendental reflection under the constraints of a radically historicized pragmatism. The answer, quite simply, lies with grasping the primacy of social *praxis* and, through that, the inescapably technological orientation of our reflection on the nature of reality. Transcendental questions may be conceded to take their familiar Kantian form: How is (this or that kind of) human cognition possible? where the querying doubt (unlike that of Rorty, Feyerabend, or Kuhn) is heuristic or methodological (Cartesian) – not skeptical or, beyond skepticism, already (transcendently) supplied with the secret knowledge of the impossibility or necessary failure of cognition. But now, transcendental questions can no longer be interpreted in terms of the cognitive competence of any singular, privately informed, or solipsistic agent[38]; they must construe cognition as the achievement of an inquiring society – as conformable with the practices and routines of a socially shaped inquiry. Inevitably, therefore, they must be brought to bear on the evolving conditions of social *praxis* under which an effective science itself evolves. Thus seen, the technological represents pragmatism's best clue linking the tacit and cognitively explicit features of whatever realism we can justifiably ascribe to our science and effective routines of life. Transcendental questions concern what – reflecting on what, within the margin of survival of the species, best appears to constitute our knowledge of the world – appears to be the necessary and enabling conditions of such knowledge. The answers are not entailed or deducible from such knowledge; they are concerned, rather, with proposed conditions that may be taken as the wider *presuppositions* on which such knowledge itself depends. So transcendental arguments are "inferences" – if one wishes so to speak – even, "inferences to the best explanation," regarding the presuppositions of human cognition. Hence, they are peculiarly occupied with the conditions of knowledge, meaning, truth, reference, validity, coherence, consistency, and the like – with just those and only those conditions that oblige us to theorize that there can, in principle, be no more than one global system within which all pertinent conceptual distinctions can be intelligibly made, that force us to concede that postulating *any* plurality of such systems is either incoherent or else self-deceptive. For example, reference to what is true-in-L and true-in-L′ presupposes a language capable of distinguishing L and L′ and of specifying what meets those conditions in a way not restricted by the qualifications of L and L′. What Feyerabend and Kuhn neglect, therefore, is just the transcendental condition under which the radical difference between two theories (any two theories) informing observation and experiment are actually recognizable as such – hence, recognizably not so radically opposed as to preclude *our* entertaining and comparing them. And what Rorty neglects is just the elementary fact that the very admission of science presupposes the possibility of grasping the transcendental conditions under which science functions as such.

Furthermore, to say that transcendental arguments are "inferences" is to say only that they are open to rational appraisal, not that there is a fixed or formulable canon by which their appraisal may be said to be governed. Transcendental arguments are *not* like formal metalogical arguments: there *can* be no relevant restriction in the scope of their premises or accessible evidence (as formally axiomatized systems presuppose), simply because they are the fruit of the continuous and undivided reflexive effort of the entire species functioning ultimately as a single cognitive community to discover what (and whatever) it can make out to be the limits and powers of its entire cognitive capacity – the same limits and powers, one must suppose, that inform inquiry at any and every object-language and metalanguage level. Again, to say that they are inferences "to the best explanation" is to say only that the power of a transcendental argument rests with the ongoing consensus of the cognitive community that weighs the dialectical advantages to itself of competing theories, not that transcendental arguments answer to a canon in any way like whatever provisional canons may obtain for inferences to the best explanation *within* the practice of an established science.

The inferences of transcendental arguments are, in this sense, *always external to a cognitive practice* as far as its purpose is concerned (that is, they are directed to its presuppositions). But they are always, and cannot but be, generated under the contingent conditions of some such particular practice; they are, therefore, speculations about conditions external to a practice made exclusively *within the historical and empirical conditions of such a practice.* This explains the temptation, the impossibility, and the unimportance of every effort to conflate the transcendental and the transcendent. That transcendental arguments, therefore, are a species of empirical argument (ii) is, broadly speaking, the consequence of Heidegger's thesis of historicity; and that our best clue about the validity of such arguments lies with the stablest technological features of social *praxis* (i) is, broadly speaking, the consequence of Marx's thesis about the relation of production and consciousness. The technological, therefore, performs a double role. On the one hand, in accord with Heidegger's and Marx's view, it signifies how reality is "disclosed" to humans – primarily because it is through social production and attention to the conditions of survival (both precognitively and through explicit inquiry) that our sense of being in touch with reality is vindicated at all; but contrary to the thrust of Heidegger's late qualification, the correction of all theories of cognition and reality thus informed is itself inevitably historicized and subject to the ideological limits of any successor stage of *praxis*. There is no escape from the historical condition, but the recognition of that fact itself is the profoundly simple result of transcendental reflection *within* the very condition of history – which obviates, therefore, the inescapability of Heidegger's various (transcendental) pessimisms and the need for his extravagant (transcendent) optimism. On the other hand, the technological signifies how the study of the whole of reality – of physical nature, of life, of the social and cultural activities and relations of human existence – is unified in terms of our own investigative interests. Hence, at the very least, not only can the theory of the physical sciences not afford to ignore the systematic role of the actual historical work of particular human investigators (for instance, against the model of the unity of science program); but also, we can neither preclude the scientific study of man nor insure that the human sciences must conform to any canon judged adequate for either the physical or life sciences. The primacy of the technological, therefore, facilitates a fresh grasp of the methodological and explanatory peculiarities that the human studies may require – for example, regarding the analysis of causality in the human sphere, the relation of causality and nomologicality, and the bearing of considerations of rationality, understanding, interpretation on the explanation of human behavior.[39]

Seen both in its transcendental role (as insuring inquiry a measure of objectivity relativized to the conditions of *praxis* and dialectical review) and in its role *vis-à-vis* the human sciences (as modelling the methodological distinction of such sciences) the technological may fairly be interpreted as helping to preserve whatever distinction bears on human freedom and dignity, the thrust and direction of human inquiry, the balance between realist and idealist components of cognition, the tolerance of plural, even incompatible, theories compatible with a common *praxis*, the provision of grounds for disclosing ideological distortion without appeal to foundationalism, the admissibility of a moderate relativism consistent with objectivity, and such similar doctrines as the recent currents of pragmatism have been advancing. But that is probably as much as one can ask of any relevant theory – and more than most can afford.

Temple University

NOTES

[1] "Die Frage nach der Technik," from *Die Technik und die Kehre* (Pfullingen: Gunther Neske, 1954); William Lovitt (transl.), *The Question Concerning Technology and Other Essays* (New York: Harper and Row, 1977). References are to the English translation.
[2] *Ibid.*, p. 4.
[3] It is perhaps useful to note that a reasonable argument can be mounted to the effect that, in a curious way, despite the influence of Marxism upon them, such authors as Herbert Marcuse and Jürgen Habermas share a good deal of Heidegger's criticism of the technological mode of thinking favored in the era of capitalism; see for instance Michael Zimmerman, "Technological Culture and the End of Philosophy" and "Heidegger and Marcuse: Technology as Ideology," in Paul T. Durbin (ed.), *Research in Philosophy & Technology*, Vol. 2 (Greenwich: JAI Press, 1979).
[4] Heidegger, *op. cit.*, pp. 4–6.
[5] Martin Heidegger, *The End of Philosophy*, trans. Joan Stambaugh (New York: Harper and Row, 1973); *Nietzsche*, Vol. 1 (*The Will to Power as Art*), trans. David Farrell Krell (New York: Harper and Row, 1979).
[6] Cf. Martin Heidegger, *What Is Called Thinking*?, trans. Fred D. Wieck and J. Glenn Gray (New York: Harper and Row, 1972).
[7] *The Question Concerning Technology* . . . , p. 24.
[8] *Ibid.*, pp. 14–15.
[9] *Ibid.*, p. 28.
[10] *Ibid.*, p. 3.
[11] *Ibid.*, pp. 10–12.
[12] *Ibid.*, p. 13.
[13] *Ibid.*, p. 14.
[14] Martin Heidegger, *Being and Time*, trans. John Macquarrie and Edward Robinson (New York: Harper and Row, 1962), pp. 95–102. Cf. Don Ihde, *Technics and Praxis* (Dordrecht: D. Reidel, 1979), p. 9.
[15] *Ibid.*, pp. 95–96.
[16] *Ibid.*, pp. 97–98.
[17] Martin Heidegger, "The Turning," in *The Question Concerning Technology* . . . , p. 37.
[18] Karl Marx, *Theses on Feuerbach*, in *Karl Marx: Selected Writings*, David McLellan (ed.) (Oxford: Oxford University Press, 1977), p. 156.
[19] Karl Marx, *Grundrisse: Foundations of the Critique of Political Economy*, trans. Martin Nicolaus (New York: Vintage Books, 1973), pp. 157–158.
[20] For a particularly touching exhibit of the Stalinist view, see Georg Lukács, *The Historical Novel*, trans. Hannah and Stanley Mitchel (Boston: Beacon Press, 1963).
[21] Peirce is noteworthy, however, in that he alone of the early pragmatists seems to have captured the imagination of Continentally oriented "pragmatists." Cf. Karl-Otto Apel, *Towards a Transformation of Philosophy*, trans. Glyn Adey and David Frisby (London: Routledge and Kegan Paul, 1980), chapter 3; Jürgen Habermas, *Knowledge and Human Interests*, trans. Jeremy J. Shapiro (Boston: Beacon Press, 1971), Pt. II.
[22] See Joseph Margolis, "Skepticism, Foundationalism, and Pragmatism," *American Philosophical Quarterly*, XIV (1977), 119–127.
[23] In Quine, the thesis appears in his insistence on testing theories only *en bloc* (which somewhat misstates the issue by favoring an epistemic idiom), in his admission of a

precognitive "quality space," and in his well-known claim that, regarding veridical choices among alternative ontologies compatible with the totality of conceded data but incompatible with one another, there is "no fact of the matter." In Goodman, particularly most recently, there is a frank emphasis on "making" worlds rather than discovering alternatively defensible ways of describing our one world (Quine's preference); but, even earlier, both Goodman's nominalism (inadequately, perhaps) and the theory of projections presuppose and are really unintelligible without, a version of the second doctrine. Goodman now describes his own position as "irrealism." Putnam's view of the work of science and its history depends essentially on an appeal to an ontological "principle of charity" with respect to diachronic reference to theoretical entities, that once again comes to the same doctrine. The most relevant texts include: W. V. Quine, *Word and Object* (Cambridge: MIT Press, 1960); *Ontological Relativity and Other Essays* (New York: Columbia University Press, 1969) – particularly, "Natural Kinds"; Nelson Goodman, *Ways of Worldmaking* (Indianapolis: Hackett, 1978); *Fact, Fiction, and Forecast*, 2nd ed. (Indianapolis: Bobbs-Merrill, 1965); Hilary Putnam, *Meaning and the Moral Sciences* (London: Routledge and Kegan Paul, 1978). In putting these specimen views forward, I do not defend them. On the contrary, I regard all three as indefensible, though not for reasons affecting the second doctrine. I have discussed these views in a number of places, including the following: "Behaviorism and Alien Languages," *Philosophia*, III (1973), 413–422; "The Problem of Similarity: Realism and Nominalism," *The Monist*, LXI (1978), 384–400; "Realism's Superiority over Instrumentalism and Idealism: A Defective Argument," *Southern Journal of Philosophy*, XVII (1979), 473–479; "Cognitive Issues in the Realist-Idealist Dispute," *Midwest Studies in Philosophy*, V (1980), 373–390. "The Locus of Coherence," forthcoming in a collection edited by Douglas Stalker, for D. Reidel.

[24] Cf. Jean-Paul Sartre, *Critique of Dialectical Reason*, Jonathan Rée (ed.), trans. Alan Sheridan-Smith (London: NLB, 1976).

[25] Cf. Ludwig Wittgenstein, *On Certainty*, G. E. M. Anscombe and G. H. von Wright (eds.), trans. Denis Paul and G. E. M. Anscombe (Oxford: Basil Blackwell, 1969).

[26] Cf. Maurice Merleau-Ponty, *Phenomenology of Perception*, trans. Colin Smith (London: Routledge and Kegan Paul, 1962); *The Visible and the Invisible*, Claude Lefort (ed.), trans. Alphonso Lingis (Evanston: Northwestern University Press, 1968).

[27] Representative texts include: Jürgen Habermas, *Knowledge and Human Interests*; Herbert Marcuse, *Reason and Revolution* (New York: Oxford University Press, 1941), *One-Dimensional Man* (Boston: Beacon Press, 1964); Pierre Bourdieu, *Outline of a Theory of Practice*, trans. Richard Nice (Cambridge: Cambridge University Press, 1977); Karl-Otto Apel, *Towards a Transformation of Philosophy*.

[28] Cf. L. S. Vygotsky, *Thought and Language*, trans. Eugenia Hanfmann and Gertrude Vakar (Cambridge: MIT Press, 1962), for an early criticism of Piaget; Roland Earthes, *Elements of Semiology*, trans. Annette Lavers and Colin Smith (London: Jonathan Cape, 1967); Paul Ricoeur, *Interpretation Theory: Discourse and the Surplus of Meaning* (Fort Worth: Texas Christian University Press, 1976). For Piaget's most recent views, see Massimo Piattelli-Palmarini (ed.), *Language and Learning: the Debate between Jean Piaget and Noam Chomsky* (Cambridge: Harvard University Press, 1980). Piaget's view of *praxis* is essentially biological but hardly social or cultural.

[29] Cf. Lucien Goldmann, *Essays on Method in the Sociology of Literature*, William Q. Boelhower (transl. and ed.), (St. Louis: Telos Press, 1980); also, *Lukács and Heidegger:*

Towards a New Philosophy, trans. William Q. Boelhower (London: Routledge & Kegan Paul, 1977).

30 Cf. Jacques Derrida, *Speech and Phenomena and Other Essays on Husserl's Theory of Signs*, trans. David B. Allison (Evanston: Northwestern University Press, 1973); also, *Of Grammatology*, trans. Gayatri Chakravorty Spivak (Baltimore: Johns Hopkins University Press, 1976).

31 Cf. Karl R. Popper, *Objective Knowledge* (Oxford: Clarendon, 1972), pp. 194–197, 261–265; *Conjectures and Refutations* (London: Routledge and Kegan Paul, 1962), chapter 3, and Ernest Nagel, *The Structure of Science* (New York: Harcourt Brace, 1961), chapter 6.

32 Cf. Margolis, "Realism's Superiority over Instrumentalism and Idealism: A Defective Argument," *Southern Journal of Philosophy*, XVII (1979), 473–479.

33 Cf. Apel. *loc. cit.*

34 Cf. Joseph Margolis, "Philosophy and Technology," in Paul Durbin (ed.), *Research in Philosophy and Technology*, Vol. 1 (Greenwich, Conn.: JAI Press, 1978), 25–37.

35 Richard Rorty, *Philosophy and the Mirror of Nature* (Princeton: Princeton University Press, 1979); Paul Feyerabend, *Against Method* (London: NLB, 1975); Thomas S. Kuhn, *The Structure of Scientific Revolutions*, 2nd ed. (Chicago: University of Chicago Press, 1970).

36 Cf. Joseph Margolis, "Scientific Methods and Feyerabend's Advocacy of Anarchism" (in translation), in Hans Peter Duerr (ed.), *Versuchungen. Aufsätze zur Philosophie Paul Feyerabends* (Frankfurt am Main: Suhrkamp, 1980).

37 Cf. P. F. Strawson, *The Bounds of Sense* (London: Methuen, 1966).

38 Cf. Apel. *loc. cit.*

39 Cf. Joseph Margolis, "Puzzles about the Causal Explanation of Human Actions," in Larry Laudan and Adolf Grünbaum (eds.), *Topics in Explanation in the Biological and Behavioral Sciences* (Berkeley: University of California Press, forthcoming).

PART V

DIRECTIONS FOR PHILOSOPHY OF TECHNOLOGY

ALWIN DIEMER

THE CULTURAL CHARACTER OF TECHNOLOGY

The word "technology" (*Technik*) belongs to a group of philosophical words for which a specific character is indicated only by getting a special, typical stamp from a real-life context. Plato gave the first formulation in the history of philosophy – a definition suitable for his time. We can discuss technology in the singular only if we feel that we all mean the same thing, the very same object (technology) – and for that reason use the same definition. If, aware of this situation, we want to take up the theme, "the cultural character of technology," we must separate our treatment from all those philosophies which pretend to provide *the* "essence" of technology – whether the definition they offer is metaphysical (Ideas), anthropological (a derivation from "forms of action" – Kapp, Sachsse, and others), sociological, or something similar. This, moreover, is true especially for all cases in which technology is simply spoken of in the singular – usually without describing and differentiating the phenomenon under discussion.

In themselves, all these views are, ultimately, oversimplified monolinear constructs which can, for the most part, contribute little to the clarification of the real problem of technology.

By contrast, a reasonable course seems to be to set out from the real problematique: technology has come to designate a phenomenon of world-wide concern today; i.e., it presents a problem in every culture throughout the world.

That technology has its origin in European cultural forms is generally and clearly recognized. This is especially true if technology is understood in the sense of *modern* technology. If this is recognized, then it appears sensible to begin with *culture* as pre-given and only then to highlight technology as a form of culture, to emphasize its cultural character. It becomes possible in this way to treat the theme not only intra- but transculturally – especially the latter. Only in this way, then, does it seem possible to apply the technology theme to Third-World culture.

The philosophical basis of the following inquiry is a phenomenology which, under the idea of a phenomenological ontology, takes the world as pre-given point of orientation. The result is a corresponding definition of culture which, in brief, lays out essential structures, elements, etc., as required

Paul T. Durbin and Friedrich Rapp (eds.), Philosophy and Technology, 313–318.
Copyright © 1983 *by D. Reidel Publishing Company.*

for a thematic treatment. The problem of the plurality of cultures is indicated in a few examples. From all this follows an "ontological culturality" of cultural givens, prior to forms-of-culture. An effort to explicate the technological culture-form then forms the conclusion – without, however, giving an essential determination, definition, or the like. Rather, technology is presented as a complex form in which it is understood as meaning modern technology which is a product of European culture. The different elements allow us to hint at the cultural character of technology in particular aspects of different cultures. A theme of especially great interest would be problems of technology in the Third World – which, naturally, must be formulated as "technology in those cultural complexes known as the Third World."

(1) *Culture* is defined, on the basis of such a pre-given, as the "totality of the horizons of a particular *Lebenswelt*, a world in which a particular human group lives, in which it gives meaning to its reality (including under this its reality as such, including as human)." This is not the place to explicate this definition more precisely. The principles are enough. But this much is true: the principles provide a frame in which to view the givens of each culture, including its intra- and intercultural reality. A characterization of the "culturality" resulting from this demands a further explication of cultural structures, cultural elements, etc. Here too only an overview can be given. A thematic treatment can then be articulated under formal and material aspects.

The pre-given *formal* horizon includes four successive components: (1) the horizon of pre-given absolutes, including such meaning dimensions as the fundamental triad: self-evidence, confidence, and familiarity (with their oppositions: strangeness, unfamiliarity, etc.); (2) the space-time unit as the precondition of the possibility of *place* (with respect to the givens of each culture); (3) "dimensioning," which in turn is further differentiated to allow for a "ground-dimension," as precondition of each particular absolute ground (e.g., God, Spirit, matter-as-absolute), and contingency dimensions grounded thereon ("creatures," things, etc.); and (4) the given "orthostructure" of each culture, including what it means by rationality and values.

Every single culture involves this sort of "formal" structuring. By means of it, among other things, it becomes possible not only to compare cultures with one another but also to make correlations among them and display their compatibility with one another, etc.

How science and technology are defined in any culture corresponds to its culture-form. We can now examine this sort of process in general terms.

The totality of horizons, once made explicit with respect to the givens of culture, can themselves be designated as "ontological culturality." (We have yet to speak of such horizons in terms of concrete examples.) Corresponding to the formal totality (above) there is, naturally (but always within the givens that belong to a culture), a *material* totality or complex. Self-evidence is something real – and in this context it highlights the fact that it pre-determines, for each culture, the meaning of the world "real." From the perspective of European culture (which is where we are located), we know how to distinguish the real in "human reality" and in the "tectonically" real (i.e., the *totality* of a cultural "form," "system," or whatever one calls it). To illustrate the problem: in a particular case, a (somehow) pre-cultural human group may in itself pre-comprehend and thus constitute what is real – including the humanly real and so what "human" means in that culture; but this does not imply that in that culture "man as totality" exists. Take "man" in our culture-form; certain members of the group know how, with others, to fashion their own culture-form; it is similar to word forms in German, e.g., words with the suffix "-ling" form the *totality* of "ling" words: *Jüng-ling, Frisch-ling, Früh-ling*, etc. Here the suffix "-ling," as contrasted with another group, say that defined by "speaking," constitutes its own (class-type) reality.

Going a step further, it should then be obvious that the collective, the "tectonic," can only be explicated on the basis of certain pre-conditions. This is so, correspondingly, for all the culture-form subdivisions within European culture, i.e., the customary "cultural spheres": linguistic, artistic, educational, economic, scientific, etc. And so it happens, respectively, with technology and "the technological." Nothing in this culture-form is *in itself* absolutely real, but only what each culture determines. There is, accordingly, no science, no art, and similarly no technology *as such*. There is, in each case, only a particular orienting type of pre-understanding.

(2) We turn next to technology, but some comments on *phenomenological method* are in order. The method involves two perspectives, one dealing with language, the other with the phenomena themselves. The first emphasizes meanings of words in both diachronic and synchronic dimensions, including cognate words. The second begins with a given phenomenon and sets out specific elements which are taken to be constitutive. Since in most cases several can be named, each culture-form can (in some cases *must*) be seen or interpreted as complex.

In such a procedure, the elements themselves naturally must then in turn be explicated in a corresponding phenomenological fashion. If we proceed in this way, it becomes obvious – just put forward as a framework or basic thesis – that there is no such thing as a single-feature essential definition, nothing like "*the* essence" of technology.

There seems, with that, the threat of a certain arbitrariness in the search for the aspects of the word-phenomenon-field seen as constitutive; on the other hand, the formal horizon-structure, with its dimension-fields, predetermines what will be considered among the elements of the culture-form. (Conversely, other cultures include still-open features of their own domain, say, with respect to Nature-versus-matter or the phenomenon of work or labor – see below, with respect to European and African culture.) It is only such a procedure that allows us so to create the presuppositions for transcending the sphere of given ideas and of a given culture and so to understand cultures and their respective culture-forms in a different way. A particular problem that may be taken up is that of technology in the Third World.

(3) We come now to technology, and two presuppositions must be noted. First, we are enclosed within European culture (assuming we put aside here a broader-ranging explication even though, in a systematic procedure, it would be in order); and second, we are enclosed within the concrete, actual situation of this culture. Here we are dealing with the historicity as well as the (ontological) "culturality" of technology. So we might perhaps say this: what we designate as features of the technological phenomenon were first identified, in terms of constitutive origins, in the second half of the eighteenth century. Since that time, for example, talk of "technology" as a (relatively) separate reality has become possible – and thus real. Given these presuppositions, as we begin our investigation, it must be stressed (again) that it is a phenomenological investigation – no more and no less. The ideas put forward here, in terms of both method and theme, are a matter for discussion. The same holds with respect to possible objections: since we are dealing with trivial or everyday matters, deciding is what will make everything hang together.

The phenomenon of technology is open to a sort of dual-character typological distinction for which the "software/hardware" designation seems appropriate. "Hardware" implies matter or the material. At times this characterization refers to a significant "distancing" feature within the Nature-dimension in the matrix, Man/matter/Nature (see the spatial "formal horizon" above, as well as the "orienting scheme," below). This differentiating feature leads logically to another: to work or labor as a specifying feature of human behavior. This phenomenon, the labor-feature, might best be characterized as "manipulation of a given matter (material) for a purpose – where the manipulation appears as *performance* or *effort*."

This basic feature of work or labor involves overarching components which are captured in the designations "construction" and "development" (in the precise sense of modern "technological-industrial development"). With this we cross over from the "hardware" to the "software" side; and this is where the (formal) orthostructure becomes especially relevant: rationality is manifested as typical "scientific-technological rationality" – or, more precisely, in terms of two-valued logic with its attendant features. In other cultures, on the other hand, there emerge rationality-forms such as magic and the like.

We are coming now to the end. (Some features of the phenomenon, technology, still have not been determined, but, with the help of the horizon formalism, they could easily be found.)

When all this is put together, the investigation in search of the totality-structure of technology can be pursued. It could be supported by still another schematic outline, involving *systematic orientations* which sketch individual elements in order to make possible bridge-building to other cultures.

We might perhaps take work or labor in the Nature/matter context. Here an outline can serve as illustration:

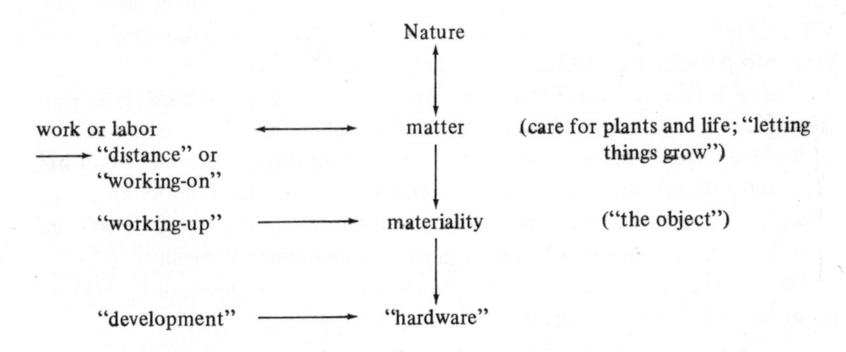

This little outline can serve as the transition to our concluding section on the Third-World problematique.

(4) The "Third-World problem," seen from the perspective of a philosophy of culture, requires first of all a phenomenological clarification from – yes, here again – a cultural perspective. It must proceed in terms of the general outline developed above. (To carry this out in greater detail here would lead too far afield; however, the author has carried out such an investigation elsewhere.) As an example we will take African culture.

Some examples should illustrate the problem: the outline dealing with work or labor (above) shows very well – above all with respect to the Nature dimension – that man is accommodated to nature. Work transforms Nature into matter, the object worked on in the process. With this there comes to be a distance between man and the (now dead) reality of Nature – however the latter is pre-determined in African culture. (This could be extended further; in the author's opinion, this whole business comes up especially in discussions of Socialism.) In this context, it must be said that there can (paradoxically) be no workers in African Socialism. There is lacking, in African culture, a materialistic distance from Nature; Man (understood categorically) is perceived as *within* Nature. This further implies that there can be no proletariat in the European sense – and so also no capitalists.

It is not by chance that in African Socialism a capitalist of this sort is always a representative of European culture. The result is that European Socialism or Communism belongs in a special way to European culture, and African Communism cannot appreciate the full meaning of specifically European elements – including European technology.

Another example, perhaps, might be rationality; here perhaps a "black philosophy" takes the place of technological thought. Wiredu, in his book *Philosophy in an African Culture* (1980), has given an example of an African VW assemblyline worker: when the European workers left, the African slid back into African ways of thinking – and, naturally, failed.

This is a task for the future: to develop a structural analysis of African culture using a pre-conditioning culture-phenomenological system – similar to the one explicated here. Such an analysis could perhaps provide a cultural pre-condition for the introduction – and eventually the assimilation – of European culture. And in turn questions about the compatibility of African with European culture should also be analyzed phenomenologically.

Only in this way could problems of so-called "aid for developing nations" be solved with respect to the Third World.

University of Düsseldorf

HANS HEINZ HOLZ

THE IMPORT OF SOCIAL, POLITICAL, AND ANTHROPOLOGICAL CONSIDERATIONS IN AN ADEQUATE PHILOSOPHY OF TECHNOLOGY

In these discussions, an emphasis on epistemological issues, to the (relative) neglect of social and political issues, runs counter to a number of well known theses.

I

(1) Technology stands to human aims in an instrumental relationship; this is the type of rationality characterized by Max Horkheimer as "instrumental." However, since this sort of rationality belongs, not to an abstract "humankind," but to individuals, instrumental rationality is always exercised for particular purposes. This stands in opposition to dialectical rationality, which postulates "reflection on reflection" in order to overcome the particularity of instrumental rationality.

(2) It is not possible to conceive of technology as just one among many of the partial systems that constitute the human world. Otherwise it would be possible to withdraw from this technologically determined region and retreat somewhere else. But technology permeates all of life (except possibly a small reserve of religious privacy). It is, like nature, universal; indeed, it is a substitute for the genuineness of nature.

(3) This expansiveness is a consequence of the so-called "scientific-technological revolution" – and it is necessary for man. Man is *Homo technologicus;* he produces for himself, according to his own purposes, the conditions of his own reproductive life. "According to his own purposes" means this: men invent means to achieve their goals.

II

These theses (and their presuppositions) require at least a minimum elaboration. Technology appears to modern man as a system which envelops and determines every aspect of life. Man's needs are satisfied by products of technological realizations. While developments in technical instruments and technological systems have come to be more and more dependent on results of scientific research, science in its turn requires new technical means to make

Paul T. Durbin and Friedrich Rapp (eds.), Philosophy and Technology, 319–322.
Copyright © 1983 *by D. Reidel Publishing Company.*

new discoveries. This basic fusion of science and technology has transformed science into a force of production.

Nuclear power, faster missiles, genetic engineering, longer life-expectancy as a result of scientific progress and the worldwide spread of an increased standard of living – all these are examples of a new quality in the relation between man and nature. There are risks inconceivable until now – including, at least in principle, the risk of human self-destruction.

An individual can master no more than a small fragment of technological capability or of the scientific knowledge needed to apply technical apparatus to production. The universality of the technological system thus appears to the individual as an alien and anonymous force which determines him. With respect to technology as a whole, man feels deprived of autonomy, subject to unknowable necessities. While, on one hand, the vast improvement in the standard of living consequent upon technological development leads to a naive belief in technological progress, on the other hand men react to the alien power of technology with an unreflective hostility – or by grasping at utopian philosophies of "alternative lifestyles," as if life without technology were possible today.

Although only in our age – an age characterized as involving a "scientific-technological revolution" – is it the case that technology determines the whole of human life, this universality of impact is not an accidental effect of contingent developments in human nature. It is an essential feature of man's special place in nature.

Man, as a natural being, reflects nature and his place within it. Objectively speaking, this reflecting amounts to this: man works on natural objects, elaborating patterns of transformation for specific purposes but, thereby, perceiving nature as a whole – i.e., natural laws – as the horizon (both limit and dream) of his transforming activities. This work on nature is mediated by tools which thus become the material realization of the reflecting-nature relations; the human hand, freed from its function as an instrument of local motion, is the first of all tools. Because of both the link of nature-reflecting to natural objects and the mediation of tools, Karl Marx speaks of man as an "object-bound being" (*gegenständliches Wesen*).

Man's relation of reflecting nature, given realization through object-bound activities, is re-presented in language, the universal system of signs. In principle, every possibility in the world can become an element in language. Language is a necessary feature in social activity, i.e., in the inter-subjective use of tools. Man's activities and their representation in language have an essential relation to the world as a whole; they are (at least potentially) a reflection of the whole world. (Recall Leibniz's monads; they mirrored the whole world, even though all their perceptions were not distinct.) Man by his activities transforms the world and the transformation in principle reaches the whole world – at least within the limits set by his instruments (the *Lebenswelt*).

It is not, however, individual men but humankind which realizes the totalizing potential of human activities; that is, it is not individual but social activity that brings about the transformation of the world as a whole. The system of instruments, their use, and the functional relationship between them and their effects – all this is what we call "technology." And from the beginning it has had a totalizing tendency precisely because, as an "object-bound being," man has always reflected the world as a whole. The subject-object relation always implies totality (cf. Jean-Paul Sartre's description of subjectivity as "totalization").

III

To return to the original theses: As means-systems get more complicated, they tend to lose their connection to their original goals and take on independent, particular aims of their own; the latter, in turn, strive to maintain themselves in competition with other partial aims. This is the source of today's anarchy of particular aims, whose dimensions no longer allow stabilization or self-regulation. For example, population growth requires a higher rate of energy production, which leads to risks in terms of safety and interference with the ecological balance, and so on. Therefore a universal, not merely a partial rationality with respect to guiding technological progress presupposes (at least the basic outline of) a hierarchical goal system – a system which is deducible in a reasonable manner if it is to be acceptable at all. In a real democracy based on majority consensus, self-determination implies the ability to argue during the decision process. If there were no agreed-upon hierarchy, society's limits on uneven technological advances might be arbitrary.

A hierarchy of goals that is persuasive might be constructed on the basis of a system of needs and wants. The basic ones will be easily agreed upon. Marginal ones – where wide disagreements as to relative importance can be expected – could be allowed to balance one another so long as vital interests were not affected.

I conclude, from these brief reflections, that – since, in terms of fulfilling human purposes, technology has only an instrumental character – technology as the universal determinant of the interaction between man and nature *must* (and can) be directed by conceptions that reflect the prevailing aims of man's present and future life. Such conceptions need not be arbitrary, as those would have it who believe in a sharp distinction between facts and values. Needs and wants can be taken as facts (biological or historical), and the systematic hierarchy among them can constitute a preference (or value) system which turns out to be persuasive in a democratic order. This means that the all-consuming character of technology is not a problem for technology itself; it suggests that an anthropological – even, in one sense, an ontological – foundation needs to be erected before the philosophy of technology can supply us with any practical "orienting knowledge" (as Hermann Lübbe calls it) that will be commonly persuasive.

University of Groningen

ELISABETH STRÖKER

PHILOSOPHY OF TECHNOLOGY: PROBLEMS OF A PHILOSOPHICAL DISCIPLINE

It has often been noticed that philosophy has come to deal with the problems of technology only in a remarkably late period. The probable reasons for this fact need not be discussed here, and the arguments for it that are normally put forward – lack of concern for *praxis* and an inappropriately high philosophical estimation of theoretical reasoning – would require detailed investigations, including a close look at the history of philosophy.

Hardly less conspicuous than the neglect of technology in traditional philosophy is, it is true, the relatively low ranking that philosophical studies of technology have up to now attained within philosophy in general.[1] It is indeed, given the impact which technology has had on the very center of nearly all areas of our life, to be noticed with astonishment that the situation of the philosophy of technology within other philosophical disciplines can at this moment only be characterized as marginal.

But the reasons for this, however various and multiple, should by no means be attributed simply to factors external to philosophy; rather, they should be looked for in the field of philosophy of technology itself. At any rate, philosophy of technology would, by considering this possibility, respond to the demand for critical self-reflection that cannot be refused by a real *philosophy* of technology. That does not exclude the possibility that there are reasons that philosophy alone is not responsible for. Can it not be that, as philosophy of *technology*, it may have peculiarities of its own that go beyond the specific subjects and methods which by tradition constitute a philosophical disipline? Probably there is only a vague possibility that philosophy of technology, as it is, is capable of a disciplinary structure analogous, for instance, to philosophy of art, philosophy of science, or philosophy of language.

It is no accident that the situation of philosophy of technology shows peculiar characteristics that have not been found and still are not to be found in other philosophical disciplines. It must be attributed to the relatively late start of the philosophy of technology – to what I would like to call the "paradox of its continual beginning." Without taking into account some of the metaphysical studies that, indeed, could never hope to resonate with lasting effect – and this is not only for reasons of growing philosophical

Paul T. Durbin and Friedrich Rapp (eds.), Philosophy and Technology, 323–336.
Copyright © 1983 *by D. Reidel Publishing Company.*

reservations about speculative global projects – and, furthermore, putting aside considerations determined more by moral pathos and political engagement than by sober philosophical description and analysis, one is still confronted at present only with a multitude of detailed philosophical studies each of which can be characterized as merely a sketch or an attempt.

Some of these sketches strike one as early skirmishes in a philosophical debate; hence they do not seek greater coherence even where attempts to deal with the same or a similar problematic already exist. Furthermore, although it occurs quite often that authors refer to each others' work – even if they disagree in their views – it is only in rare cases that a criticial discussion of the same subject is carried out. Apart from that, the preparation of bibliographical materials tends towards documentary completion;[2] however, systematic elaborations and assessments have yet to be entered upon.[3]

These may all be external symptoms of the lack of structure that still exists in the philosophical problematic of technology. Such a state, a sort of "anthology philosophy," seems to be a literary reflection of the situation; it is a project involving the presentation of a variety of new – and they are indeed new – problems and attempted solutions, which project makes remarkable progress in individual subjects but which nevertheless is stagnant in terms of systematics.

It could be assumed that this situation in the philosophy of technology is due to the enormous complexity of its field of research. To be complex is, however, no privilege of technology. Neither is "complex" a predicate belonging to an object itself; it is ascribed to an object only according to the distinctiveness of the methods of investigating it. On the other hand, could it be that technology displays a peculiar kind of complexity, and that is what is novel for philosophy? If this were true, philosophy might not be able to fall back on those traditional approaches which have so far structured the problematic of philosophy of technology.

As is generally known, technology has multiple ties to other fields of research: science, society, the economy, politics. This requires interdisciplinary efforts for a philosophical investigation. With this, philosophy is soon faced with limitations on its competence; judgments of specialists in non-philosophical areas are required. But if it were only a matter of interdisciplinary difficulties, these would not, in principle, be any greater for philosophy of technology (nor, on the other hand, any smaller) than for other cooperative ventures. However, there is a further point: the philosophical problematic of technology is complex in a twofold respect.

I. First, not only are the relations just mentioned a matter of mutual interdependence, so that the relations between technology and the economy, technology and science, technology and ethics, etc., must be dealt with from both sides if the philosophical problematic of technology is to be expounded in an appropriate way. But there is also mutuality in areas external to technology, where the mutuality is mediated by technology. The often cited "totality of modern technology" can be given a precise interpretation only through a deeper penetration into this closely woven net of mutual relations.

Does this mean that to examine the connections between, for instance, science and ethics, or the economy and politics, inevitably involves technological implications? It cannot be maintained that these implications must *always* be part of the matter of investigation. According to the aims of a particular investigation, it may well make sense to focus on pairs of relations and their permutations as not being primarily technological, to set aside their technological implications.

Philosophy of technology represents quite a different case, however. It has to be part of its aim to get in touch with the phenomenon of technology even where it does not show itself immediately; accordingly, philosophy of technology has to make use of methods capable of penetrating into that network of relations if its numerous threads are to be disentangled. But have the analytical procedures of philosophy, as so far applied, been adequately applied, or has the network become so impenetrable that philosophical procedures, traditionally proved satisfactory, may no longer have "bite"? Could it be that, in the end, philosophy of technology is not only the newest philosophical discipline, but already too late – the effort to structure a field and to find an identity hopelessly doomed to failure?

II. Furthermore, the problematic aspect of philosophy of technology shows itself in still another respect. Difficulties of the kind just mentioned – as they come to haunt this philosophical discipline as it strives for self understanding – seem to be the lesser part, to be only a reflection of a more fundamental difficulty that arises as soon as its subject is identified.

While it is not open to question, for example, that philosophy of art is concerned with art, or that philosophy of language deals with language, by contrast it seems to be a critical question what the subject matter of philosophy of technology really is. Is it something "technical" (*Technik* in German) or something "technological" (*Technologie* in German)? (The collective singular noun *Technologie*, analogous to "natural science" in the singular form, may stand for a set of technological sciences.)

It seems no accident that the German *Philosophie der Technik* or *Technikphilosophie* has been translated into English as "philosophy of technology";

it is no less accidental that the translation poses no obvious questions for either side. Of course, there is a close connection between *Technik* and the technological sciences (*Technologie*). But as close as this connection might be, it is not an identity. That this differentiation is philosophically significant can already be seen; a distinction between *Technik* and *Technologie* is increasingly being made within the philosophy of technology.[4]

Neither can the distinction in question be minimized by analogies with situations in other philosophical disciplines. Philosophy of language, for example, in certain respects, also deals with both language and the science of language (or linguistics); whether it sets language off against linguistics to further characterize its own problems and aims, or whether the philosophy of language in its explicit reflections upon linguistic problems gains new questions for itself.

There is an even more illustrative example in the philosophy of nature. It has developed to such an extent, with nature transformed into scientifically investigated nature, that the old characterization has been given up in favor of a philosophy of science with new features for which there was no pattern in the old philosophy of nature.

Could we draw the inference that philosophy of technology, too (at least nowadays), will have to investigate technology (*Technik*) only in its scientific forms – and so will have to be understood exclusively as a philosophy of technical sciences? Even if the analogy would lead to an affirmative answer to the question, the distinction between technical and technological issues would, nonetheless, be neither eliminated nor rendered unimportant. If philosophy makes this distinction (philosophy of science, interestingly enough, fails to mention the distinction between nature and the science of nature), this is already an indication of the fact that the differentiation between technology (*Technik*) and technical science (*Technologie*) is categorically different from the one that exists between nature and natural science, between language and linguistics, etc. But if it is such a philosophically relevant distinction, it cannot in the long run be neglected without shortcomings, inadequacies in the conception of problems, and even confusions. If, indeed, despite this important distinction between technology and technical sciences, philosophy of technology in most of its detailed studies seems to be uninfluenced by it, the situation is once again symptomatic – not necessarily of a fundamental confusion in distinguishing its subject, but of that state of "continual beginning" that was hinted at earlier. For, according to what objective is aimed at, philosophical studies can be carried out which, in preliminary stages, permit elimination of the distinction in question without precision being impaired. Nevertheless, the distinction is very important for the philosophy of technology.

Considering the special nature of the distinction between technology and technological sciences, it might be suggested that we make a comparative study – especially with respect to the similar distinction between nature and natural science. But such an analogy limps. This is obvious if only from the fact that nature exists independently of and beyond any science of nature; or from the fact that nature, despite scientific investigation of it, continues to carry features of the Greek *physis*. As such, it has never, up to now, stopped being important for us in shaping our lives, and at the present time it exerts genuinely new impacts in terms of the scientific research devoted to it. Nor is it an accident that, today, nature has become the focus of interest in its non-scientific form as well – above all as nature worth preserving and protecting as it becomes badly endangered and severely damaged.

If, by contrast, one should maintain that technology correspondingly existed before all technological sciences and independent of them, the claim might, at best, have historical meaning; and even then it would run the risk of anachronistic distortion. Technology, rooted in the ancient *technē*, doubtless existed as a distinct ability to produce things to be put to use in daily life, to bring forth artificial objects – i.e., those not pregiven by nature. This could be witnessed long before anything like technological science arose, late in the eighteenth century. When it arose, technological science not only investigated scientifically the methods for producing the earlier artificial technical objects under various specific conditions; it also started to systematically examine the prerequisite know-how and thus helped pave the way for the development of what were later called engineering sciences.

The technological sciences, however, have not left the pre-scientific *technē* untouched, as if the latter continued to exist side by side with them, in the same way nature exists side by side with natural science. It is not just a matter of opinion whether, on the one hand, we consider the field of research on technology to be the technological sciences, or whether, on the other hand, we regard technology as the set of practical production procedures that have remained as independent of the technological sciences as they had always been. Wherever pretechnological forms of producing artificial objects have continued to exist, it is not as prescientific technology but as separated from it – as arts and crafts regarding themselves as sharply opposed to technology.[5] In the course of historical development, the more technology has progressed, the more it has been adopted by the crafts; it has even found its way into the production of works of art. This process quite obviously is continuing. It proves, however, that the technological sciences have gotten hold of technology's object (technical *praxis*) in a way that has not happened in the supposed analogous case of natural science and nature. The development of the technological sciences is marked by a trend towards a radical, technological penetration of every kind of technical production even where the latter still shows some remaining features of the old *technē*. Thus there is – at least in terms of tendencies – nothing about technology that retains effect and validity outside scientific practice.

This state of affairs makes the very close interconnection between modern technology and technological sciences understandable. Step by step there is a progressive development, a change of the one along with the other, with technological know-how at work in technical *praxis* at the same time it gains new scientific problems from this *praxis*. Such a symbiotic relationship offers some justification for those problems of the philosophy of technology in which the sharp distinction between technology and technological sciences is disregarded.

Another philosophically relevant distinction – quite different from that between technology and technological sciences, far older and more systematically comprehensive – is the distinction between natural science and technology. It is not a new distinction for the philosophy of technology, and had already, as is known, been given considerable attention in the philosophy of science. It has, moreover, been the philosophy of natural science that has pointed out the difference between natural science and technology most distinctly; it has done so, indeed, more than it has emphasized their mutual relationship. Varying claims of natural science and technology favored a conceptual distinction to the effect that (from the viewpoint of natural science) technology came to be customarily characterized as mere applied natural science.

It is true, necessary corrections of this conception have been made. With more attention given to the technological preconditions for the application of instruments with which to achieve scientific knowledge of nature, the problematic within the philosophy of natural science of the relationship between understanding and action had to become salient. In this case, the meaning of "theory" in a search for knowledge that is dependent on instrumental *praxis* has become the object of discussion. Philosophers must take into account, too, more precise differentiations between technical acting and the experimental activities of science.

However, there has been little progress within analytical research in this area. The philosophy of technology, for the most part, still takes as its orienting guidelines such indefinite (and hardly valid any more) opposites as "theory and *praxis*" or "knowledge and action." This problematic is common to both theory of science and theory of action. As the theory of action (at least within European philosophy) has made notable progress – although only in the last few years – it is to be expected soon that it will give impulse to the clarification of the term "action" in natural science as well.[6] One should, however, beware of underestimating the special philosophical difficulties that affect the relations between natural science and technology in a particular way.[7] These are difficulties based on a deep epistemological problematic that has so far not only remained unsolved but has hardly even been appropriately expounded. So the philosophy of technology is confronted with a mystery. For, how is it to be understood that modern natural science arrives at a kind of knowledge in which, on the one side, it makes use of ideal mathematical structures to grasp real nature, while in doing so it is successful only by intruding into nature by way of experiment? What, indeed, is to be made of the fact that natural science engages mathematical science – which is in principle independent of reality – for the study of real nature? It even happens that it comes across regularities whose hypothetico-deductive connections in certain respects correspond to the axiomatic-deductive system of mathematics. These hypothetical connections, however, come to be conceived neither on the basis of mathematical science nor on that of merely pre-given nature; the mathematical regularities of nature arise only when nature is at the same time instrumentally reorganized in experimentation.

Generally, for this functional relationship between mathematicization and instrumental structuring, we use such phrases as "the application of mathematics" or "experimental application." "Application" has more than merely a double meaning here; normally we fail to reflect on how nature can be conceived as "real" nature if it is subject to such a twofold – i.e., mathematical as well as experimental – manipulation. In whatever way the meaning of reality may have to be more explicitly expounded here, in any case nature, as recognized in natural science, without doubt has a sense of something that is made, produced. Its reality is not only pre-given; it is a produced reality. It is, nevertheless, not the reality of the artificial object of technological production, for experimental *praxis* is not technological *praxis*. The activities are structurally different in both cases.

The fact that modern technological production requires the help of natural science knowledge must not lead us to forget that results obtained from experiments cannot automatically be regarded as instructions to act concerning the technological treatment of nature – as an instrumentalist philosophy of science all too rashly maintains. The argument against this is not only that there is knowledge of nature which excludes any technical transformation; it must furthermore be emphasized that something like the technological sciences is required to make possible such a transformation, which natural science of its own cannot bring about. What, then, are the relations between experimentally established facts of nature and their technical utilization for new artificial objects of technology? For an answer to this question, an important epistemological role can be attributed to technological science, although it is not intrinsically its own but must be assigned to it by philosophy. It seems to be necessary to make fundamental statements about the relationship between technical science and natural science, if only because, in analyzing this relationship, the question of the reality of both scientific and technical objects can far better be dealt with. This distinction remains unclarified as long as the technological sciences are not taken into account.

There is an additional fact that can just be hinted at here: Natural science, preoccupied with the so-called "pure case" of a natural process – one that can be reproduced as exactly as possible – downgrades deviations that occur to the minor level of "disturbing" conditions. And it does so not infrequently with methodological legitimacy; for its own purposes science can rightly regard itself as exempt from any duty to view such disturbing factors in any other way. Technological science, however, takes such disturbing factors to be real factors; and, in order to discover the how-to of a particular technical production process, it must count on these as real factors of natural processes. If one were to ask what, in fact, it is to "disturb" such factors, the question in a technological sense would probably be meaningless. Natural science, on the other hand, with good reason could give the answer that what might be disturbed is the "mathematically idealized reality."

Now what is this paradox of an idealized reality supposed to imply? What is meant here by "idealization"; and in what sense is reality in science "really" conceivable? For the technological sciences do not get hold of nature simply in a receptive way either; they also mathematicize nature and reorganize it in their own way. This, however, is not done in the technological sciences by idealization but by simulation. What, then, is the relationship between the idealization of nature in natural science and the simulation of nature in the technological sciences? Here we are faced with a number of unsettled basic epistemological questions the problematic nature of which has not been given adequate attention.

Still another factor makes the relationship between natural science and technological science an important subject for philosophy of technology. The relations between the two fields of research are not determined merely by the fact that there are technological implications of experimental natural science or that there are repercussions of the latter on technological projects. Rather is it the case that the modern technological sciences are, in many cases, ahead of developments in natural science; they secure technological knowledge that is not yet justified by sufficiently corroborated natural laws. In doing this, moreover, they give rise to the impulse to search for such laws of nature.

But even this does not complete the picture of the correlation between scientific and technological development; rather, more peculiar relations become visible – if only vaguely so: relations involving scientific and technological regularities. For natural laws of science prescribe necessary conditions for technological research insofar (and indeed only insofar) as they exclude technical effects which would contradict them. Technological science, however, is quite capable of leading the way to the discovery of new natural laws. In view of this it can be said that with certain problems technological regularities play a methodological role in the reverse direction not unlike the results of natural science experiments on technology. These regularities, moreover, can take on the function of corroborating instances for theoretical scientific laws; they can play the role of the much discussed "empirical basis" of science insofar as technological regularities have already stood up to tests in technical *praxis*. Scientific theory would, accordingly, be indirectly related to processes of technical *praxis*. It is true, however, that scientific theory in these cases does not yield any explanation in the strict sense for technical processes; this is supplied in such cases by the technological sciences which consider a number of additional real but random conditions which are of no concern to science. Here, too, it becomes apparent that some form of the relationship between natural science and technological science should be closely and analytically examined.

These reflections have so far neglected philosophically relevant problems mentioned earlier, correlations that exist between technology and the economy, ethics, politics, and society. My main concern, however, has been to show – in its outlines as it were – the central area within which the philosophy of technology has to do its research. A one-sided concentration on *the* view of technology was avoided, with questionable attempts to define the term "technology" deliberately left unconsidered. On the contrary, the attempt here has been to present in place of a definition a kind of basic *matrix* of those four elements – nature, science, technology, technological science – which make up the *structure* of the problematic of the philosophy of technology. This is meant to imply that it is not the single elements of this matrix as such, but their *mutual relations* that primarily determine the direction of philosophical research.

However, with the level of problems thus fixed, there is, so to speak, a third dimension, a wider social problematic. For, according to which of the six relations is, at a given time, one's theme, specific perspectives on the other disciplines related to technology are opened up. Thus, for example, questioning economic implications is quite different from examining the relationship between technology and natural science. And ethical questions, as well as those relating to sociology and/or politics, take on quite a different appearance and impact if one's theme is, for instance, nature and natural science or technology and nature.

This is no proposal, however, for a rigid schematic for technological-philosophical questions; neither is a definite localization of individual problems aimed at that might be systematically settled once and for all. The primitive matrix proposed is to serve a heuristic function; it is only to serve the purpose of raising questions, of locating positions. In this way, philosophical questions about technology can perhaps be protected against abstract one-sidedness. Moreover, different proposals that apparently contradict one another and that are proposed to solve the "same" problem can, in a more discriminating perspective, often be reconciled; it is probably merely different but not clearly formulated contexts of the problem in question that have caused the apparent contradiction. To take just one example, the question of "technology assessment," of some urgency at the present time, will need to be seen in quite different ways according to whether either technology (*Technik*) or technological sciences (*Technologie*) is taken as the object of assessment. And for its various answers it cannot be unimportant which concepts of nature or of reality, more or less inexact as they may be, are taken as starting points. The question of natural science and ethics, nowadays much discussed, would likewise get a precise meaning from reflection on its exact starting point: Is it the relation mentioned earlier that is intended, or is not the discussion rather often based on a quite different (yet hardly clearly explained) relationship between technological science and ethics? There is a clear risk that by confusing such problem-relations one may be led to infer "ethical consequences of science" which natural science can not in fact have.

There is no point in extending the catalogue of sample questions any further. No answer, however, has yet been given to the aforementioned question whether the procedures of the philosophy of technology that have been tested so far are sufficient to penetrate into the close network of problems deeply enough so that a philosophy of technology capable of systematic progress can be founded. The field of research has turned out not to be structurally homogeneous. For its basic structure is not provided by just one other already established philosophical discipline like philosophy of natural science, but depends on still another science (namely technological science), and in its "third dimension" still more sciences demand their share of attention. This is certainly a peculiarity that makes the philosophy of technology stand out among other philosophical disciplines. A consequence (in the strict sense) is that the philosophy of technology is to a special degree dependent on interdisciplinary cooperation if it is not to get lost in the mists of unrealistic speculation. But this does not mean that a better philosophical systematization is impossible. (A certain scepticism, it is true, is aroused as to whether philosophical analysis in the same way as has been tried can succeed in arriving at larger contexts that exceed mere beginnings.)

A promising way seems to be to break up the texture of philosophically relevant relations by examining the historical conditions of technology's emergence. What from a historical point of view actually presents itself as all too blurred and only slightly penetrable in its synchronic existence, may in its diachronic historical genesis offer a possibility of conceiving the successive relations philosophy of technology has to deal with. Not only can it be seen that in historical development the complexity of the philosophical problematic has been steadily but gradually increasing (which enables an easier penetration); but history also gives evidence of the range of assessments and the varied prevalences of diverse factors in the network of philosophical problems. Not least, however, does history – and history alone – provide all those concepts that form part of the repertoire of the philosophical analysis of technology. It may be no accident that the phrase "historicity of technology" has become popular just in our times. This implies more, though, than just the trivial fact that technology, like all other cultural achievements, does have its history. The philosophy of technology would be ill advised if it left the question of historicity to the historiography of technological development alone.

The historical reconstruction proposed here does not mean, then, that the history of technology should be substituted for the methodological analysis of technological problems. Nor is the latter to be conceived as a mere add-on to historiographical studies of technology (of which a great number are available). At present nothing but this should be worth considering: whether the philosophy of technology in historical perspective ought to limit its attention to its own relatively short history – as has been explicitly dealt with so far – or whether the essential point should not be to acquire a genuine history through historical reconstruction of its area. This should, to be sure, be done in such a way that historical philosophy of technology displays its analytical instrumentarium – until now almost exclusively tested on present-day questions – so that systematic insights may arise from the elaboration of historical contexts.

Only when it has truly acquired such a history, it seems, can the philosophy of technology effectively arrive at the development of its systematics. But only as a systematic discipline can it take on the rank among other philosophical disciplines that it deserves (today more than ever) if one judges by the importance of its tasks.

The starting point of this whole discussion has, of course, been that the philosophy of technology might consider it necessary to obtain a more definite structure. Could it be, however, that the philosophy of technology sets too much store by greater systematization? It might be argued that the question whether or not philosophy of technology is a systematic philosophical discipline is relatively irrelevant, a matter of indifference. For its structure seems to be a merely academic problem, whereas at present and quite obviously it must be a matter of greater importance to find pragmatic solutions for actual problems under consideration. For instance and above all, there are complexes of pragmatic questions especially about the relations between technology and society; and it is not a matter of pure chance that these questions are attracting particular attention in our day.[8] They are, indeed, important enough to be dealt with in detail. It must not be overlooked, however, that every practical problem has a theoretical dimension which cannot be overlooked. It could turn out, here as elsewhere, that downplaying theoretical questions about technology has its risks; neglect of philosophical discussion might hide contradictions and prevent equivocations from becoming apparent, and erroneous arguments might pass unnoticed. Thus the discussion of the pragmatic problems of technology would in the long run be affected too. Only a heightened consciousness of theoretical problems can provide the precautions that the philosophy of technology needs.

University of Cologne

NOTES

[1] I refer here to philosophy in Germany; German philosophers of technology are aware of this state of affairs. It seems, nonetheless, that the situation with philosophy in the United States is not essentially different.

[2] See especially the elaborate, annotated *Bibliography of the Philosophy of Technology*, by C. Mitcham and R. Mackey (special number of *Technology and Culture*, 1973; available in book form from the University of Chicago Press). There is also plenty of bibliographical material in H. Lenk and S. Moser (eds.), *Techne, Technik, Technologie* (Pullach, 1973), and in F. Rapp (ed.), *Contributions to a Philosophy of Technology: Studies in the Structure of Thinking in the Technological Sciences* (Dordrecht, 1974). Special emphasis is given to the Western Christian tradition in H. Beck, *Kulturphilosophie der Technik: Perspektiven zu Technik-Menschheit-Zukunft* (2d ed., revised and enlarged, Trier, 1979; original edition, 1969, was entitled *Philosophie der Technik*).

[3] This also makes it difficult to write an introduction to philosophy of technology; introductions should not be too onesided, but normally we can deal with only selected topics. One example: H. Stork, *Einführung in die Philosophie der Technik* (Darmstadt, 1977). For introductory purposes, another type of book might be better. Much of the literature is organized and discussed in F. Rapp, *Analytische Technikphilosophie* (Freiburg and Munich, 1978), and in A. Huning, *Das Schaffen des Ingenieurs: Beiträge zu einer Philosophie der Technik* (2d ed., enlarged; Düsseldorf, 1978).

[4] Despite a commitment to certain neo-positivist assumptions, H. Rumpf's "Gedanken zur Wissenschaftstheorie der Technikwissenschaften," in Lenk and Moser (eds.), *Techne, Technik, Technologie* (see note 2, above), pp. 82–107, is useful. See also H. Lenk, "Zu neueren Ansätzen der Technikphilosophie," in the same volume, pp. 210ff. There is certainly justification in Lenk's contention that the link between technology and the technological sciences has not yet been examined in a systematic way (p. 222).

[5] Although there is a long tradition of using the term "technology," that should not deceive us; since the rise of modern science, its meaning has changed fundamentally. Various proposals that we reserve the term "technology" for its modern sense, that is, that we apply it only to technology after its marriage to science, have been unsuccessful – although this would aid in clarifying the discussion. Using the traditional term in dealing with modern issues, it must be admitted, has contributed to the historical inadequacy of many views.

[6] For the first detailed analysis, see T. Kotarbiński, *Praxiology: An Introduction to the Science of Efficient Action* (Oxford and Warsaw, 1965). On this question, see also G. Ropohl, *Eine Systemtheorie der Technik: Zur Grundlegung der allgemeinen Technologie* (Munich and Vienna, 1979); Ropohl distinguishes three dimensions of technology: natural, human, and social; since systems theory is the general framework, technology must be embedded in a general system of action.

[7] From a different perspective, J. Agassi has thoroughly dissected "The Confusion between Science and Technology in the Standard Philosophies of Science"; see *Technology and Culture* 7 (1966): 348–367.

[8] In the long tradition of German philosophy, it is only very recently that this problem has been taken up. In the early decades of this century, strangely enough it was only someone coming from a Hegelian and neo-humanistic tradition who emphasized this

problem and saw its treatment as being of the first importance for academic instruction: see T. Litt, *Technisches Denken und menschliche Bildung* (4th ed.; Heidelberg, 1969). For a critical discussion of Litt's argument, see E. Ströker, "Naturwissenschaft und Technik als geschichtsbildende Mächte," in J. Derbolav *et al.* (eds.), *Sinn und Geschichtlichkeit: Werk und Wirkungen Theodor Litts* (Stuttgart, 1980), pp. 206–222. For something closer to recent questions about technology and its influence on culture, see (for instance) H. Beck (note 2, above) or H. Sachsse, *Anthropologie der Technik: Ein Beitrag zur Stellung des Menschen in der Welt* (Braunschweig, 1978).

NAME INDEX